Minor Fruit Crops of India
Tropical and Subtropical

THE AUTHOR

Prof. Bibhas Chandra Mazumdar had been in the profession of teaching and research of post-graduate students of doctoral fellows in Indian Universities for more than 34 years. He had also served in fruit development, extension education, farm management and research positions for 7 years. Singly or jointly with others, he has authored 14 books, 163 research papers and 53 non-research articles in horticulture, human nutrition, medicinal plants and plant physiology.

Minor Fruit Crops of India
Tropical and Subtropical

2nd Revised and Enlarged Edition

– Author –

Dr. Bibhas Chandra Mazumdar

M.Sc., (Agri.); Ph.D.
Ex-Professor in Horticulture,
Institute of Agricultural Sciences,
Calcutta University, Kolkata;
Reader in Horticulture, B.C. Krishi Vishvavidyalaya, Mohanpur, W. Bengal;
Lecturer in Horticulture, Allahabad Agricultural Institute, Allahabad, U.P.;
Fruit Development Officer, Govt. of West Bengal

2024

Daya Publishing House®
A Division of
Astral International Pvt. Ltd.
New Delhi – 110002

ISBN: 978-93-5461-871-0 (HB)

Published by : **Daya Publishing House®**
A Division of
Astral International Pvt. Ltd.
– ISO 9001:2015 Certified Company –
4736/23, Ansari Road, Darya Ganj
New Delhi-110 002
Ph. 011-43549197, 23278134
e-mail: info@astralint.com
Website: www.astralint.com

Printed at : **Replika Press Pvt Ltd**

Dedication

To the venerated memory of William Roxburgh (1751- 1815), the Scottish surgeon and botanist who came to settle in India and explored plant wealth in this country. Regarded as "Founding father of Indian botany, this legendary biologist delved into plentiful species of plants which he had envisioned to be of worth in fundamental botany, agriculture, horticulture, pharmacobotany, forest science and various plant based industries in this country in future.

Preface

Organized farming of crop plants is presumed to have been started in the world thousand years ago in the fertile land of Mesopatamia. But fruits as satiable food to the human beings was brought to the notice of the human beings long before that and we owe to our ancient mothers for this remarkable discovery in the area of food wealth.

Some examples of fruits and nuts that were tasted by the primitive people were date, grape berries, pomegranate, almond, bael, coconut, jamun, apricot, peach etc. and the present archaeological evidence indicates that the fig tree was domesticated in the Near-East some 11,400 years ago.

Ancient people had learnt from experience which fruits could be considered fit for consumption by the human beings, which would taste better, improve their toiling potential, efficacious to cure or defend against illnesses and also those which might bring about harmful effect to them on consumption. Backed by this knowledge, they conceived the idea to make distinction among the various categories of fruits. The fruits which were considered highly acceptable and useful were given prominence. The trees bearing such fruits were planted by them along their dwelling places, farm lands and other places which could be accessible to them. The necessity of applying dung of pet animals as manure, irrigation water and stirring of the soil to enhance yield and quality of the harvested produce same to their knowledge in course of time.

The process continued and those fruits which were not given much emphasis were recognized as of less, i.e., minor importance. While those which were given prominence continued to grow and produce on receiving care, those of lesser importance did not halt their multiplication and they began to strike roots here and there, at odd places, inferior and uneven land and in the forests. In the struggle for existence, these materials learned how to survive in adverse edapho-climatic situation and survive through progenies and in their efforts, they received co-operation of wild animals. Arraying fruits into principal, i.e., *major* and the less important, i.e., *minor* ones came into being in this way.

It is worth mentioning that such discrimination of fruits into major and minor groups however, had not called attention to the people of ancient India. The sages and scholarly people of the bygone days in this country who had a persistent leaning to plant fruit trees had the apprehension that the fruits which were not highly palatable or the trees which bore them might had been bestowed with medicinal wealth or other attributes and hence, it would not be a logical approach to set them at defiance. This is the prime factor for existence of so vast multitudes of varieties, hybrids or types of naturally grown wild fruits in this country, in the land, odd places and mountains.

With passage of time, fruit farming in India gained momentum among the kingly and affluent people who were much inclined to set fruit trees that would bear produce of high organoleptic appeal as compared to those which could not stand in comparison with them. Such discrimination of fruits got impetus in the later period in this country.

At the present time when the worth of biodiversity in plants is comprehended, the horticultural scientists in this country are paying heedful attention to explore the less utilized in other words, minor fruits that grow in different parts of the country instead of passing over them.

Plentiful species and varieties of such less-exploited, minor fruits are found to grow in various parts of India and these are attuned to produce in tropical, sub-tropical and temperate environments. Among them, only a few species of pertinence having adaptability to thrive and give satisfactory production under tropical and sub-tropical situations have been chosen and considered for discussion in the present text.

A title of the same subject had been brought out by the author earlier in 2004 and the present one has been attempted to make that revised and enlarged. Evidently, he will be amply rewarded if the present title meets the requirements of the students and researchers in horticulture, pomology, forestry, applied botany, ayurveda and related subjects. The author gratefully acknowledges Dr. N. K. Dutta, M.Sc., Ph.D., Dr. Kaushiki Mazumdar Dutta, M.Sc., Ph.D. and Mrs. Purabi Mazumdar, M.A., B.Ed. for various types of assistance received in preparing the book. Lastly, the author welcomes valued suggestions from the readers such that the monograph could be further improved in future.

Bibhas Chandra Mazumdar

Contents

CHAPTER–1

INTRODUCTION

I. Concept of Fruit

One of the highly choicable items of food to a human being is fruit. In botanical sense, it is defined as an organ belonging to a species of higher plant (= plantae) as a mature or ripened ovary. It may also contain one or more seeds but seed may be absent if it is developed parthenocarpically, whether by vegetative means or by exogenous stimulation of ovary.

In horticulture on the other hand, the term *fruit* is conceived in a broader sense. In this science, the following three criteria are in fact, commonly taken into consideration to designate a plant organ as a fruit. It may however, be mentioned that the conditions are flexible and are not rigidly followed (Mazumdar, 2004).

(i) Botanical Part

In most cases, the horticulturally designated fruits are same as with the botanically defined fruits. Thus, mango, guava, papaya, banana, citruses, grape, apple, pear *etc.* that are developed on maturation and ripening of ovaries are grouped under the purview of fruits in botanical sense. In horticultural science also, these food-stuffs are recognized as fruit.

On contrary, tomato, brinjal, pumpkin, gourds, okra, beans *etc.* which are also matured and ripened ovaries and accordingly, are grouped as fruits in botany, these are considered as vegetables in horticulture and not as fruits.

In regard to the concept of fruits in horticulture, cashew-apple may be cited as an example. Botanically, this organ of the cashew tree (*Anacardium occidentale* L.) is the swollen peduncle with the thalamus and does not arise from the ovary.

Hence, this organ is not considered to be a fruit in botany. But paying attention to the attractive appearance and colour, succulent texture, pulpiness, high content of juice and sweet-sour taste which resembles it to be fruit, it is considered as fruit in horticulture and has been given a name, *i.e.*, cashew-apple.

(ii) Savouriness

Savour, in other words, palatableness of a fruit to the human beings is an important consideration in horticulture for its recognition as a fruit. Thus, mango, guava, papaya, banana, citruses, apple, pear *etc.* are regarded as fruits in horticulture as they are highly palatable to the human beings in fresh form. But examples of karonda (*Carissa carandas*), dillenia (*Dillenia indica*), hog-plum (*Spondias pinnata*), wild citruses *etc.* may be cited which are not palatable in fresh form and yet they are placed under the group of fruits in horticulture.

On the other hand, examples of water-melon (*Citrullus lanatus*) and musk-melon (*Cucumis melo*) may be cited in this connection. These are not only grouped under fruits in botany as they are formed from the ovaries but are also highly palatable when taken as fresh. Thus, although they meet all the requirements to consider them as fruits in horticulture, many horticultural scientists of India have considered them in the area of vegetables and not in fruits.

(iii) Mode of Consumption

An important criterion to recognize a botanical fruit as a horticultural fruit is, whether it could be taken in fresh form, in other words, without cooking or processing. Keeping this in view, botanically recognized fruits like mango, banana, apple, pear *etc.* are relished as fresh and hence, these are well considered as fruits. But cucumber may be cited as an example which is a botanically recognized fruit although it is universally taken as fresh and without any sort of cooking. Nevertheless, horticulturists consider cucumber as a vegetable and not a fruit. In this connection, it may also be stated that some fruits like papaya, jackfruit *etc.* are consumed by cooking in their immature condition but when they are ripe, they are relished as fruit.

Thus, it could be opined that definition of fruit is horticulture is not rigid and hence, a fresh review on the subject appears to have necessity.

II. Classification of Fruits

Classification of fruits is made by the botanical and horticultural scientists in different ways. The primary distinction between these two types of classification lies in that, while the horticultural scientists consider fruits to those food-stuffs only which have edible value to the human beings, the botanical scientists consider those to the group of fruits without regard to whether they are edible or not to the human beings. Each classification is extensive and some of the principal criteria considered in each are stated below in brief.

(a) Botanical Classification

A number of considerations are made by the botanists to classify fruits and among these, the following are of pertinence.

(i) Morphology and Ontogeny

A fruit is considered a true fruit in botany when it is developed from a single ovary of a single flower and is considered a false fruit (= spurious fruit) when this does not happen so as in case of aggregate and multiple fruits. All fruits are however, classified into three broad groups which are recognized as simple, aggregate and multiple fruits.

Simple fruits may be dehiscent, *e.g.,* pod, follicle, siliqua, capsule and schizocarps or, indehiscent, *e.g.*, caryopsis, achene, cypsela, nut, drupe, berry and pome. Aggregate fruits may be etaerio of follicles, achenes, drupes and berries while multiple fruits of the two types are sorosis and syconus.

(ii) Taxonomic Position

Systematic position of a fruit and the botanical family to which it belongs is given importance by a section of botanists to classify fruits.

(iii) Forms of Stem

Whether the stem of the fruit plant is erect, *i.e.*, herb, shrub or tree or of weak form, such as creepers, trailers or climbers is considered as a criterion to make classification.

(iv) Ecological Types

To which eco-types a fruit plant belongs, *e.g.*, hydrophytes, xerophytes, halophytes *etc.* is given importance by some botanists to make classification.

(v) Centre of Origin

Some botanists classify fruit plants according to their centre of origin. For example, mango, sweet orange *etc.* come under Assam-Myanmar centre of origin, banana, shaddock, coconut *etc.* under the south-east Asian centre, litchi, longan, pear *etc.* under the Chinese centre, fig, apple, pear, cherry *etc.* under the Middle-east centre, cashew, pineapple *etc.* under Brazilian centre and the like.

Apart from the above, botanical classification of fruits or plants is made in many other ways.

(b) Horticultural Classification

In horticultural science, quite a large number of classifications are made and some of the common ones may be stated as follows.

(i) According to Climate Adaptation

Horticulturally designated fruits are most commonly classified in accordance with their adaptability and productive capacity both in terms of yield and quality of the produce in climatic zones of Earth (Köppen climate classification). The zones are separated according to the latitudes both in the north and south of the equator (0°). Plants are found to grow in three zones, which are tropical, sub-tropical and temperate zones. Tropical zone (also known as torrid zone) extends on Earth between the tropic of cancer, *i.e.*, 23.5° North and the tropic of capricorn, *i.e.*, 23.5° South of the equator. Sub-tropical zone extends from 23.5° to 35° North and 23.5° to 35° South of the equator and temperate zone extends from 35° to 66.5° North and same latitude in the South of the equator. The latitudinal lines and zones are however, not exact and divisions are made on approximation basis.

Climate in the tropical zone is hot (megathermal) and humid. Average temperature in this zone is 25° to 30° C and average precipitation is 60 mm in wet season, for which relative humidity is high. Day length (= photoperiod) in this zone ranges from 10–13.5 hours. Seasonal fluctuation in climatic constituents is very much less in tropical zone and is least between 5° North and South of the equator which is referred to as equatorial calm zone. In tropical zone, rain forest, monsoon and savanna types of climate are also existent.

In sub-tropical climate, both winter and summer are distinct. There is no snowfall in winter but frost may occur. Day length differs widely between summer and winter. Sub-tropical climate may be humid where rainfall is high or dry (Mediterranean type) where it is low. Temperate zone is characterized by snowfall and mean temperature is (–) 3°C but less than 18°C. Difference in day length between summer and winter is much wider than in the sub-tropical climate.

Fruit plants whose productiveness both in yield and fruit quality go to the highest level when grown in tropical zone are called tropical fruits. Similarly, those which are most adaptable and productive in yield and quality, respectively in sub-tropical and temperate zones are recognized as sub-tropical and temperate fruits.

In the above context, two aspects come to light. These are firstly, variations in climatic condition in some pockets in tropical and sub-tropical zones may exist due to the presence of mountains, ocean current, lake *etc.* and thus, some sort of temperate type of climate may prevail in the tropical and sub-tropical zones. Secondly, varieties, hybrids or types of a tropical fruit plant species may adapt equally well in sub-tropical climate and similarly, species of temperate fruit plant may exist that thrives in sub-tropical climate.

Thus, classification of fruit plants into tropical, sub-tropical and temperate groups is made and followed in a general manner only. In fact, any sharp line of distinction is difficult to be made for all species of fruit plants. It may further be mentioned that climatic requirement in respect of the tropical and sub-tropical fruit plant species has not been widely different and hence, many horticultural scientists prefer to consider species of tropical and sub-tropical fruit plants into the same category.

It may however, be mentioned in this context that species of some fruit plants may be grouped as highly tropical, *e.g.*, banana, cashewnut, coconut, carambola, mangosteen, hog-plum, pineapple, sapota, karonda *etc.*, species of fruit plants which are highly sub-tropical, *e.g.*, grape, litchi, sweet orange, pomegranate, loquat, ber, kiwi, *etc.* and species which are highly adaptable to temperate climate may be named as apple, plum, cherry, pear, almond, apricot *etc.*

(Broad agro-climatic zones making fifteen groups based on physiography and climatic condition has however, been accepted by the Planning Commission of the Govt. of India in 1985-90. The agro-climatic zones have later been brought into a total of 120 sub-zones by the National Agricultural Research Project).

(ii) Texture

In consideration to texture, very hard-textured and one-seeded fruits are called nuts, soft-textured ones as pulpy fruits while fruits in which juice is only taken, *e.g.*, citruses, pomegranate, phalsa *etc.* are grouped as juicy fruits.

(iii) Consumers' Demand

Fruits which have much fondness to the consumers for desirable blend of sweetness and sourness and other traits are regarded as dessert or choice fruits.

(iv) Season of Availability

In accordance with season of availability, fruits are grouped as summer season, rainy season, winter season fruits.

(v) Ripening Behaviour

Fruits in which respiratory rate suddenly increases before the onset of ripening are called climacteric fruits and in which fruits this is not apparent are regarded as fruits of non-climacteric respiration.

(vi) Seedlessness

In accordance with presence or absence of seeds, fruits are grouped as seeded and seddless fruits.

Other considerations are also made to group horticulturally recognized fruits, *e.g.*, shelf-life, medicinal value, existence of phytochemical compounds *etc.*

III. Concept of Minor Fruits

Minor fruits are also known by other names, such as less important fruits, less utilized fruits, less exploited fruits, under-utilized fruits *etc.* The terms denote that fruits, belonging to these species are of lesser importance as compared to those of other species which are of greater importance with respect to a locality, zone or country.

However, the above criteria of making any explicit difference between major and minor groups of fruits is difficult to be made. It could be reasonably said that logically, the terms, minor and major fruits are in fact, relative and not definite. The distinction between minor and major fruits is primarily based on the following two considerations.

i. Requirement of the consumers.
ii. Area and production of fruits.

Requirement of any particular fruit or fruits for all people in a locality or country is not same. It depends on a number of factors, among which the following are of importance.

Palatability of fruits is beyond doubt, an expressive factor in regard to their acceptance. Fondness in acuity however, differs among people. It is a common observation that fruits which have both sweetness and sourness to a desirable degree are of much preference to the people in north India and also in most parts of the world. In the eastern India on the other hand, people have a greater preference of such fruits which are highly sweet in taste while fruits of sour taste do not gain favour. Fruits which have relatively high sourness, *e.g.*, dillenia, carambola, karonda, cicca *etc.* are found to be highly popular to the people of south India.

Texture of the edible part of a fruit is much a determining factor in regard to its acceptance to many people. Mango fruits of sucking varieties are not found to be liked by the people of western countries and these fruits are considered as of minor importance in those countries. Mango fruit of the Alphonso variety is choicable for desirable texture of its pulp, *i.e.*, mesocarp. However, in the north and south Indian condition, sucking mango fruits are highly liked by many people.

Flavour of a fruit exerts great influence for its acceptability to many people particularly in the western countries. Thus, tropical fruits which have fine flavour, *e.g.*, scented varieties of mango, banana, queen group of pineapple are highly favoured by them. In fact, fruits of many species which are characterized by pleasant flavour are categorized as fruits of major importance. An annonaceous fruit, *viz.*, cherimoyer may be cited as an example. This fruit is highly favoured in western countries for its excellent flavour and is even considered as one of the best sub-tropical fruits in the world. Under Indian condition on the other hand, the related fruit of the same family, *i.e.*, custard apple has much greater importance for its high sweetness although it has no flavour and the fruit is considered to have major importance. Cattley guava fruit which though has nice flavour due to caryophyllene content is considered as a minor fruit in India while the common guava is ranked as a major fruit.

Despite having many desirable attributes, existence of astringent or bitter principles in a fruit debars its status to be a fruit of major importance. Mention may be made to the grapefruit which is a highly popular fruit in the U.S.A. and other western countries has not gained favour to most of the Indian people for slight bitterness of its juice due to the presence of limonin and naringenin. The related fruit, shaddock is given high importance which is free from bitter compounds. A fruit, *Citrus limettioides* has not been considered to be a major fruit due to existence of limonin in its juice although it has many other attributes.

It may be mentioned in this context that improvement made in the quality of minor fruits belonging to many species has however, turned them to consider as fruits of major importance. Under Indian condition, examples may be cited to the Banarasi and Narendra varieties of bael, Umraon, Gola and Kaithli varieties of ber, Ra jamun, Bedana, Paras and Narendra varieties of jamun *etc.* which of late, have high demand in the markets and tend to assume as major fruits unlike other varieties that have been grouped as minor fruits.

Area and production of a fruit crop belonging to a species in a locality, region or zone in a country are given high emphasis in assigning it to be a major or a minor fruit crop. Accordingly, in India, mango which is extensively grown to an area of about 2400 thousand hectares, yielding a production of 12,800 thousand tonnes is regarded to be a major, in other words, principal fruit crop. The same fruit which is now grown to a small extent in many other countries is regarded there as a minor fruit crop.

Even in the same country, area and production of a fruit crop may differ in agro-climatic zones. For example, a number of temperate zone fruit crops have covered extensive area now. Apple is grown to about 78,000 hectares in Jammu and Kashmir giving a production of about 714900 tonnes. Temperate fruits comprise as major fruit crops in Jammu, Kashmir, Himachal Pradesh and adjoining areas. Similar examples may be cited in Nasik, Pune, Ahmednagar, Satara and other parts of Maharashtra where area under viticulture is about 0.1 million hectares and production is 0.80 million tonnes. Some other examples may be cited as Nagpur region in growing mandarin orange, Khandesh area in growing Cavendish banana, Varanasi and Pratapgarh area in Uttar Pradesh in growing choice emblicas, Muzaffarpur area in north Bihar in growing litchi a Jalpaiguri area in West Bengal and eastern States in growing pineapple, Kalimpong district in West Bangel in growing strawberry, South 24-Paraganas district and southern West Bengal in growing Baruipur variety of guava and the like.

It may be brought to notice that area and production of a particular fruit crop in a zone may substantially change with time. Increase in area of any specific fruit crop may take place for a number of reasons, such as demand by the consumers, requirement in the processing factories, export to other countries and so on. Mention may be made to the fact that cashew tree was unknown to the Indians until the Portuguese people planted the trees in the Malabar coast of the Arabian sea in the years, 1560 to 1565. They had planted the trees there only to check erosion of the soil by the sea. (Nobody in the world knew at that time that a cashew tree could produce delicious kernels inside their nuts and the cashew-apple).

Quite a number of fruit crops have been introduced in India from other countries. Introduction of temperate fruits and nuts and their culture in Kashmir and Kulu valley is a classic example. This had become possible by the efforts of the

Frenchman, M. Pychard, the then Director of Agriculture in Kashmir (1910-1920), Captain A.A. Lee, S.C. Stokes and other European experts in fruit science who came to India.

In the recent years, viticulture in northern India which was unknown before 1960 has covered over 5000 hectares of land in Punjab, Haryana and the adjoining areas by introduction of early cultivars of grapes from the Mediterranean countries. The tangor citrus variety, Kinnow which was introduced in the Punjab six decades ago is now growing there in about 90 thousand hectares and giving an annual production of 3 thousand tonnes of fruits.

Area and production of some fruit crops which were once considered as of minor importance have now markedly stepped up in some parts of India. Examples may be cited of ber, jamun *etc.* in Punjab, Haryana and other parts of north India, longan in West Bengal, bael in Uttar Pradesh, West Bengal and Odisha and the like. It may also be pointed out that apart from increase in the area, decrease in the area of some fruit crops is also not uncommon in some parts of India. Example may be stated of guava in the eastern part of Uttar Pradesh and particularly the famous cultivar, Allahabad Safeda. Area under this fruit crop tends to squeeze in that zone due to serious threat by the soil-borne fungal pathogen of *Fusarium.*

It is apparent from the above discussion that area and production of a particular species of fruit crop could not be said to be static in a zone or a country. Again, a fruit crop which is considered to be a minor fruit or a less utilized one in an area may assume the status of a major or principal fruit in the same area with time.

Therefore, in assigning a species of fruit crop to be a major or a minor one, some other aspects are needed to be specified. These particulars should include variety belonging to the species of the fruit crop, zone or locality of its growing, area and production as compared to the other fruit crops grown there, period of assessment besides other relevant informations. Necessity of these specifications cannot be disregarded for the fact that minor fruit crop is a relative rather than an exact term.

Under Indian condition however, at least 150 species of tropical and sub-tropical fruit crops could in general, be assigned as of minor importance considering the entire country as a whole.

Out of that long list of tropical and sub-tropical minor fruit crops, only 42 have been considered for discussion in the present text.

IV. Importance of Minor Fruits

All fruits have plentiful attributes and minor fruits are of no exception. However, some of the particular merits of the fruit crops that are designated as of minor importance under Indian condition are stated in the following.

(i) Stepping up Fruit Production

India's total foodgrain production is estimated to be 295.67 million tonnes during the year, 2019-20, which is higher by 10.46 metric tonnes achieved in 2018-19.

Production of fruits has been 97.38 million tonnes in the country in 2018-19 as compared to 97.36 million tonnes in the previous year. Although there has been some increase in the area and production of fruit crops during the past five years but the increase is meagre as compared to foodgrains.

Significant increase in the area under fruit crops is impeded for many factors. An important factor is aversion of the crop growers in taking up fruit growing as a farming enterprise for the possible risk factors that are associated in it, such as long gestation (juverite) period for most of the fruit trees, high cost of initial investment, cultural operations, harvesting and also other factors.

The above encumbrances usually do not come in the way in growing minor fruit crops. Being hardy and tolerant to adverse edapho-climatic situations, they are able to grow and produce in land of low agricultural potential. Thus, these fruit crops do not come into competition with other crops in respect of land.

Land is in fact, a finite, *i.e.*, determinate resource and the geographical area of 328.73 million hectares that exists in the country is not possible to be increased for the purpose of food crop production.

(ii) Utilization of Waste Land

In India, 63.85 million hectares which comprise 20.17 per cent of the total geographical area are designated as waste land. From agricultural point of view, these are recognized as degraded lands. Waste lands in the country are grouped into categories, *e.g.*, gullied and ravinous land, upland with or without scrub, waterlogged and marshy land, saline and alkaline land *etc.* Percentage of waste lands to the country's geographical area is high in some States, *e.g.*, Jammu and Kashmir (65.5 per cent), Manipur (58 per cent) and Himachal Pradesh (56.9 per cent).

Reclamation of waste lands in wider scale in a large country like India has many practical difficulties. Hence, securing food from these lands by raising agricultural crops has limitations. Many of the minor fruit plants however, thrive well in such lands and give rise to production. Thus, productive utilization of waste lands in securing food has its possibility in growing minor fruit plants.

Many of the fruit crops have been studied in Florida in the U.S.A. to be salt resistant. Minor fruit plants which could withstand soil salinity have been stated as *Carissa* species, *Phoenix sylvestris, Blighia sapida, Annona reticulata, Feijoa sellowiana, Casimiroa edulis, Chrysophyllum cainito, Psidium cattleianum etc.*

(iii) Securing Food from Drought Prone Area

Drought, *i.e.*, dryness of soil due to lack of rain may be of several categories, *e.g.*, agricultural, meteorological, hydrological, economic *etc.* Agricultural drought is a condition when there is prolonged water deficit to an abnormal level. This is recognized when the crop plants are in greater necessity of moisture.

Some of the minor fruit plants, *e.g.*, ber, phalsa, some relatives of guava, wild date, palmyra, monkey jackfruit, karonda, emblica *etc.*, could be named which thrive successfully and produce under drought condition. Hence, securing food from drought prone zone gains support by growing these minor fruit plants.

(iv) Protection against Storm

The coastal areas of India in particular, are at greater threat due to violent storms which are referred to as tropical cyclones. These appear 5 to 7 times during the end of summer and advent of south-west monsoon. Hazards associated with cyclone are storm surges, gale wind and squall besides heavy rain that causes to bring flood and tornadoes.

Many species of minor fruit trees are deep-rooted and have highly spreading roots. Thus, they are able to resist high velocity of wind without being uprooted. Mass planting of such trees can reduce wind velocity to a high extent. These fruit trees protect against damage of various types. They are also highly effective to serve as wind-break rows around the orchard boundaries.

(v) Protection against Soil Erosion

Minor fruit trees belonging to many species have deep and highly spreading roots. They ramify and cover large area of soil. To control erosion of soil in undulated area and river banks, many of such trees are highly effective. It has been mentioned earlier that the Portuguese settlers planted cashew trees along the Malabar coast of India to check soil erosion by the Arabian sea.

(vi) Afforestation

Many species of the minor fruit trees are highly suitable materials for large-scale planting in forests. Trees which are tall, bushy with large conopy and hardy such as jamun, monkey jackfruit, ber, elaeocarpus, diospyros, sterculia, hog-plum, palmyra, elephant apple, madhuca *etc.* are highly suitable materials for planting in tropical rain forests. It may be mentioned that many species of minor fruit trees and shrubs make rapid growth and are suitable materials for planting in social and agro-forests.

(vii) Use in Industries

Many of the minor fruits serve as excellent materials in making varieties of food products, *e.g.*, pickle, preserve, morabba, dried product, beverages, jelly, jam *etc.* Accordingly, many cottage and small-scale industries have been developed by making their use. As stated in the subsequent Chapters, cattley guava, karonda and many other minor fruits are highly suitable materials to make jelly and jam. Calamondin and many wild citrus fruits are found to be suitable materials to make excellent quality of marmalade.

Juices obtained from palmyra, madhuca, wild date *etc.* offer as suitable materials to make fermented beverages and industries in the rural sector have been developed to make these drinks in many parts of India.

Apart from fruits, leaves of diosphyros, dillenia, palmyra *etc.* are also used to make many products and serve as materials in running cottage and small-scale industries in many parts of the country.

(viii) Securing Timber

A large number of species of minor fruit trees have high timbėr value and are used to make building materials, furniture and structure for building up industries.

(ix) Providing Food to the Rural People

A general survey done in the recent past has revealed that consumption of fruits by the rural people in India is lesser than what is consumed by the urban people. Many fruits which are supplied from distant places are also not supplied to the village markets or are scantily supplied. Even considerable proportion of local fruits that are produced in the rural areas are sold in the urban markets for high economic return.

Owing to the above trend, rural and the tribal people are regularly deprived in gettling fruits in adequacy. However, many of the minor fruits produced in the rural areas are not much disposed to the urban markets for difficulty in transportation, very soft skin as in mulberry and hence, these fruits reach the hands of the rural people.

(x) Nutritional and Medicinal Value

Minor fruits belonging to a number of species are of high nutritive value and have plentiful medicinal properties. These have been stated in the subsequent Chapters.

(xi) Use as Breeding Materials

To effectuate genetical improvement of fruit crops, species of minor fruits are explored for hybridization by the fruit breeders.

(xii) Use as Rootstocks

For hardiness and ability to thrive in adverse soil conditions, a number of species of minor fruit trees and shrubs are used as effective rootstocks to graft superior varieties over them.

(xiii) Providing Food to the Wild Animals

Minor fruit trees that grow as wild or in forests are of great benefit to the birds and wild animals in getting food and materials to build nests. Thus, these fruits are of great help in wild life preservation.

(xiv) Eco-environmental Improvement

Minor fruit trees grow here and there, in odd places, undulated and hilly areas, forests *etc.* They grow and produce satisfactorily without requiring inputs from human beings. Apart from providing fruits and other valuable materials, these trees and shrubs play great role in eco-environmental sustenance in a locality.

CHAPTER-2

LIST OF SPECIES

Species: *Anacardium occidentale* Linn.
Taxonomic Serial: 28793

Common Name

In English, the commonly used name of the species is Cashew or Kaju. The word, Kaju is the pronunciation in Portuguese language. The generic name, *Anacardium* has been given in consideration to the shape of the swollen peduncle, *i.e.*, cashew-apple. Cashew-apple looks like an inverted heart. Ana of *Anacardium* means upwards, cardium means heart and thus, *Anacardium* means inverted heart. In Spanish language, cashew is called Maranón. In Portuguese, the name is both Kaju and Kajueiro.

In Hindi, cashew is known as Kaju, in Kannada Geru beeja, in Malayalam, Kasuandi, in Gujarati, Kash, in Tamil, Mundiri paruppu, in Telegu, Jeedi pappu, in Oriya, Lanka ambu and in Bengali as Hijli badam or Kaju badam or Kaju.

Botanical Family

Anacardiaceae which is also known as Sumac family comes under the botanical order, Sapindales. The family had been treated as a series of five tribes by Engler and subsequently, into sub-families by Takhtajan as Anacardioideae and Spondiadoideae. In the sub-family, Anacardioideae has been included the tribe, Spondiadeae.

The family, Anacardiaceae consists of 77 genera and 700 species, which are trees, shrubs or woody vines. They are mostly tropical and few species are temperate. They are evergreen or deciduous and have resin ducts in the bark. Some species of the family of commercial importance are mango, cashew, pistachio, spondias, sumac, smoke tree, poison ivy, marula, pink peppercorn *etc.*

Leaves of the plants are simple or often pinnately compound with leaflets in various arrangement. Stipule is none or very small. Stem is woody with resin duct in bark. Flowers are inconspicuous, regular, actinomorphic, perfect or mostly imperfect. They are poisonous. Sepals are 3, generally 5 and often fused at the base. Petals may be upto 5 and are small. Stamens are 5 or 10 in two series, alternating with petals. Carpels and styles are usually 3. Ovary is superior. Fruit is a berry or berry-like drupe and is usually oily.

Genus: Anacardium

The genus, *Anacardium* of the family, Anacardiaceae comes under Magnoliopsida and has 20 species, although according to some taxonomists, it is still less. *Anacardium occidentale* is most recognized species of the genus. Some other species of importance are *A. amapaense, A. excelsium, A. brasiliense, A. amilcarianum A. humile, A. giganteum, A. nanum etc.*

Leaves of the species are alternate, petiolate or sessile, simple, entire and shape is oblong, oval or obovate. Inflorescence is corymbose panicle, terminal or sub-terminal. Sepals are 5, imbricate and are basally connate. Petals are caducous. Stamens are 7-10 and unequal in length. Pistil is obcordate or obovoid, sessile with free ovary. Fruit is obliquely renal-shaped on swollen peduncle. Seed is dicotyledonous, unequally reniform and has prominent testa.

The plants have secondary metabolites in leaves, branches and fruits which have use in commerce. Some species have medicinal importance. Adhesive property of some plants is also known. The plants are adaptable to thrive in inferior soil.

Importance

Among the various types of nuts that are consumable to the human beings (N.B. : Nut is not consumed as such but the kernel which is present inside the nut is the edible part), cashew-nut, *i.e.*, cashew-kernal has great popularity around the world. The tree is highly tropical and although it is grown to some extent in mild sub-tropics but the performance is not satisfactory.

India is a leading country where cashew is grown in about 820,000 hectares, yielding a production of about 540,000 tonnes of nuts. The tree is extensively grown in the southern and western parts in the country and also to a high extent in the coastal parts of Odisha, West Bengal and other tropical parts of the eastern States.

In a cashew tree, there are two parts which are consumable to the human beings. These are, (i) kernel of the nut and (ii) succulent peduncle (= fruit-stalk) with the thalamus which remains attached to the upper end (= proximal end) of the nut. The latter part assumes highly attractive yellow or orange colour, becomes succulent, has good flavour and the pulp is sweet-sour in taste. For this reason, common people consider the peduncle as a fruit when it is fully grown and a catchy name, *i.e.*, "cashew-apple" is ascribed to it. At present, in commerce also, the name, cashew-apple is common. However, it should be remembered that botanically, cashew-apple is not a fruit. Apart from the above two parts, sometimes the tender leaves of the cashew tree are also cooked with rice in Java to add flavour.

In commerce, the kernel which is located inside a nut is most popular around the world. On breaking the hard shell of the nut, the kernel is removed out. This is dried in sun, or roasted in suitable drier or fried in oil and then taken. It is taken either directly or with addition of salt, with or without spices. Dried or fried kernels are also used as ingredients in confectioners, chocolates, sweetmeats, cakes *etc.* for value addition of the products. The kernels have nice taste, pleasurable texture and flavour for which they appeal all sections of consumers.

Cashew-kernels are of great worth from the nutritional point of view as well. They are enriched with high amount of protein, fat and carbohydrate which is 21, 47 and 22 per cent respectively. They are also rich sources of vitamins of the B group.

Although kernels of the nuts of cashew have high commercial importance, it needs mentioning that the swollen peduncle, *i.e.*, stalk of the fruit which is known as cashew-apple is of no less importance. Although botanically, it is not a fruit but the horticultural scientists consider cashew-apple as a fruit and is in fact, recognized as a "minor fruit" (Mazumdar, 2000 C).

Cashew-apple is relished like a pulpy fruit and is popular particularly in Brazil and other south American countries. Preserved products like jelly, jam, squash *etc.* are also made from cashew-apple.

However, in many cashew growing countries, juice of cashew-apple is extracted and then it is fermented and distilled to make a liquor which is known as "feni". In India, Goa and other parts of Konkan region in India have high reputation to make this liquor in large-scale. The liquor which contains 40-45 per cent alcohol has great demand in many countries.

Nevertheless, it may be commented upon that the cashew-nut which contains the delicious kernel has far greater importance in the world than the cashew-apple or the feni made from it. Exhaustive research studies have been done in many cashew growing countries in the world with the objective to enhance yield and quality of the nuts. These studies have not been considered in the present text.

Textual matter of the present Chapter being cashew-apple, studies and knowledge on *cashew-apple have only been considered for discussion and not the cashew-nut, nor an oily substance present in the shell of the nut which is called cashew-nut shell liquid (CNSL).*

Description of Plant Parts

Cashew is a strong and robust tree (Plate-1) which attains a height of 10-12 metres. The tree is evergreen and produces 3-4 growth-flushes in the year. Trunk of the tree is of short height and irregular in shape. Branches arise at low height and are of spreading nature. Leaves are in terminal clusters, simple and have prominent veins. They are sprially arranged, leathery in texture, elliptic to oval or obovate in shape, 5-20 cm long, 2-10 cm wide and have smooth margins.

Flowers appear from September to February. Inflorescence is indeterminate panicle and is polygamo-monoecious. Flowers are small, pale green at first and later became pinkish. Petals are 5, slender and 8-15 mm long.

Fruit of cashew is a one-seeded nut, which is kidney-shaped and botanically is a hard drupe. It encloses kernel which is a large dicotyledonous seed. Kernel is surrounded by hard shell. The shell has three layers and these are, exocarp at the outer side, mesocarp in the middle and endocarp at the inner side. Mesocarp contains an only liquid which is known as cashew-nut shell liquid (abbreviated as CNSL) and this liquid has many industrial uses.

At the distal, *i.e.*, lower end of the nut is fruit-stalk, known as peduncle. With progress of growth, the peduncle becomes highly swollen along with thalamus and very much pulpy. This assumes attractive yellow or orange colour according to varities and its taste becomes sweet-sour. For these attributes, this part resembles a seedless fruit and hence, in horticulture, it is called cashew-apple and is usually considered as a minor fruit. However, botanically this is a pseudo-fruit (= false fruit) and is not a real fruit. The real fruit is the one-seeded nut which is present at the distal end of the nut.

(N.B.: It is brought to light in this connection that an unusually large cashew tree is known to be grown in Natal in Brazil. The tree has occupied an area of 8.5 square metres. It is supposed to be a genetic mutation and had possibly been present at that site since 1888).

Origin

Nativity of cashew is Brazil and many taxonomists are of the opinion that the tree has actually originated in the north-eastern part in that country. It may be pointed out in this connection that very old fossil of cashew has provided indication that the tree had its existence in Europe and also in north America.

Geographical Distribution

Cashew is grown in tropical countries in plantations, forests and also grow here and there as stray trees. Nowadays, cashew tree is also grown in countries having mild sub-tropical climate.

Important cashew growing countries may be named as Brazil, Peru, Bolivia, Venezuela, Guatemala, Caribbean Islands and other countries in America and Atlantic Islands, Nigeria, Sudan, Niger, Ethiopia *etc.* in Africa, southern, western and eastern India and Sri Lanka in south Asia and all the islands in south-east Asia. Significantly high production is obtained in Vietnam, Nigeria, Côte d'Ivoire and some south-east Asian countries.

India occupies a significant position in the cashew scenario of the world and the country contributes to more than 15 percent of global production. Cashew tree was in fact, unknown in India and it was the Portuguese settlers who had first brought cashew trees in Goa and Konkan region in India between the years 1560 and 1565, a territory in India which was then under the possession of the

Portuguese. The Portuguese settlers planted cashew trees along the Malabar coast in Konkan region of the Arabian sea. They planted the cashew trees only to check soil erosion which regularly occurred there.

However, it is surprising that neither the Portuguese settlers nor anybody in the world knew at that time that each nut of cashew enclosed in it a large, kidney-shaped, white kernel, *i.e.*, a dicotyledonous seed which on drying or frying would have a delicious taste and one day, it would have a large demand in the world.

It should be noted that in cashew, it is the nut which is highly valued in commerce and not the cashew-apple. Hence, in most of the cashew growing countries, much higher attention is paid in securing the nuts of cashew for commercial purpose and not the cashew-apples. Further, it may be noted that cashew-apples start to develop when the nuts have sufficiently grown. Thus, harvesting of both the nuts and the apples together could not be done or has practical difficulty. Hence, the cashew growers prefer to harvest the nuts from the trees for commercial demand, ignoring plucking of the cashew-apples.

Nevertheless, some countries like Brazil, Malagasy, Mali *etc.* may be named which pay adequate attention to harvest cashew-apples from the trees. Brazil could be ranked first which produces about 1.78 M and thus, contributes to more than 92 per cent of global production of cashew-apples. The countries, Malagasy and Mali contribute to 3.9 and 3.7 present of global production. Goa and Konkan region in India have also great importance where cashew-apple is harvested to a fairly high extent for making the fermented wine, known as feni and also the unfermented juice for use as a non-alcoholic drink.

Climatic Requirement

Cashew is considered to be a strictly tropical tree and is best grown between 25° North and South of the equator. However, the tree is found to grow and produce in mild sub-tropical climate depending on varieties, although the performance is not so good in respect of productiveness and quality of the kernels and cashew-apple.

Hot and humid atmosphere throughout the year is preferred by the cashew trees. Minimum annual rainfall should be about 300 mm for the tree to grow. Low winter temperature is unfavourable and temperature lower than 20° C damages the trees. In the tropical climate, the tree grows upto an altitude of 700 metres but the yield declines in high altitude. For colour development and quality of the cashew-apples, bright sunshine is necessary.

High summer temperature during early stage of growth of the nuts cause heavy incidence of nut drop. Bhattacharyya *et al.* (1989) observed in a large cashew plantation in the coastal area of West Bengal that about 29 per cent of nuts borne by about 7 per cent trees were cracked at their proximal ends extending upto 1.6 cm length and 1.3 cm width. According, to the growers, cracking of nuts occurs in years due to high summer temperature. But this is not confirmed as nutritional deficiency may also be a factor and it may be mentioned that fruit cracking in guava has been stated (Vasudeva and Raychaudhuri, 1954) to be due to zinc deficiency.

Soil Requirement

Soil is not an important factor for growing cashew but drainage should be very good. Tree becomes tall and vigorous when it is grown in deep soil. Slightly acidic soil is better for vigorous growth of the tree which is moderately salt resistant as well. Yield performance of the tree increases with higher soil fertility.

Propagation

Cashew is propagated by a number of methods. Seed propagation is by far, very common in India. Seeds germinate easily but mechanical scarification by rubbing the nuts between two sand-papers for sometime quickens germination. In some parts of India, seedlings are raised in nursery-beds or in perforated polythene bags and then transplanted. But direct sowing of the seeds *in situ* in manure-filled pits is better as the tap-root is liable to be broken during transplanting and thereby, causing stunted growth of the plants. However, true characters as with the mother tree are not maintained in a tree when raised from seed. Hence, vegetative propagation is advisable for superior varieties or types.

A number of vegetative propagation methods are successful in cashew and these are practised in many countries, Air-layering gives high degree of success when it is done in rainy-season and root regeneration is quicker and higher by application of auxin. It is practised in many parts of Kerala, south India and coastal parts of West Bengal. Mound-layering is also done in cashew in some countries and ringing the shoots ensures quicker protruberance of roots in them.

Among budding methods, patch-budding, Forkert-budding and chip-budding give high degree of success. Grafting is done in cashew in south American and south-east Asian countries but not common in India. Stem-cuttings are successful when invigorated side-shoots that appear on beheading thick branches are used with application of auxin. Soft-wood cuttings are more successful.

To secure rooted plants within a short time, *in vitro* regeneration of roots in gaining ground. Thimmapaiah and Samuel (1999) at the National Research Centre for Cashew in Puttur reported that axillary shoot-bud proliferation, *i.e.*, 1-13 shoot buds per explant had been observed when shoot tip/nodal explants from the *in vitro* grown cashew seedlings were cultured in MS medium and supplemented with thidrazuron or with auxins such as NAA, IBA or BAP. Maximum proliferation of roots had been seen to occur when thidrazuron was used at 0.1 mg per litre concentration, in combination with NAA or IBA solution at a concentration of 0.1. mg per litre.

Culture

Cultural aspects in cashew have a great impact on productiveness and quality of both nuts and apples. The influence of laminar level of nutrients and quality of cashew-apples had been observed by Kumar and Sreedharan (1987). The said authors had applied 150-450, 50-150 and 50-160 gram of nitrogen, P_2O_5 and K_2O per tree per year respectively and observed that while a negative correlation had existed between the nitrogen, phosphorous and potash content of the leaves with sugars and juice reaction of apples, there had been a positive correlation with juice and vitamin C content of apples.

Egbekun and Otiri (1999) studied on the effects of different levels of nitrogenous, phosphatic and potash fertilizers as well as plant growth-regulators on the yield and quality of cashew-apples. It was observed by them that the effect was best when nitrogen, phosphorous and potash were applied, at the rate of 500, 250 and 250 gram respectively per tree. Among growth-regulation, ethylene releasing compound, *i.e.*, Ethephon at 50 ppm solution proved better than with 2,4-D or NAA solutions.

A study done with the objective to determine the effect of tree spacing, Akinwale and Aladesua (1999) in Nigeria maintained three spacings and observed that spacing had not any significant effect on the performance of cashew-apples. Training the trees by *open-centre* (= vase)system at early stage of growth to make the branches more spreading is observed to produce apples of bigger size and brighter colour as they receive more sunshine.

Top-working– Many cashew trees which are grown in various parts of India are found to be poor yielders of nuts and apples and their quality is inferior. These trees may be transformed to superior varieties or types by the horticultural practice, known as top-working. (The method of top-working has been described in other Chapters).

Pests and Diseases

Cashew trees are damaged by a large number insect pests and diseases. Among insect pests, tea mosquito, leaf miner, borers, webbers *etc.* are very common. Tea mosquito causes heavy damage by sucking sap from young shoots, inflorescences and apples. Control of the pest may be done by spraying 0.05 per cent Endosulfan during appearance of new shoots. Leaf miners attack new shoots which appear in rainy-season and the caterpillars suck the sap. The pest may be controlled by spraying 0.05 per cent Endosulfan or Fenitrothion. Caterpillars of leaf webbers which web inflorescence with shoots may be controlled by the same insecticides or by spraying 0.1 per cent Carbaryl.

Among diseases, drying, *i.e.*, blight of inflorescence is serious, which is a fungal disease. Other fungal diseases are anthracnose which attacks fruits and twigs. Die-back of shoots is sometimes serious which is common in rainy-season. The fungal species causes the bark to split and the shoots dry up from the tips, gradually downward. To control fungal diseases, Bordeaux mixture (3: 3: 50) or other copper fungicides, *e.g.*, copper-oxychloride (0.2 per cent), Dithane M-45, Carbandazin (0.1 per cent) *etc.* may be applied. Animal pests like birds, bats and squirrels also damage the nuts and apples.

Harvesting

Cashew trees grown from seeds start flowering after 4 to 6 years of growth. Vegetatively propagated trees flower a year or two earlier.

Growth of cashew-nut and cashew-apple does not occur simultaneously. That is, growth of the nut takes place at a quicker rate than the growth of the apple and this continues upto a month from fruit-set. After that, *i.e.*, when the nut has grown upto 50 to 60 per cent, the apple grows. It grows very quickly such that it overgrows the nut. In fact, after fruit-set, ovary grows much but the ovule grows less and hence, the kernel does not fill. The trend changes after that and kernel grows much faster. During the first month after fruit-set, the nut grows very fast and attains almost maximum growth and size. During the next week, the nut becomes hard in texture but its size becomes somewhat less during the next week.

Interestingly, the apple starts growing when growth of nut almost stops and the apple grows at a very fast rate. Thus, the apple is ripened after about two months from fruit-set. Unfortunately, the nuts become over-mature at this stage, *i.e.*, they surpass the stage of harvest. Therefore, harvesting of the nuts and the apples at their proper maturity is difficult to be done together. As a matter of fact, if nuts are desired to be harvested, apples have to be forgone and *vice versa*.

It may however, be stated that the trend of growth of apple and nut as stated above is sometimes not found in some varieties or types. Further, it may be possible to manipulate the growth of apples and nuts by foliar application of plant growth regulating chemicals to some extent.

Performance of the cashew-apples is best when they become fully ripe in the tree. But at full ripeness, they become very soft and harvesting should be done with great care. As stated in the Chapter on *Dimocarpus longan*, a sickle-shaped, movable blade attached to a cloth-bag and fastened to a bamboo pole may be used for harvesting the apples without making damage to them.

Yield

Yield of cashew-apples varies on many factors and has practical difficulty to determine the yield precisely. However, according to Madhava Rao (1967), a tree may produce 35 kg of cashew-apples.

Varieties

Although not recognized as varieties, a large number of types of cashew are grown in India and many eco-types are found.

Moura *et al.* (2001) made an evaluation of cashew varieties. Chadha (2012) has stated in his book that about 33 varieties of cashew have been released in India and most of them have a mean yield of 8-10 kg of nuts per tree. He provided a list of varieties that are recommended in different parts of India. Some of these varieties are, BPP-4,6,8 and VRI-2 for Andhra Pradesh, Chintamani-1, Selection-1, 2 *etc.* for Karnataka, Madakkathara-1,2, *etc.* for Kerala, T. No. 40 and Vengurla-4 for Maharashtra and Goa, VRI for Odisha, VRI-1,2 and 3 for Tamil Nadu and Jhargram-1 for West Bengal. Singh and Saxena (2008) presented a list of varieties developed in India and mentioned high yielding varieties.

Mode of Consumption

Fully ripened cashew-apples are best relished as fresh, with or without mixing salt and spices. Use of the apples in fruit salad is also common. Drying and grinding of the apples for prolonged use is done in some parts of south India. Akinwale and Aladesua (1999) stated that steaming the apples is done in Nigeria for higher palatability but the practice reduced Vitamin C content in them.

Quality and Nutritive Value

Morton (1987) stated in her book that 100 gram of cashew-apples contained moisture, protein, fat, carbohydrates, fibre and ash, respectively of 84.4 - 88.7, 0.101 - 0.162, 0.05 - 0.5, 9.08 - 9.75, 0.4 - 1.0 and 0.19 - 0.44 gram. Calcium, phosphorous, iron, carotene thiamine, riboflavin, niacin and vitamin C had been stated by her to contain 0.9–5.4, 6.1–21.4, 0.19–0.71, 0.03–0.742, 0.023–0.03, 0.13–0.539 and 146.6–372.0 milligram respectively in 100 gram of cashew-apple.

From the National Institute of Nutrition (I.C.M.R., Govt. of India) in Hyderabad, it has been reported that 100 gram of cashew-apples containing 86.3 per cent moisture has the following nutritive constituents (Gopalan *et al.,* 2018).

Protein: 0.2 g.; Fat: 0.1 g.; Carbohydrates: 12.3 g.; Crude fibre: 0.9 g.; Minerals: 0.2 g.; Calcium: 10 mg.; Phosphorous: 10 mg,; Iron: 0.2 mg.; Carotene: 23 mcg,; Thiamine: 0.02 mg,; Riboflavin: 0.05 mg.; Niacin: 0.4 mg,; and vitamin C: 180 mg. It has been stated that 100 grams of cashew-apple could provide in the body 51 K calorie (213.384 K Joule) of energy (Index: g = gram, mg = milligram, mcg = microgram).

Maikup *et al.* (1997) determined some physico-chemical constituents of cashew-apples of red and yellow colour, which were borne by trees growing in the coastal area in the Medinipur district of West Bengal. It was observed that in general, most of the constituents were found to be at higher level in the red-coloured apples.

Cashew-apples borne by 9 early dwarf clones were compared by Moura *et al.* (2000) and they put forward that in general, the tannin content of the pulp had been at a lower level than that which was stated in earlier reports. Olveira *et al.* (2000) in Brazil determined alanine, serine and some other *free-* amino acids in the pulp of cashew-apples. Volatile compounds are important constituents of cashew-apple and Pino (1997 a) in Cuba made a review of such compounds. Some compounds such as esters, terpenes, alcohols, hydrocarbons, lactones, ketones *etc.* which play influence on the quality of cashew-apples have been determined by Bicalco and Rezendo (2001).

Medicinal Value

Cashew-apple has a great use in traditional medicine. Being a very rich source of vitamin C, it is an antidote against scurvy disease as well. High content of tannin in the apple saves the vitamin C against loss. In chronic dysentery, cashew-apple is much used in Cuba and Brazil.

The apple is known to have diuretic and sudorific properties. Topical application of cashew-apple paste relieves rheumatic pain. In containing high carotene, vitamin C, anthocyanin, anacardic acid, cardanols, cardols and other compounds, cashew-apple is a potent antioxidant.

Anti-microbial activity of cashew-apple juice has much been studied by a number of investigators. Kubo *et al.* (1999) in California in U.S.A. studied that anacardic acid (E)–2 hexanal derived from cashew-apple exhibited antibacterial activity against the gram-negative bacterium, *Helicobacter pylori* which has been known to be responsible for gastritis. The said authors (Kubo *et al.*, 1993) in earlier works reported anti-tumoric agents from cashew-apple.

Phytochemicals

Cashew-apple contains a large number of phytochemicals. Tannins and flavonoids are rich in the pulp. Existence of vitamin C in cashew-apple which amounts to 180 to 250 mg or more per hundred gram is a great attribute.

Anacardic acid is present in considerable amount in cashew-apple. This is used in the industries for production of cardonol required for resins, coatings, frictional materials *etc.* Among other compounds, 6–pentadec (en) yl salicylate is obtained from the apple and the nut which has anti-oxidative property.

Constituents in Cashew-apples during Development

With cashew trees grown in the coastal area of West Bengal, Roy and Mazumdar (1989) experimented by spraying the trees with 1, 3 or 5 per cent aqueous solutions of zinc sulphate when the nuts were 5–7 days old. The apples and nuts were physically and chemically analyzed after 20, 30 and 40 days. Acidity (equivalence of citric acid), vitamin C and sugars in the apples were found to have increased by treatment with the solutions at all stages and directly with concentrations. In the same experimental site, Bera and Muzumdar (1992) sprayed cashew trees with 200 ppm solutions of IAA, IBA or NAA. It was observed that vitamin C and acidity level of the apples increased by the auxin treatments but not sugars.

A study undertaken with the objective to make possible enhancement of protein content in the cashew kernels mediating through penicillamine Bhattacharya and Mazumdar (1993) sprayed aqueous solutions of 0.1, 0.5 or 0.7 per cent benzyl penicillin to the cashew trees. Vitamin C content of the apples had though somewhat increased by application of 0.1 per cent solution, higher concentration decreased it.

In Andhra Pradesh, Naidu *et al.* (1998) determined dry matter content of cashew-apples with progress of their development. They observed a slow increase of dry matter in the apples upto fourth week and the increase had occurred to a marked extent at the subsequent stage of development. Constituents of cashew-apples during their development had been studied by Egbekun and Otiri (1999) and Figueiredo *et al.* (2001, 2002).

Gordon *et al.* (2012) determined vitamin C content, phenolic compounds and antioxidant activity of cashew-apples at three maturity stages of them. The results revealed that phenolic compounds in the apples were highest in unripe apples and decreased during ripening. Vitamin C has been stated to have influenced anti-oxidant activity.

Undesirable Effect

Despite having high vitamin C, cashew-apple produces some undesirable effect. It causes allergic reaction to some people though everybody is not allergic to it. The astringent compound, uruciol which is present in cashew-apple cause oral irritation. It may be stated that this compound is present in mango peel also. Steaming the cashew-apple however, removes the irritant from it. Sobhana and Mathew (2015) from the Cashew Research Station in Thrissur in Kerala stated that for preparing jelly from cashew-apple juice, astringency may be removed by boiling 250 ml of extracted juice with 500 ml of water for half an hour.

Post-harvest Storage

Cashew-apples are very soft and are easily rotten physiologically or by microbial infection. When stored in ordinary atmosphere, these are readily infected by yeasts and other fungi besides some bacteria. Morton (1987) quoted in her book that researches undertaken by the food technologists in India provided evidence that cashew-apples could be stored for upto 5 weeks at a temperature of 0°–1.67^{o} C with maintenance of 85–90 per cent relative humidity in the storage chamber.

Perishability of cashew-apples under post-harvest condition was studied by Filgueuras *et al.* (1999). They emphasized that suitable handling of the apples had important role to extend their shelf-life. Kumar *et al.* (2015) at Bagalkot in Karnataka observed that treatment of cashew-apples with 0.5 per cent calcium chloride caused minimum percentage of physiological loss which was 5.62 per cent by weight and less rotting, *i.e.*, 10 per cent during their fourth day of storage.

Utilization

Cashew-apple has greater importance for the products that are made from it than its use as a fresh fruit. Many products are in fact, made from cashew-apples and some commonly made ones are as follows.

High quality jelly could be made from cashew-apple juice and standardized procedure to make jelly had been put forward by Shobhana and Mathew (2015) from Thrissur. Jam is also made from cashew-apple and usually by mixing with other fruits.

Squash is made from the juice of the apples and this is done by steaming the apples to remove the astringency. In some cashew growing countries and also in India, canning of cashew-apple is done preserving in syrup. In India, dried or oily form of pickle of cashew-apple is made with addition of spices.

In some countries, bottled juice of cashew-apple is made in large-scale for high demand. Leite *et al.* (2015) from Brazil stated that high-pressure homogenization (HPH) might be used to improve the physical condition of the juice.

Cashew-apple juice is used for many industrial purposes as well. On converting the juice to make ethanol by reaction with the microbe, *Saccharomyces cerevisiae*, preparation of vinegar has been described by Thiripurasundari and Usharani (2011) from the Annamalai University in Tamil Nadu. Lactic acid production from the cashew-apple juice as a low-cost substrate has been reported by Silveira *et al.* (2012) from Brazil.

Ethanol production from cashew-apple juice and optimum environmental condition to produce that have been put forward by Devi *et al.* (2015) from Tamil Nadu. Ethanol, *i.e.*, ethyl alcohol could be obtained from cashew-apple bagasse also.

Preparation of Feni

Whatever products that are made from cashew-apple, it is "feni" of Goa which could be said to surpass all products. It is said that "Portuguese people brought cashew trees in India and Goan people's gift in the country is feni made from the juice of cashew-apple." Feni which is also called and spell as fenny, fenno, fenim *etc.*, is in fact people's a fermented and distilled liquor which is country-made. This is produced in large-scale in Goa and Konkan region and to a lesser extent in some parts of Tamil Nadu and south India and in Gujarat also.

However, feni is made not only from the juice of cashew-apple but also from the toddy derived from coconut palm. Accordingly, when feni is made from cashew-apple juice, it is called cashew-feni and when it is made from coconut toddy, it is called, coconut-feni.

The feni was in fact, innovated by the Goan people. But why the name, feni was given to this fermented beverage is not clearly known. A great many people is of the opinion that, the term, feni has derived for the fact that when a bottle of feni is vigorously shaken or it is poured in a glass, frothing occurs and gush out. In Konkani language, the frothing is called feni and hence, the name was given. It is likely that the word, feni is derived from the Sanskrit word, fena which refers to froth.

It may be stated that in olden days when the Portuguese people had not brought cashew trees in India, the Goans used to prepare only coconut-feni. But when the Portuguese people planted large number of cashew trees in Goa and adjoining areas (1560–1565 A.D.) along the Malabar coast, the Goans tasted the kernels and apples of cashew. The idea whether feni could be made from the juice of cashew-apple came to their mind. They translated the idea into action and observed that feni made from cashew-apple was actually of very high standard and better quality than that made from coconut toddy. People's acceptance of cashew-apple-juice-feni had been very high thus, effectuating high demand of this feni in the local and distant markets.

For the above reason, a large part of Goan people gradually switched over to make feni from cashew-apple juice, leaving aside feni production from coconut-toddy. Nevertheless, coconut-toddy-feni has its importance in that, this liquor is possible to be made all throughout the year because coconut-toddy is obtained in all seasons but cashew-apples feni is possible to be made only when ripe cashew-apple are available. That is, it is made only in one season. However, nowadays the juice of cashew-apple is preserved for months and feni is made in other seasons also.

It may be mentioned in this connection that some consumers have a greater liking to coconut-toddy-feni than cashew-apple-juice-feni. Mention may be made that cashew-apple-juice-feni is produced to a large-scale in the northern part of Goa while coconut-toddy-feni is largely produced in the southern part of that territory.

Feni making is in fact, a traditional practice among the Goans and it runs in their families. Intelligently done experiments made by them have brought about tremendous improvement in making this beverage. They prefer to confine within themselves the improved techniques which they have gained through many years of experience.

Feni made by Goans from cashew-apple juice has now assumed the status of a world-class alcoholic beverage and has been awarded the Geographical Indication Registration (GIR) by the Govt. of India in 2009. This liquor has been recognized as a heritage beverage by the Govt. of Goa in 2016. Cashew feni is valued as a symbol and glory of Goa such as saké in Japan, Vodka in Russia, Champagne in France, Makkoli in Greece *etc.* (It may be mentioned in this connection that some parts of India also have the fame to make some special types of country-liquor of high popularity, *e.g.*, mohua-liquor in Jharkhand and Purulia district of West Bengal, palmyra-sap-liquor or tari in West Bengal, verjuice-liquor in western India.)

Cashew-apple-juice-feni has the attribute in that, no chemical or biological material is added in any stage of its preparation. Many people have a fondness for this brew for this reason and this is honoured by calling eco-friendly wine.

(a) Mode of Preparation

Preparation of good quality feni requires care and technique at every stage and this has been well-standardized at the Central Plantation Crops Research Institute at Kasargod in Kerala.

However, the following steps are in general, involved in preparation of feni as is practised in Goa and other parts of Konkan.

(i) Selection of Raw Materials

To prepare feni, only fully ripe cashew-apples should be used. It should be remembered that cashew-apple is non-climacteric and ripens best in the tree. Pre-ripened fruits if harvested do not ripen any more after harvest.

(ii) Extraction of Juice

To extract the juice from the apples, they are needed to be thoroughly crushed. It is a conventional practice of the producers in Goa and Konkan to crush in a rock which is at first carved out to make the shape of a basin. It is called there as columbi. At present however, many producers crush the apples in mechanical pressing device instead of columbi. Conventionally, extraction of juice is made twice. That is, small mound of the pulp is made and is crushed with a heavy weight boulder. After second extraction of the pulp, the juice that is obtained is used to make feni. The juice after second extraction is called neero. The neero is also used as an appealing but, non-alcoholic drink.

(iii) Fermentation

The juice obtained after second extraction, *i.e.*, the neero is then fermented. For fermentation, the juice is transferred to a large-sized earthen pot. The pot is not fully filled and adequate space is left at its top. The pot is buried half in the ground and left for 2–3 days to get the juice adequately fermented. Instead of earthen pot which is breakable, many producers use plastic drum.

For fermentation, no enzyme preparation or microbial culture or any extraneous material is added to the juice from outside. In other words, fermentation is allowed to take place by the action of the naturally occurring yeasts which are tropically grown species of these microbes.

When about 4–5 per cent of the juice is fermented, sugars that are present naturally in the juice are enzymatically catabolized, *i.e.*, broken down to simple sugars by microbial action. Due to this catabolism, alcoholic, esters and other compounds are formed as the end-products and these are responsible to impart flavour of the feni when it is prepared. Besides catabolic end-products, a thick layer of scum appears on the surface of the juice during fermentation as the bubbles float on the surface.

(iv) Distillation

After fermentation of juice, it is collected in earthen pot which is known as bhatti, or preferably in copper pot and then distilled. The fermented juice is heated and the vapour of alcohol which gives out is collected by condensation process, *i.e.*, under cold condition.

Distillation of the fermented juice is done for three times. After first distillation, the brewage which is obtained is called, *arrack* (= urrac). The strength of arrack is mild (14–16 grao) and this is almost non-toxic. Those who prefer non-toxicating beverage, drink this arrack usually mixing with juice of orange, sour lime, pomegranate *etc.* The arrack is then distilled for second time. The product which is obtained after second distillation is called, *cazulo.* The cazulo is then distilled for the third time, *i.e.*, charged for final distillation. After final distillation, the beverage becomes of higher strength than cazulo (20–24 grao). *This product is actually feni.* However, the second distilled product, *i.e.*, cazulo is also sometimes called feni, or low-strength feni.

(v) Bottling

The final product, *i.e.*, feni is colourless and of sweet, alcoholic flavour. It is stored in sterilized bottles, which are then tightly stoppered and labelled giving necessary particulars. It should be noted that although wines of different brands are usually stored for years and then drunk as it is believed that their strength increases with ageing, it is not so in the case of feni. Feni is best relished as fresh and it is not stored.

(b) Juding Quality

Quality of feni that is prepared is assessed on many factors. Alcohol content is given much importance and a high quality feni should contain 43–45 per cent ethyl alcohol (= ethanol). The marketable feni should contain about 43 per cent ethyl alcohol. Aprat from alcohol, aroma of the feni is also given much importance in judging quality.

(c) Mode of Consumption

Feni can be taken as neat or over ice. But as it contains fairly high amount of alcohol than many other liquors, many consumers prefer to drink on blending feni with other beverages, *e.g..*, lemonade, orangeade, squashes, coconut water lime-juice cordial, aerated water *etc.*

(d) Medicinal Value

Feni is known to have therapeutic value also. It is believed to be efficacious in intestinal worms, gum infections, body ache, fever, influenza attack *etc.*

General Remarks

Large-scale marketing of cashew-apple-juice-feni as well as coconut-toddy-feni in different countries needs encouragement. Problem-oriented researches on feni also need attention in India to prepare better quality of the product. Mechanization at every step in preparation has its necessity. Preparation of feni on mixing juices of cashew-apple and coconut-toddy or with other fruit juices needs attention.

In India, more than 20 lakh tonnes of cashew-apple are estimated to be produced and a tree yields 25–35 kg of apples. Unfortunately, most of the apples are wasted. Critical attention is needed to be paid in regard to utilization of cashew-apples to make fruit products and feni in wider scale for economic progress and employment in the rural sector.

Family: Anacardiaceae

Species: *Spondias dulcis* Forst.

[Synonym: *Spondias cytherea* Sonn.]

Taxonomic Serial: 825213

Common Name

In English, the fruit of *Spondias dulcis* is known as Ambarella, Golden apple, Polynesian plum, Otaheiti plum, Jew plum *etc.* In Malaysia and Indonesia, it is known as Kedondong, in Thailand, 'Makok-farang', in Camodia, Mokok, in Vietnam, Pomme, cytherea in, Myanmar, Gway in Bangladesh, Bilati amra in, Tanzania, Embe kizungu, in France, Pomme cythere, in Germany, Goldpfame, in Jamaica, June plum, in Colombia, Hobo de racimos, in Venezuela, Mango jobo *etc.*

The names of the fruit in India are Bilati amra in West Bengal, Omoro in Assam, Adavi mamidi in Andhra Pradesh, Amtekai in Karnataka and Ama bazhanga in Kerala. Other names of the fruit in India are Mampuli, Mambu lichi, Ranamba, Amba lamu, Marima, Amate, Pundi *etc.*

Genus

The genus which is known as mombin is magnoliopsid and contains 17 species. The trees of the genus are mostly deciduous or semi-evergreen and their height may be 20–25 metres or more. Leaves of the trees are pinnate, simple and have spiral phyllotaxy. Fruit is a drupe, yellowish when ripe and the length is 5–10 cm, which contains one hard seed. Fruits belonging to about 10 species of the genus bear edible fruits. Fruits are sour, moderately sour or somewhat sweet-sour in taste.

Importance

The fruit has much popularity in south-east Asian countries, Bangladesh and Sri Lanka. Popularity is also high in many parts of India although the trees are not in high abundance in this country. The tree thrives well and produces satisfactorily in inferior soil. Planting of the tree in home gardens is common in Tripura, West Bengal, Maharashtra, Kerala and other parts of south India. In some countries, it is planted in crop fields to provide shade. The amount of pulp in the fruit is much greater than that of the allied species, *Spondias pinnata*.

Description of Plant Parts

The tree is deciduous. It grows fast and attains a height of 10–12 metres but in deep soil, it becomes more tall. Bark of the trunk and main branches is smooth and light grey in colour. Leaflets are 6–12 cm long with finely toothed margins.

Leaves are 20–50 cm long, imparipinnate and their colour turns brownish in autumn season.

Inflorescence is a long penicle. Flowers are very small, whitish and have fragrance. Male, female and bisexual flowers are borne in each cluster. Fruits are borne in bunches of 8–15 on long stalks. They are 6–10 cm long, oval, yellowish when ripe and thin-skinned. Morton (1987) stated that in South Sea island, a fruit may weigh upto 0.45 kg. A fruit contains one seed which is soft in early stage and later, it becomes hard and fibrous.

Origin

Indonesia and Malaysia have been stated by many as the origin of *Spondias dulcis*. According to Morton (1987), the species had originated from Melanesia to Polynesia and from there, it spread to other countries.

Geographical Distribution

The tree grows abundantly in Malaysia, Cambodia, Vietnam, Celebes, Thailand and other south-east Asian countries. It is also much grown in Bangladesh, Sri Lanka, Kenya, Mexico and south American countries. In India, the tree is mainly grown in Assam, Tripura, Eastern States, West Bengal, Odisha, Kerala, Maharashtra, Tamil Nadu and Karnataka.

Climatic Requirement

The tree grows satisfactorily with high fruiting performance in tropical climate and humid tropics is considered ideal. Full sunshine has necessity for the tree for optimum flowering. Frost damages the tree if continues for many days.

Soil Requirement

Soil is not an important factor for the tree, except that drainage should be good. Growth of the tree is vigorous in rich soil. In Florida in the U.S.A., the tree is found to thrive in oolitic limestone.

Propagation

Ambarella is easily propagated by seeds. Germination takes place in 3–4 weeks and temperature of 25°–30° C is favourable for germination. The seeds lose viability within a short period and hence, those stored for long period should not be used. Soaking the seeds in lukewarm water for 4 to 5 days with regular change of water facilitates germination by leaching out germination inhibitors.

Stem-cuttings are successful for propagation, using hardwood stem. Applications of auxin greatly helps in protrusion of adventitious roots. Pinto *et al.* (2012) reported success of cuttings by treatment with sugar solutions of different concentrations. Air-layering is much practised for propagation of the tree in many south-east Asian countries. Grafting is also done in some countries on rootstock of same species or on *Spondias pinnata*. Budding is usually not done and when it is done, chip-budding method is mostly practised with success.

Culture

If mass planting of ambarella trees is desired at a site, the trees should be spaced to grow at 7–12 metres. The tree is grown without use of agro-inputs but its growth is vigorous and productiveness is high by application of organic manure. Pruning the branches encourages new growths and higher yield of the trees. Liberal irrigation in hot-weather is greatly helpful to improve size of fruits and yield of the trees.

Pests and Diseases

Mites are often found to attack leaves which feed on undersurface of leaves and produce a velvetty layer, known as erinose. To control, dusting sulphur or spraying 0.05 per cent wettable sulphur is effective. In Indonesia, leaves of the tree are found to be attacked by the larvae of the beetle, *Podontia affinis*. Necrosis of leaves due to attack of a wasp is observed in Costa Rica. Bark of the trunk and branches are also attacked. Gummosis, *i.e.*, gummy exudation from the trunk or branches is found to occur in Costa Rica due to fungal infection. To control, Bordeaux mixture–1 per cent or copper oxychloride–1.25 per cent or any copper fungicide may be applied. Aurofungin–0.1 per cent is also effective. Sooty mould is found to appear in many trees. Throwing water on affected branches partly removes the mould.

Harvesting

Ambarella trees bear fruits after four or five years of growth. Vegetatively propagated trees produce fruits a year or two earlier. Fruits are best ripened in the tree and do not ripen under post-harvest condition if harvested before they are fully ripe. Acidity of the fruits goes to the minimum level and concomitantly, sweetness to the maximum level when the fruits become fully ripe. Ripening of the fruits start from September and continues upto November.

Varieties

Varieties of ambarella are not known but many types are found which differ in tree and fruit characters. Trees which are much dwarf are sometimes found. Dwarf trees are usually heavy yielders. Bartoo and Badrie (2005) studied on the effect of adding nectar of the flowers on physico-chemical and sensory quality of yoghurts made from cow's milk.

Quality and Nutritive Value

Morton (1987) put forward a table on analytical reports of ambarella fruits done in other countries. She quoted that the total solids, moisture, protein, fat, sucrose, crude fibre and ash in the fruits were 14.53–40.35, 59.65–85.47, 0.50–0.80, 0.28–1.79, 8.05 - 10.54, 0.47, 0.85–3.60 and 0.44–0.65 per cent respectively and calorific value of 100 gram of pulp had been stated as 157.30.

Analytical studies done by Mandal and Mazumdar (1995) with ambarella fruits borne by trees grown in the central part of Wets Bengal revealed that the weight of fruit, specific gravity, length, total sugars acidity as equivalence of citric acid and vitamin C had been 78 gram, 1.04, 6.2 cm, 2.58 per cent, 0.85 per cent and 50 mg per 100 gram of pulp. Appreciable amount of pectin in the fruits has been stated by Majumdar and Mazumdar (2001).

Medicinal Value

Various parts of ambarella trees have many medicinal properties. Roots have abortifacient properties and the extract is helpful to control itching. Leaf extract has anti-oxidant property and has anti-microbial, cytotoxic and thrombolytic activity. Chewing leaves cures mouth ulcers. Fruit is helpful in urinary trouble, type-2 diabetes, constipation and in softening skin. Anti-cancerous activity of the fruit has been stated.

Phytochemicals

Fruits are known to contain a number of terpenes and these give rise to flavour in them. They are also rich sources of pectin and other polysaccharides.

Constituents in Developing Fruits

Changing responses of some physico-chemical constituents in round as well as oval shaped fruits of ambarella at three stages of their maturity had been assessed by Ishak *et al.* (2005). It has been reported by them that the soluble solids, insoluble matter and *p*H increased with maturity of fruits. Ripe fruits were significantly less acidic than the green and the half-ripe samples. With maturity of fruits, total phenolic content had decreased. Vitamin C content ranged from 4.65 ± 0.27 to 5.86 ± 0.73 mg per 100 gram of pulp. Water-soluble pectins in fruits decreased with advancement of their maturity. They had also determined the amount of phosphorus, magnesium, calcium and zinc in the developing fruits.

With fruits borne by trees grown in the central part of West Bengal, Mandal and Mazumdar (1998) determined weight, specific gravity, stalk-stylar length and diameter at maximum bulge of the fruits as well as total, reducing sugars, non-reducing sugars and the vitamin C content of the pulp at four equally spaced stages of development, such that the fruits sampled at the last stage coincided with harvest-maturity of the fruits under local condition. Their findings revealed that while acidity had consistently decreased in fruits with progress of their development, other constituents made steady rise. Sinha Roy (1996) and Mandal (1997) studied the changes in different constituents of ambarella fruits during their progress of development, maturation and ripening.

Utilization

A number of products are made with ambarella fruits in south-east Asian and other countries. Fruit juice is very common which is drunk with spices, condiments and sometimes by mixing juices of other fruits. Jelly and jam are also made with the pulp or in most cases, by mixing with pulp of other sweet fruits (Mazumdar, 2019).

A thickened sauce is made using the pulp of the ambarella fruit in Thailand. In West Bengal, Tripura, other eastern States and in Bangladesh, a preserve is made with the fruit. In some parts of India, fruit pulp is dried, ground and used as a condiment.

Family: Anacardiaceae

Species: *Spondias pinnata* (Linn. f.) Kurz.

[Synonyms: *Spondias mangifera* Willd.; *S. paniculata* Roxb. ex Wight and Am.; *S. amara* Lam; *S. acuminata* Roxb.; *Mangifera pinnata* L. f.; *Wirtgenia decandra* Jungh]

Taxonomic Serial: 895118

Common Name

In English, the fruits of *spondias pinnata* is commonly called Hog plum, Indian hog plum, Yellow plum, Indian mombin, Libas, Spanish plum *etc.* In Japanese, it is called Amura tamagonoki, in Chinese, Bin lang quing, in Nepalese, Amaaro, in Thailand, Makok pa, in German, Mangoflaume, in Spanish, Mango jobo, in Portuguese, Cajamanga *etc.*

In Sanskrit, the fruit has been referred to as Amrataka. In Hindi, the fruit as called Bhiringi phal, Amra, Metula, Jangli aam, Ambara *etc.*, in Tamil as Ambalam, Ambazham, Pulima *etc.*, in Telugu as Ambalamu, Adhavamu *etc.*, in Marathi as Ambada, Khatamba, Ranamba *etc.*, in Malyalam as Marahunsi, Ambazham *etc.* and in Bengali, the fruit is known as Desi amra, Amra, Tok amra, Piala, Amratak *etc.*

Importance

The tree is a suitable material for planting along avenues for its graceful appearance. Fruits are very sour and not eaten as raw but are much used in culinary preparations in south India. In West Bengal and the eastern States, the fruit is very common in making pickles, preserves and in preparation of sour juice which is known as *omlo, ombol*, chutney *etc.* of late, the tree is given importance for planting in tropical forests, primary and secondary and tidal forests.

Description of Plant Parts

Hog-plum tree is deciduous or semi-evergreen. Height of the tree attains 10–15 metres but may be upto 20 metres or more in deep soil. Tap-root of the tree is deep. Bark of the trunk is smooth with irregular cracks and is greyish to pale reddish brown in colour. Leafltes are in 4-5 pairs, opposite or sub-opposite and pointed at the base.

Leaves are crowded at the end of branchlets, spirally arranged, alternate, imparipinnate, 20–50 cm long with 5–15 cm long petioles. Inflorescence is an axillary panicle. Flowers are sub-sessile, bisexual, small and creamy white in colour, Sepals are 5, petals are 5, glabrous and ovate to oblong in shape. Stamens are 10 and are inserted below the disk. Disk is deeply 5-lobed, thick and fleshy. Ovary is 5-celled, sessile and superior. Fruit is a drupe, 4–10 cm long, ovoid-oblong, fleshy and yellowish. Pulp is soft, finely flavoured and acidic. Endocarp is woody and is surround by fibres which are interwoven. Seeds are 1–3 of which one is perfect and oblong in shape.

Origin

Exact origin of the species is not known but India or south-east Asian countries have been considered by many to be the origin (Singh and Saxena, 2008).

Geographical Distribution

The tree is much grown in moist sub-tropical parts in many countries, *viz.*, China, Japan, Korea, Laos, Thailand other south-east Asian countries, Bangladesh, Sri Lanka, South Africa, Brazil, Columbia and other south American countries.

In India, the tree is found to grow in West Bengal, Assam, Odisha and the eastern States, Bihar and also in Tamil Nadu, Karnataka, Andhra Pradesh and Kerala.

Climatic Requirement

Tropical climate with high rainfall is most favourable for vigorous growth and high yield of the tree. Frost is a damaging factor when it prolongs for several days. Branches of the tree are highly breakable and high wind causes breakage of branches particularly in case of tall trees.

Soil Requirement

Soil is not a limiting factor for the tree but drainage should be good. The tree is deep-rooted and anchorage of the tree is better when grown in deep soil.

Propagation

The tree is most commonly propagated by seeds. Seeds remain viable for upto a year. Mechanical scarification of the seeds by rubbing between two sand-papers loosen the hard seed-coates and facilitates germination. Among vegetative propagation methods, air-layering gives success when done in rainy-season and particularly by the use of auxin. Stem-cuttings with hardwood or semi-hardwood stem may produce roots when done in rainy-season. Dipping lower ends of cuttings with 200–300 ppm of IBA solutions helps in protrusion of roots in cuttings.

Culture

The tree grows well and produce without care but irrigations during summer is helpful for higher production and improving size of the fruits as well as in retarding fruit drop. Training of the trees at early stage of their growth to ensure wider crotch angle between branches make them to assume better appearance in mature condition and also make them more productive. Pruning during dormant season by removal of dead and decaying shoots has neccessity to produce new growths and admit more sunshine in the inner parts of the trees. Fruit thinning at early stage of fruit growth may be necessary in some trees to increase size and quality of the fruits.

Harvesting

Inflorescences appear in the trees before emergence of new leaves in June–July. Fruits start ripening and become flavescent, in other words, assume yellowishness from end of September to October. It is observed that in most parts of India, semi-ripe fruits which have not assumed yellow colour are harvested. This is because, fruits of such maturity appear to be more suitable materials in cooking, use as condiments and in making pickles. Semi-ripe fruits also emit a pineapple like flavour. However, quality of the fruits is best when they ripe in the trees and pre-ripened fruits do not ripen any more after harvest. Harvesting of the fruit bunches could be efficiently done with the implement as described in the Chapter on longan.

Yield

Exact yield which is obtained from a tree is not known. In general, a mature tree is found to produce 300–500 fruits.

Mode of Consumption

When eaten fresh, salt and spices are added to the fruits as they are very sour. The crispy skin in ripe or semi-ripe fruits can be easily removed to take out the white pulp. Acidity of the fruits decreases with ripeness and fully ripe fruits are preferred by many.

Quality and Nutritive Value

Harish and Vijay (2010) determined nutritive and mineral constituents in ripe fruits of *Spondias pinnata* which were procured from trees of two locations in the western Himalayas. The findings of their study revealed that the crude fat, crude fibre, total carbohydrate, sodium, calcium, iron and copper content were 12.23–12.54, 3.13–4.03, 23.54–16.30, 1.38–0.96, 0.93–0.15, 1.5–1.3 and 0.9–1.23 per cent respectively.

It has been reported from the National Institute of Nutrition (I.C.M.R., Govt. of India) in Hyderabad that 100 gram of fruit containing 90.3 per cent moisture have the following nutritive constituents (Gopalan *et al.,* 2018)

Protein: 0.7 g; Fat: 3.0 g; Carbohydrates: 4.5 g; Crude fibre: 1.0 g; Minerals: 0.5 g; Calcium 36 mg; Phosphorus: 11 mg; Iron: 3.9 mg; Carotene: 270 mcg; Thiamine: 0.02 mg; Riboflavin: 0.02 mg; Niacin: 0.3 mg and vitamin C: 21 mg. Energy value of 100 gram of fruit has been stated as 48 Kilo calorie (=200.8 Kilojoule)

(Index: g = gram, mg = milligram, mcg = microgram).

Nahar *et al.* (1990) studied that *Spondias pinnata* fruits grown in Bangladesh contained appreciable amount of dry matter, lignin, starch and dietary fibres. They pointed out that glucose and fructose comprised as the major components of fruits. Majumdar and Mazumdar (2001) reported pectin contents of the fruits.

Medicinal Value

Hog-plum has greater worth for its medicinal value than as a fruit. All parts of the tree are known to have one or more medicinal properties. In folk medicine in Bangladesh in particular, use of different parts of the tree is very common.

Decoctions made from root, bark and leaves of the tree have been experimentally found to have remedial effect in cholera, liver ailments, diarrhorea, dysentery, stomach troubles and are also helpful to lower blood sugar and blood pressure. Paste made from the bark is known to be efficacious in skin ailments, ringworm and gonorrhoea and in burns. Extract made from the leaves is appetizing and astringent. Fruits have anti-oxidant activity and are also known to have anti-cancerous property.

Phytochemicals

Plant parts of the hog-plum tree contain a number of chemical compounds (Bora *et al.,* 2014). Among many compounds, different parts of the tree are known to contain flavonoids, saponin, tannin, terpenoids, sterols *etc.* Beta-amyrin, oleanolic acid and other compounds are found to be existent in fruits.

Constituents in Developing Fruits

With fruit samples obtained from trees grown in the central part of West Bengal, Bhattacharyya *et al.* (1992) determined some constituents at three stages of development, *i.e.,* later part of April, early part of May and middle of June. It was observed that the total soluble solids, total, reducing and non-reducing sugars and vitamin C content in fruits had gradually increased. Acidity level of the fruits as equivalence of citric acid had increased initially but declined later insignificantly.

Utilization

In most parts of India where hog-plum trees are grown, the fruits are generally utilized in making pickles. For this purpose, the fruits are at first halved or made into splits with knife and then dipped in weak emulsion of lime water for 15–20 minutes to reduce the acidity. Then the materials are repeatedly washed in clean and preferably running water to wipe out the lime particles.

The washed materials are dried in sun for few days to make them half-dried and then processed to make pickles. Pickling is done either using much of mustard oil or using very little of it. In both methods, adequate amount of ginger paste, various types of spices and condiments, green chilli and if required, little amount of sugar, lime juice and black pepper are added. After mixing thoroughly, the product is heated in slow heat or dried in sun for 1–2 weeks and the pickle become ready for use.

Apart from pickling, other products of hog-plum are made in different countries. The fruit pulp is mixed with pulp of other fruits and made into jelly or jam. A thickened type of sweet-sour sauce is known to be made in south-east Asian countries.

Apart from making pickles, in West Bengal, other eastern States and also in Bangladesh, the fruits are boiled with molasses after removing the peels. Sometimes immature fruits are broken into pieces and thoroughly mixed with pieces of green chilli and salt and preserved for short duration. In this zone, making juicy type of sour preparation known as *ombol* is very common and this is taken with rice.

Family: Annonaceae

Species: *Annona reticulata* Linn.

Taxonomic Serial: 18100

Common Name

In English, the species *Annona reticulata* is most commonly known as Bullock's heart. However, other names that are used in English are Sweet sop, Ox heart, Bull's heart, Honey apple, Sugar apple *etc.*

In other languages, Bullock's heart is known by other names. For example, in French, it is termed as Bois cachiman, Coeur de boeuf, Corossol sauvage *etc.*, in Portuguese, Aondesa and Coracao de boi, in Spanish, Anona de seso, Anona roja, Anona colorada *etc.*, in Malayan, Nona kapri, in Vietnamese, Qua na *etc.*

In Sanskrit, Bullock's heart is known as Seetaphala, Sudha, Subha *etc.* In Hindi, the fruit is called Ramphal, Seetaphal, Luvun, Nonai *etc.*, in Malayalam as Atha, Athachakka, Seetapazaham, Manilanilam *etc.*, in Kannada, Marathi and Gujarati as Ramphal, in Tamil as Ramchita, in Oriya as Ramopholo and in Bengali as Nona on or Nona ata. Other names that are used in Indian languages are Krishabeeja, Ramaseeta, Mamapandu, Anoda *etc.*

Botanical Family: Annonaceae

Annonaceae is the largest family under the Order, Magnoliales and consists of 130 genera or more and about 2500 species. Among the species, about 900 are Neotropical, 450 are Afrotropical and the rest are Indo-Malayan. Edible fruits are obtained only from few species and notably, from *Annona, Anonidium, Rollinia* and few other genera. Most of the species abound in tropical and subtropical parts and few are grown in temperate climate.

Genus: Annona

The name of the genus, Annona has derived from anón, which is a Hispaniolan Taino word for the fruit. Annona is a very large genus which is next to Guatteria in the family, Annonaceae. The genus contains about 166 species which are Neotropical and Afrotropical trees or shrubs. Only a few species have importance in producing human consumable fruits and there are some species which have commercial importance.

Importance

Fruits belonging to the genus, Annona are merited in that, in accordance with species, they give fruits all throughout the year. Thus, while the fruits of bullock's heart (= *Annona reticulata*) are available in summer season, those of *A. muricata* in rainy-season, *A. squamosa* in autumn while the *A. cherimolia* gives fruits in the winter season.

As regard quality, fruits of custard apple (*A. squamosa*) are usually given most important status at least under Indian condition although on global basis, fruits of cherimoyer (*A. cherimolia*) which are characterized by nice flavour are considered to be the best of the genus, *Annona*. Fruit of this species is even regarded by many as one of the best subtropical fruits in the world.

Evidently, the fruits of bullock's heart do not stand comparison with custard apple or cherimoyer in respect of quality but compactness of the pulp and much less number of seeds are the special attributes of these fruits. Among annonaceous fruits, bullock's heart is grown next to custard apple in India (Hayes, 1970).

Popularity of the fruit of bullock's heart is particularly high in West Bengal where plantations of this fruit crop are even raised in soils of low agricultural suitability.

The fruit is valued and appreciated in some other countries as well and notably in Maxico and the West Indies. Apart from the fruit, the trunk, leaves, bark and seed of the tree have usefulness in commerce and medicine.

Description of Plant Parts

Trees of bullock's heart are more vigorous than those of other species of Annona. Tree is erect, crown is rounded or spready, shape is irregular and not very attractive in appearance. Life-span of the tree is short. Habit is deciduous or sometimes semi-evergreen. Height of the tree is 6–10 metres and the crown spreads from 5 to 10 metres. Trunk of the tree is 25–35 cm thick.

Leaves are simple, alternate, arranged opposite to one another and mildly green in colour. They are oblong, narrow, lanceolate, 10–20 cm long and 2–7 cm wide. They emit ill smell.

Flowers are yellowish green and in drooping clusters of 3 or 4. They have fragrance. Petals are slender with 3 long outer ones which are fleshy and 3 inner petals which are very small. Flowers never fully open.

Fruit is compound and is heart-shaped (hence the name, bullock's heart) to cylindrical or somewhat spherical in shape and has a depression at the base. It is lopsided or irregular. A fruit (Plate-2) measures 10–15 cm in length and 7.5–12.5 cm in width. A pointed, fibrous, central core is attached to the thick stem which extends about half or more of the fruit pulp. Two types of bullock's heart fruits are in fact, recognized, which are squamosa, *i.e.*, heart-shaped in appearance and the other is of reticulate form, which has distinct finger-prints on the surface (Sham Singh *et al.*, 1967). Peel of the fruit is thin, tough and brown or yellowish in colour when ripe. Pulp is creamy white and somewhat granular. The carpels in the fruit are fused together to form a rind, marked with hexagonal areoles (Hayes, 1970). Pulp of the fruit is much less sweet than custard apple but compactness of the pulp and lesser number of seeds, which may be 50 to 70 in a fruit are its merit. Seeds are like those of custard apple, *i.e.*, deep brown or black, glossy, oblong, smooth and each seed measures 1–1.5 cm in length.

Origin

Bullock's heart is originated in West Indies, although some authors have stated to be of Mexican origin. However, from West Indies, it has been stated to be migrated to Mexico, central America and then to south America in very early time.

Geographical Distribution

Bullock's heart grows in almost all tropical countries in the world. Important countries where it is largely grown are those in West Indies, Mexico, Guatemala, Ecuador, Peru and other south American countries, southern parts of the U.S.A., India, Sri Lanka, Bangladesh, south-east Asian countries, French Polynesia, Guam, many parts in Australia *etc.*

In India, the fruit is grown in all tropical parts and the trees are most abundantly grown in West Bengal, Odisha and Assam. In West Bengal, it is a highly popular fruit.

Climatic Requirement

Bullock's heart is primarily a tropical tree and its performance is best in the warm and humid tropics. However, the tree is cold-resistant to a considerable extent and grown in higher altitude in tropical climate. For example, it grows well upto 1500 metres in coastal lowlands of Ecuador, upto 1220 meters in Guatemala while upto 915 metres in Sri Lanka. In India, the tree is found to grow and fruit upto an altitude of 800 metres. At higher altitude, the trees shed leaves and remain dormant in winter season.

It may however, be said that although the tree is capable to grow at higher altitude in tropical countries, its performance is not so high when grown at lower elevation in tropical situations. As regard rainfall, 1200 mm annual rainfall is considered optimum for the tree. Yield and quality of fruits are also high when the trees are grown in open sunny situation.

Soil Requirement

Soil is not an important factor for growing bullock's heart and it is found to grow even in inferior soil. The tree is resistant to drought as well. In the southern part of Florida in U.S.A, bullock's heart is grown even in oolitic limestone and in light soil. Nevertheless, yield and quality to the fruits are very high when grown in deep, rich and well-drained soil. In West Bengal, fruiting performance of the trees which are grown in alluvium soils is found to be very satisfactory than when grown in inferior lateritic soil.

Propagation

Seed propagation is easy and very common method in bullock's heart. Seeds remain viable for more than a year, if stored in air-dry storage at 5° C. Improvement in seed germination and growth of seedings especially by application of gibberellic acid has been stated by Cartagena Valenzuela and Barreto Osoria (1998). Scarification of the seeds is a good practice to improve germination. It may be done by rubbing the seeds between two sand-papers for sometime. Dipping seeds in concentrated sulphuric acid for 1–2 minutes may also be done to scarify seeds.

Bullock's heart is also propagated by budding and grafting on rootstocks raised from seedings of bullock's heart, or custard-apple or pond-apple. In Mexico, Marcelino Ponce (1986) studied the performance of side-grafting and oblong wedge-grafting methods in bullock's heart by grafting on eight rootstocks. It was observed that percentage take in the grafting was maximum (100 per cent) and growth of scions was also best when side-grafted on *Annona muricata, A. reticulata* and *A. lutescens* rootstocks.

In a study aimed at making comparison of three budding methods in bullock's heart, Gholap *et al.* (2000) at Akola, Maharashtra observed that highest success (91.5 per cent) was achieved in softwood-grafting, followed by patch-budding (89 per cent) and shield-budding (80.5 per cent).

Apart from budding and grafting, air-layering is also done in bullock's heart with high rate of success. Satisfactory performance of air-layering has been observed by Chovatia and Singh (2000) in Gujarat by the use of auxin in suitable concentrations.

Culture

If plantation is desired to be set up with bullock's heart, a spacing of 4–6 metres should be maintained between the trees. No culture is usually done for the fruit crop but it is observed that yield performance and quality of the fruits are markedly increased by manuring, mulching and irrigation. Sharma *et al.* (2012) have stated in their book that for bullock's heart and other *Annona* fruits when grown at a spacing of 5 metres should be manured with farmyard manure @ 10 kg and fertilized with nitrogen, farmyard manure @ 10 kg and fertilized with nitrogen, phosphorus and potash respectively as 250, 125 and 250 gram per tree per year.

In studies done with determination of leaf sampling technique, Dhandar and Bhargava (1993) observed that sampling tissue from the 5th youngest leaf gave the best index for nutrition in bullock's heart. With pot-culture experiments done in China, Zang and Xu (2000) stated that nitrogen fertilization was best in bullock's heart and this was followed by potassium and calcium while phosphorus and magnesium application had not much effect. Mandal, U. and Mazumdar, B.C. (unpublished) observed that fertilization had greatly enhanced yield and quality of fruits and the effect of potassium had been very high. Spraying zinc sulphate solution (0.5–1.5 per cent) was also found by them to have made increase in size of fruits and their sugars content. Pruning is an important operation in custard apple and pollarding gives rise to production of new shoots.

Pests and Diseases

Fruits of bullock's heart are heavily attacked by the insect, fruit fly when the fruits are ripe. Bats also cause much damage to the fruits and as protective measure, fruits or the trees may be covered with suitable nets or bags.

Mummification, *i.e.*, drying and hardening of fruits due to fungal infection is a common occurrence. Long ago, Chowdhury (1947) from the Sylhet district in the then Bengal reported dry rot of the fruits which had spoiled 15–20 per cent of the crop in a plantation. The causal organism had been stated to be the fungal species, *Glomerella cingulata*. According to him, this had been controlled by application of Bordeaux mixture (2-2-50) when the symptoms appeared or during late December.

Harvesting

Bullock's heart requires very long time to develop and ripen fruits from their set. The trees flower in August to October and fruits are ripened and harvested in April to May and thereby, 7–8 months are required from the time of flowering to harvest of fruits.

Fruits are best ripened in the trees but in ripened condition, they become very soft and are difficult to handle. Hence, Morton (1987) has suggested to harvest the fruits slightly earlier than when they are ripe, *i.e.*, when the green colour of the fruits has disappeared and without splitting. However, ripening of fruits does not take place any more if they are harvested too early. Fruit-stalk of bullock's heart should be clipped with a suitable implement and preferably with a clippers as stated under the chapter on longan.

Yield

Yield of the fruit varies greatly and is very high if agro-inputs are used. Normally a tree gives rise to 25–50 fruits and each fruit may weigh 250–350 gram. With a tree and row spacing of 5 metres, 1.2–1.5 quintal fruits may be obtained in a hectare area of land.

It is generally observed that the fruits which are of reticulate form bear much more than those which are of squamosa form, *i.e.*, heart-shaped. Yield is higher in some hybrids.

Varieties

Although much variation is found in the character of trees and fruits, named variety of bullock's heart is hardly known. Fruits of two types are however, recognized, which are squamosa form and reticulate form. Reticulate form of fruits are usually more productive. Colour of the fruit peel is considered by many to make groups and it is claimed that fruits which have yellow peel are of superior quality than those which are of brown colour. Some hybrids of bullock's heart have been developed. Pino *et al.* (2003) in Cuba had mentioned four varieties, *viz.*, Cenizo, Rojo, Verde and De Ojo with which they had conducted studies aiming at isolating volatile compounds.

Mode of Consumption

Bullock's heart fruits are almost exclusively eaten fresh. Taste of the fruit is not so good as compared to the custard apple and hence, to make the pulp sweet, many people in West Bengal add sugar to it, make a paste of the pulp and then take.

In Mexico and central America however, although the fruit is taken fresh, but many people suitably process the pulp and take. Scooping out the pulp and blending with light cream and sugar is a practice in those countries. The pulp is also pressed through a sieve and added to milk shakes, custard or ice-cream. Mashing the pulp with banana pulp is also a practice in Mexico.

Quality and Nutritive Value

Many people do not like this fruit perhaps for the fact that the taste of custard apple and cherimoyer comes to their mind while taking the bullock's heart fruits and they attempt to make a comparison. However, it should be kept in mind that every fruit has a characteristic taste of its own and a tamarind cannot be compared with a ripe mango fruit of the Alphonso, Chowsa, Himsagar or Langra variety as regard taste.

Bullock's heart is predominantly a starchy fruit. The pulp of the fruit contains more than 18 per cent starch of granular characteristics and is comparable to that of starch derived from corn.

The proximate principles per 100 gram of the fruits as put forward by the National Institute of Nutrition, (I.C.M.R.) Govt. of India at Hyderabad (Gopalan *et al.* (2018) presented below:

Moisture: 76.8 g; Protein: 1.4 g; Fat: 0.2 g; Minerals: 0.7 g; Crude fibre: 5.2 g; Carbohydrate: 15.7 g; Calcium: 10 mg; Phosphorous: 10 mg; Iron: 0.6 mg; Carotene: 67 mcg; Riboflavin: 0.07 mg; Niacin: 0.6 mg; Vitamin–C: 5 mg; Energy derived: 70 Kilocalories (= 292.88 K Joules).

(Index: g = gram, mg = milligram, mcg = microgram)

Medicinal Value

All parts of the bullock's heart tree including root, bark, leaves and fruits have been on use in medicines since ancient time. As regard fruit, it may be said that ripe and unripe fruits of bullock's heart are used as medicines for various ailments and some of these are intestinal worms, diarrhoea, dysentery, constipation, vomiting, cough, anaemia *etc.* Kaladhar and Rayavarapu (2014) stated phytochemical profile, antioxidant and antimicrobial activities of raw fruit peel of bullock's heart, collecting samples in Jami mandal, Vizianagram.

Phytochemicals

A number of phytochemical compounds are known to occur in fruit, seed, root, leaf and bark of bullock's heart and few of these may be mentioned as acetogenins, anonaine, bullacin, carvone, corydine, linalool, motrilin, muricine, muricinine, phenolic acid, reticulatacin, roemerine, squamocin, squamocenin *etc.*, among many other compounds. Chang *et al.* (1998) reported acetogenins and other compounds from chromatographic studies done with seeds of bullock's heart. From four cultivars of bullock's heart fruits, Pino *et al.* (2003) could be able to isolate volatile compounds in Cuba. Kaladhar and Rayavarapu (2014) reported alkaloids, cholesterol, coumarins, flavonoids, saponins, phenols *etc.*

Constituents in Developing Fruits

Studies have been done by Bhattacharya *et al.* (1992) on the changing pattern of some constituents in fruits of bullock's heart during their development, borne by the trees grown in the South 24–Paraganas district of West Bengal. Fruits were sampled at three stages of development, which were later part of April, early part of May and at harvest-maturity and the total soluble solids, titratable acidity as citric acid equivalence, total, reducing and non-reducing sugars and the vitamin C were determined at each stage. Increase of all the constituents had been observed by them with progress of development of the fruits except the acidity level which consistently declined. The rate of increase of the constituents was higher from early April to early May.

Undesirable Effect

Many toxic substances are found in bullock's heart. Morton (1987) stated that the seeds of bullock's heart were very hard and although they might be swallowed whole with no ill effects but the kernels were very toxic. According to her, "The seeds, leaves and young fruits are insecticidal. The leaf juice kills lice. The bark contains 0.12 per cent anonaine. Injection of an extract from the bark caused paralysis in a rear limb in an experimental toad. Sap from cut branches is acrid and irritant and can severely injured the eyes. The root bark has yielded three alkaloids, *viz.*, anonaine, liriodenine and reticuline (muricinine)."

Post-harvest Storage

Bullock's heart fruits cannot be stored for more than a few days under ordinary atmosphere. In containing high amount of starch, the pulp becomes very soft and rot due to catabolism of starch although sweetness is slightly increased in them. Fruits may however, be stored for 2 to 3 weeks, maintaining a temperature of 4°–6°C and 85–90 per cent relative humidity in the storage chamber.

Utilization

Although for enhancing taste, the fruit pulp of bullock's heart is sometimes processed with addition of sugar as has been stated earlier, making preserved products from the fruits is not a common practice. Sometimes juice is extracted by squeezing the fruit pulp through a sieve. The juice is added with sugar, permitted amount of chemical preservative, flavouring and colouring agents and bottles. This may be used as a drink or used in desserts of various types, pudding, ice-cream and sweetmeats. But the preparation does not keep well for long time.

Drying of the pulp in home-scale is a practice which is done in some parts of West Bengal. Pulp of bullock's heart is blended with soft pulp of other fruits, *e.g.*, banana, sapota, jackfruit *etc.*, made into a paste with addition of sugar and on making small pieces, dried in sun for several days until they become hard. These are taken like Banana figs and are used as snacks.

Family: Apocynaceae

Species: *Carissa carandas* Linn.

[Synonyms: *Carissa congesta* Wight; *C. carandas* Auct.; *Arduina carandas* (L.) K. Schum.]

Taxonomic Serial: 30164

Morton (1987) quoted that *Carissa carandas* was formerly believed to have 2 distinct varieties (= sub-species), *viz.*, *Carissa carandas* var. *amara*, having oval, dark-purple, red-fleshed fruits of acid flavour and *C. carandas* var. *dulcis* having round, maroon, pink-fleshed fruits of sweet-subacid flavour. However, David Sturrock, the horticulturist in Florida, U.S.A. ruled out this and opined that those were in fact, variations in the seedlings.

Apart from *C. carandas*, some other species of Carissa are also of horticultural importance. These are, *C. inermis* Vahl. (*C. macrophylla*), *C. grandiflora* D.C., *C. spinarum* L., *C. edulis* Vahl., *C. spindrum* D.C., *C. ovata* etc.

Common Name

The common names of the species in English are Christ's thorn, Bengal Currant and Karanda. In Thailand, it is called Namphrom and Namdaeng and in Malaysia, Karenda.

In India, there are several names in different languages and some are as follows:

Sanskrit: Kamrad, Kamardaka and Sushena. Hindi: Karonda. Assamese: Karja tenga. Bengali: Karamcha. Oriya: Kerendokuli. Telugu: Vakkay, Peddakalavi. Tamil: Kalakai, Kilakkai, Karamcha, Kalakkay. Kannada: Karjige, Karekaya. Gujarati: Karamdaa. Malayalam: Karakka.

Botanical Family

Apocynaceae which is also called Dogbane family of the Order, Gentianales consists of more than 424 genera and about 1500 species which are herbs, stem succulents and vines having nativity to Europe, Asia, Africa, Australia and tropical and sub-tropical parts of America. The former family, Asclepiadaceae is included in this family according to the APG III system. The family is divided into several sub-families. There are about 1500 species under this family and many of them have economic importance. Most of the species exude milky latex when injured and some species are poisonous if taken by mouth.

Leaves of the species belonging to this family are simple, usually opposite and decussate or whorled and usually without stipules. Inflorescence is terminal or axillary and the flowers are showy, actinomorphic and bisexual. Calyx is 5–lobed and united into a tube at the base. Petals are 5 and united into a tube with 4–5 epipetalous stamens. Style is expanded at the apex into a clavuncle just below stigma. Ovary is superior, bicarpellary and apocarpus. Fruit is a drupe, berry, capsule or follicle.

Genus: Carissa

The genus, *Carissa* consists of about 8 species and many species may be synonyms or are assigned to other genera, *e.g., Acokanthera*. The species belonging to the genus, *Carissa* are shrubs or trees which are 2–8 metres tall and consist of sharp thorns, *i.e.*, modified stem and not to be considered as spines. The thorns may be simple, forked or dichotomously branched. Leaves of the species are 2–8 cm long, green, smooth, glossy, simple, entire and elliptic to ovate in shape. Flowers are 1–5 cm wide with a 5-lobed corolla and are solitary or are borne in clusters in umbel or corymb. In many species, flowers have a pleasant fragrance. Fruit is a prolate or spheroid shaped berry and is of red colour when ripe in some species. Seeds are brown and 1–4 but in some species may be more. Ripe fruits are sour and are mostly used to make preserved products. Some species are used as highly protective, strong hedge plants and some are planted in gardens for ornamental value.

Importance

Karonda finds favour mostly as a protective and strong hedge plant. The plants contain large and strong thorns and are bushy in nature. Hence, it makes a highly protective hedge plant which is impermeable to the animals and knaves. The plants also produce fruits which are consumable to the human beings. Therefore, it is a dual purpose plant.

Fruits are utilized in many ways. In southern India, West Bengal and the eastern States, fruits are used in culinary purpose and cooked to make sour preparations. Excellent jelly could be made from the fruits as they contain high amount of pectin which is of high jelly-grade. The tree is graceful in appearance, leaves are showy, flowers are fragrant and the fruits are of attractive reddish and purplish colour for which, it is planted in decorative gardens and also used in making topiary. In West Bengal, a product is made from the berries which has a resemblance of cherry and has a high demand. Karonda trees are also effective in conservation of soil against erosion.

Besides the above attributes, fruits and leaves of karonda have medicinal properties and have valuable phytochemicals (Morton, (1987); Devmurati *et al.*, 2010; Itankar *et al.*, 2011; Kumar *et al.*, 2013; Mehmood *et al.*, 2014; Merai and Jadhav, 2014).

Description of Plant Parts

Karonda is a straggly, woody, semi-vine or upright small tree which is evergreen (Plate-3) and attains a height of 3–5 metres if not trimmed. Roots of the plant are highly branched. Branches are numerous, rigid, spready and produce straight, simple or forked thorns. These arise on axil and sometimes may be very long of upto 5 cm. Bark is light grey in colour.

Leaves are shiny, opposite, dark green, leathery, 4–6 cm long, glossy on the upper surface and dull on the underside. They are simple, exstipulate, elliptic or broadly elliptic or obovate. Base of the leaves is obtuse to rounded, margin is entire, tip acute and often shortly mucronate and unicostate. If leaf or stem is injured, milky sap oozes out. Flowers are 2–3 cm long, tubular, white, borne in clusters of 2–3 and fragrant. Corolla is twisted the left in the bud. Fruit (Plate-5) is 3–5 cm long with tough skin which is smooth and glossy and taste of the fruit is sour or sometimes semi-sweet.

Origin

Origin of karonda is believed to be India or the region from India to Java (Singh and Saxena, 2008) while other species of *Carissa* of horticultural importance might have been originated in South Africa, Malaysia or in other countries in the south-east Asia.

Geographical Distribution

Karonda grows widely in all tropical and subtropical parts of India (Mazumdar, 2004 a) and is much grown in south India, West Bengal and the eastern States. Regular plantations of karonda are found in Varanasi in Uttar Pradesh (Chadha, 2012). Outside India, it is extensively grown in Thailand, Malaysia and other south-east Asian countries, Myanmar, Bangladesh, rain forests in Sri Lanka and also in east African countries and South Africa.

Climatic Requirement

The plant thrives in tropical and sub-tropical climates and tolerates low temperature of upto 5° C for some period. It grows in the Nilgiri hills in south India but the performance is not good. Full sunshine is required by the plant for better quality of fruits.

Soil Requirement

Soil is not a very important factor for karonda and it grows in a wide range of soil. In India, it is grown as a hedge plant in many types of soil including dry, sandy and rocky soil and in Florida in the U.S.A., it is found to grow in sandy soils and limestone. However, productivity of the karonda plant is high and fruit quality is better when grown in deep, fertile and well-drained soil. In wet soil, vegetative growth is high and productiveness is low.

Propagation

Karonda is propagated from seeds in most parts of India. Seeds remain viable for a year or more. Germination of the bold seeds is fairly good and is normally 70 per cent or more but treatment with gibberellic acid has been stated to enhance germination percentage and vigour of the seedlings (Banker, 1987). Seeds are sown in the nursery bed at the end of rainy-seasons and seedlings are transplanted after 3–5 months or more. Direct sowing of seeds *in situ* is also common in India to build up rows of hedges. Seed propagation is an easy and cheap method and one kilogram may consist 500–600 seeds.

However, seed propagation of karonda is confronted with two problems. These are, firstly large variations are found in the resulting plants in their fruit and plant characters and secondly, growth rate of the seedling plants becomes slow (Hayes, 1970). Enhancement of seedling growth by gibberellin treatment had been stated by Banker (1987). Subsequently, Misra and Jaiswal (1998) studied the effect of applying 250, 500, 750 or 1000 ppm of GA_3 solutions on the growth pattern of 9-month old seedlings. It has been observed by them that both height and stem diameter of the seedlings had increased with increasing concentrations of the growth-regulator solution applied.

Stem-cuttings do not produce roots ordinarily but treatment with auxins in suitable concentrations have been reported to promote rooting (Bandopadhyay *et al.* 1983, Bandopadhyay *et al.,* 1999). Sanjay Tyagi *et al.,* 1999). Singh and Saxena (2008) have stated that cutting of 2–3 month old terminal shoots treated with 800 ppm of IBA in lanolin paste resulted in higher extent of rooting. The author (Mazumdar, B.C., unpublished) at Baruipur in West Bengal observed that the leafy shoots which were obtained on heading-back the thick branches struck adventitious roots in high propagation when maintained in a crude type of mist-chamber after application of NAA (100–200 ppm) or IBA (200–300 ppm).

Mound or stool layering as is done in guava is found to give success in karonda and many rooted plants are obtained by the method. Misra and Jaiswal (1993) studied stool-layering by application of varying concentrations of IBA and NAA and they observed best results by treatment with IBA at 10,000 ppm. As compared to other methods, root regeneration takes place more easily in karonda by air-layering, particularly when thicker branches are girdled. Misra and Singh (1960) reported success in air-layering using IBA and NAA and obtained maximum rooting by the use of 5000 ppm of IBA. It has been observed by Ghosh *et al.* (2011) under West Bengal condition that rooting success in air-layering was significantly influenced by season. Highest rooting success of 90 per cent was observed by them when the operation was done on June 15 and this was followed by June 30.

Grafting by inarching has also been stated in karonda but to a limited extent (Sham Singh *et al.*, 1967). Ghosh *et al.* (2011) studied the effect of season on the success of chip-budding and wedge-grafting. According to them, highest budding success of 60 per cent was observed when it was done on June 15 and grafting was most successful when carried out from the middle of July to the middle of September.

Micro-propagation of karonda has been studied with success by some workers, Rai and Misra (2005) cultured shoot tips on MS basal medium supplemented with BA and IBA during different seasons. Maximum sprouting was observed by them with 1.5 cm long explants collected in spring seasons followed by those collected in summer. Shoot proliferation was highest on MS basal media supplemented with 3 mg/$^{1-1}$ BA. Micro-propagation of Asam karonda was studied by Hasmah *et al.* (2013), in Malaysia. The results revealed that MS medium supplemented with 2–4 mg/litre of BA inducted an average of 3.6–5.5 shoots per explant but the number of multiple shoots formed was not significantly different. Addition of IBA with BA had however, shown a reduction trend in the number of shoots being formed. The MS medium supplemented with 2 mg/litre of IBA was found to be effective for rooting of micro-shoots. After the acclimatization process, 84 per cent of the *in vitro* plantlets were found to have survived but there was a reduction in the number of survived plantlets after 4 weeks of incubation in green-house.

Culture

If hedge-row is necessary to be set up with the plants, it is better to dig a trench of 15–20 cm depth and 10–15 cm width, fill it with soil manure mixture, irrigate to settle the soil and then sow the seeds or transplant the seedlings in the filled in trench. If seeds are sown, the seedlings that will emerge should be thinned out to maintain a spacing of 90 cm to 1 metre. Spacing may be increased upto 1.5 metre if the hedge plants have been planned to be regularly trimmed. Similar spacing should be given in case of transplanting of seedlings.

Where exclusive planting is necessary to be done, it is better to dig a pit of 50–60 cm cube for each plant, fill it with manure and soil and sow the seed or plant the seedlings in the filled in pit. Sowing many seeds at one place and thin out the resulting seedlings in instalment retaining ultimately the most vigorously growing one would give better result.

Karonda is a hardy plant only little care is necessary to the plant but regular removal of suckers and trimming of dead and dried shoots are of much help before growth of the plants. At early stage of growth, training the plants by single or double stem maintains them in good shape and annual pruning of the plants by the act of heading-back of thick shoots and thinning-out of shoots which are overcrowded should be done to admit more sunlight in the inside part of the tree which increases productivity of fruits. Lopping of branches and young stems leads to production of new shoots in karonda. Manuring and irrigation at early stage of growth are helpful to boost up growth of the plants as they are slow growing in nature when young. Efficient floor management of karonda plantations markedly steps up yield and quality of fruits. The level of nitrogen, phosphorus and potash in the leaves of karonda at monthly interval from December to May have been analyzed by De *et al.* (1996) under West Bengal condition. Sharp increase had been observed upto the month of March in case of nitrogen and potassium and upto February for phosphorus. The components steadily declined thereafter and intensively from April to May.

Pests and Diseases

A number of fungal diseases are found to appear in karonda. Anthracnose caused by the fungus, *Colletotrichum inamadarii* had been stated by Lal and Singh (1953) long ago in Uttar Pradesh. According to them, the fungus invades through the wounds of leaves, produces spots on the leaves and result in their death. Die-back of shoots due to infection by the fungal species, *Diplodia natalensis* and stem canker by the attack of *Dithiorella* species are observed in karonda at some places. A type of fruit rot in karonda has been reported by Kumar and Tandon (1982). Kumar and Patel (1982) reported a bacterial disease in the plant.

Harvesting

Karonda plants start bearing after 2 to 3 years of growth. The plants flower in the month of February and the fruits are ready for harvest from August to October. Many flowers are produced before February and the fruits produced by these flowers ripen in rainy-season. Ripe fruits assume dark purple, maroon or reddish purple in colour. Harvesting of the ripe fruits is done in 2–3 instalments. Fruits are also picked before they are fully ripe when the colour is pinkish. In some parts in south India, flowers and fruits are produced in 2–3 times in a year. Harvesting is done by manual clipping of the fruits. Yield of fruits greatly varies on many factors.

Varieties

Karonda has no established varieties. Plants bearing fruits of two types are commonly found, one having dark purple fruits and the other of creamy colour with pink brush (Mazumdar, 2004 b). Diversity in plant characters of karonda had been stated by Mehra and Arora (1982) in plants grown in Mount Abu, Khandala and other parts of north-western India. Karale *et al.* (1989) identified some promising types in the plants grown in Maharashtra. Improved types of the species are found to grow in the karonda plants that are much grown in Varanasi area in Uttar Pradesh. (Mazumdar, B. C. unpublished) had observed fruits and leaves of varying shape in the plants that grow to a high extent in the Baruipur area under the South 24-Paraganas district in West Bengal.

However, at the G.B. Pant University of Agriculture and Technology at Pantnagar in Uttaranchal, some selections of karonda have been made. These have been named at Pant Manohar, having dark pink coloured fruits on whitish background, Pant Sudarshan, having pink coloured fruits on white background and Pant Suvarna, fruits of which are light brown with green background and of upright and bushy nature of growth of the plants.

Mode of Consumption

Karonda fruits being sour are not taken as fresh and hence, are used for culinary preparations and for making preserved products. However, fresh fruits suitably mixed with sugar, salt, black pepper or cumin are also found to be sold by road-side hawkers in many parts of India and these are purchased and relished by the children.

Quality and Nutritive Value

From the National Institute of Nutrition (I.C.M.R. of the Govt. of India) in Hyderabad, it has been stated that fresh fruits of karonda contain 91 per cent moisture, 1.1 per cent protein, 2.9 per cent fat, 0.6 per cent minerals, 1.5 per cent crude fibre, 2.9 per cent carbohydrate, 0.021 per cent calcium, 0.028 per cent phosporous and 42 K calorie (= 175.7 K Joule) of energy is obtained on consumption of 100 gram of pulp (Gopalan *et al.,* 2018). The composition of dried berry of karonda has been stated by Sham Singh *et al.* (1967) and Gopalan *et al.* (1993).

As regard suitability of karonda fruits in making jam, jelly, chutney, pickle and preserve, Singh and Singh (1998) from their analytical studies opined that the white-yellow coloured fruits which were less sour in taste were more suitable than the red-green fruits. The studies done by Sinha Roy and Mazumdar (1996) and Majumdar and Mazumdar (2001) under West Bengal conditions revealed that the fruits contained significantly high amount of pectin of high jelly-grade. Oil and some other compounds from the fruits, leaves and flowers of karonda have been reported by Chandra (1985), Naim *et al.* (1988) and Sekar and Francies (1998).

Medicinal Value

Karonda is known to have many medicinal properties. Kumar *et al.* (2013) have mentioned in their review that all parts of the plant are of medicinal value and are used in traditional medicine. Fruits are rich in iron and thus, is helpful in anaemia. They are also of help in biliousness and have anti-cancerous activity. Anti-diabetic potential of the extract made from unripe fruit extract of karonda has been reported by Itankar *et al.* (2011). Decoction of the leaves is helpful in fever, diarrhoea, oral ulcer, ear ache and microbial infections in throat. Root is stomachic, vermifuge and cardiotonic. Anti-ulcer activity of the root extract has been studied by Merai and Jadhav (2014).

Constituents in Developing Fruits

The changing pattern of constituents taking place in developing fruits of karonda had been studied by Uthaiah (1988). Sinha Roy (1996) undertook a study by spraying karonda plants grown in West Bengal with solutions of varying concentrations of nutrient salts, *e.g.*, urea, potassium chloride, magnesium sulphate, copper sulphate, zinc sulphate, sodium tetra-borate and plant growth regulators, *e.g.*, IBA, GA_3, Ethephon and kinetin after 7–10 days of fruit-set and the treated and untreated fruits were analyzed at four equally spaced stages of their development till attainment of harvest-maturity condition. Changes in the content had been seen to have taken place for a number of fruit constituents among which calcium pectate content increased particularly with potassium chloride application. Results of the changing pattern of physico-chemical constituents of the fruits by spraying the trees grown in the same agro-climatic zone with solutions of nutrient salts and growth regulator had also been stated by Mandal (1997) and Mandal and Mazumdar (2008).

Undesirable Effect

Experiments have shown that root extract of karonda have the property of releasing histamin (Joglekar and Gaitonde, 1970; Kumar *et al.*, 2013). As the fruits of the plant are consumable to the human beings, existence of the compound in fruit needs in fruit need to be examined as it acts as a strong allergen to them.

Utilization

A number of preserved products are made from the berries of karonda. In containing high amount of pectin of high jelly-grade, high quality jelly is made (Sinha Roy and Mazumdar, 1996; Mazumdar, 2009) from them that have importance in commerce. Jam is also prepared from the berries and shelf-life of the jam under ambient temperature has been studied by Wani *et al.* (2013). Making pickles from the unripe fruits and drying of the ripe fruits are very common in India. Johar and Anand (2003) have studied in regard to brining of the fruits. In West Bengal, slightly immature berries are suitably processed, dulcified and imparted red colour and a cherry like product is made (Plate-4) which is used in sweetmeats, puddings and confectionary (Mandal *et al.*, (1992). Many cottage- industries have developed particularly in the rural areas of the South 24-Paraganas district in West Bengal, centering on making the product (Plate-6). In some countries, squash and partially fermented wine are made from the water extract of karonda fruits.

Family: Araceae

Species: *Monstera deliciosa* Liebm.

[Synonym: *Monstera lennea* Koch.]

Taxonomic Serial: 503872

Common Name

The plant is known by various names in English and other languages. Some of the more commonly used names are as follows.

Ceriman, Swiss cheese plant, Cheese plant, Monster fruit, Monsterio, Monsterio delicio, Fruit salad plant, Monstera, Mexican breadfruit, Locust and wild honey, Delicious monster, Balazo, Banana-pineapple, Penglai-banana, Window leaf, Split-leaf Philodendron (N.B.: Philodendron is an incorrect name since the genus, *Philodendron*, although is the species of both genera have some resemblance).

In India, there is no specific name of the species but many people call it, Amar phal.

Botanical Family

The genus *Monstera* comes under the Family, Araceae which is known as the Arum family and belongs to the Order, Alismatales. In this Family, *Monstera deliciosa* comes under the sub-family, Monsteroideae and the tribe, Monstereae.

Araceae consists of monocotyledonous phanerogams and the flowers of the plants are borne on spadix. The inflorescence, spadix is accompanied by or covered in leaf-like bract (spathe). The family is known to contain more than 110 genera and 3700 species. Most of the species are rhizomatous or tuberous and contain calcium oxalate crystals (raphides).

The plants are climbers, aquatic, epiphytic, marshy and may be similar to trees. Roots are fibrous, adventitious, two-aerial or epiphytic. Velamens are present in aerial roots. Stems are underground or subterranean as corm or rhizome. Inflorescence is made of a spadix, which is mostly surrounded by spathe, *i.e.*, modified leaf.

Plants are both monoecious or dioecious. In monoecious species, female flowers are located at the lower part and the male flowers at the upper part. As the stigma is not receptive when pollens are made free, self-pollination does not take place. Interestingly, many plants of the family are thermogenic, *i.e.*, they produce warmth. Their flowers become somewhat warmer than the surrounding air.

Genus: Monstera

The genus is named Monstera and the word actually comes from the Latin word, monstrous, *i.e.*, abnormal. This is so named as the plants of the genus have very unusual types of leaves. They are very large in size and have peculiar *window-like perforations, i.e., holes on them.*

The species are herbs and are evergreen vines. The vines climb with the help of aerial roots which tend to hook on trees, walls or rough surfaces. The roots also grow underground in soil and hold the vines firmly in position. The leaves measure upto 100 cm in length or even more and 70–90 cm in width. In the species, *Monstera dubia*, the leaves are extremely large and may be more than 1.2 metre in length.

Leaves of *Monstera* species are leathery, dark green in colour and are alternately arranged. They have long and large perforations looking like windows. The plants produce inflorescences which are spadix and flowers are borne in the spadix. The fruit consists of a cluster of white berries.

The genus contains more than 60 species which are mostly tropical. Among different species, *Manstera deliciosa* is of much importance as its fruit is edible to the human beings. The fruit has a taste like a blend of banana and pineapple. Some other species of importance are *M. adansonii, M. acuminata, M. obliqua, M. dubia etc.*

Importance

Monstera deliciosa is a dual purpose creeping plant for its foliage as well as for producing human consumable fruit. The creeper has unusual appearance of leaves in having window-like holes, leathery texture and very large size and hence, it is grown inside houses for interior decoration. Some types have variegated leaves which are highly preferred for interior decoration. It is an excellent creeper for training on arches and pergolas.

Fruit of the vine looks like a cob (as a cob of maize) and when it becomes fully ripe, it gives a taste which resembles a blend of the fruits of banana, pineapple and mango. In some parts of Australia, Mexico and the U.S.A., the vine is cultivated in field to harvest the ripe fruits for edible purpose. Processing of the fruit is also done to make jam and drinks, shakes *etc.* Some medicinal properties of the fruit are known. Besides fruit, the aerial roots of the vine are used in making ropes in some central and south American countries.

Description of Plant Parts

Monstera deliciosa is a vigorous, straggling and evergreen vine. The vine is epiphytic and it clings to its support on trunks and branches of trees, walls and rough surfaces with the help of cord like, heavy aerial roots produced by it. The roots act as hooks to enable the vine to climb upwards. Although the roots are aerial, many of the roots also move downwords and on touching the ground, they ramify and provide extra support to the vine.

The vines reach a height of 10–20 metres. Height is found to be more in tropical rain forests. When it comes in contact with tall and spreading trees in the forests, it creeps easily by means of aerial roots which cling to the trees.

Leaves are the characteristic features of the vine as they are peculiar in shape. They are alternate, pinnately arranged and are typically dissected with deep slits and thus, they are conspicuously perforated with oval, oblong or rectangular holes. The holes resemble windows of leaves and hence, the plant has been jokingly named as *window-leaf vine*. When the leaves are young, they are heart-shaped, dark green and holes are not present in them. But as the leaves grow, their greenness is somewhat faded, petioles become thicker and they start to produce perforated, scalloped holes on them. Size of the holes increases in length and width with further growth of the leaves.

When fully grown, a leaf may be 100 cm long and 80 cm wide. Leaf size becomes very large when the plant grows in hot and highly humid climate. (Mention may be made in this connection to another species of *Monstera, viz., M. dubia*, in which species a leaf may be more than 1.2 metre long). Leaves of *M. deliciosa* overlap and closely cling to the trunk of a tree and in such a stage, they are called *shingle* plants.

Morphology of the leaves of *M. deliciosa* has been studied by Sanabria *et al.* (1999) in order to develop an identification system for the ornamental species of the genus, *Monstera*. As the number and size of leaves determine commercial demand of the plant, Lojo and Bernedetto (2014) studied the effect of spraying various concentrations of 6–benzyl aminopurine on the vines. Most important result obtained was early appearance of perforated leaves by spraying with 50–200 ppm solution of the chemical.

Flowers of *M. deliciosa* are borne in specialized inflorescence which is called spadix. Spadix is creamy white in colour, fleshy, upright and measures upto 18 cm in length. It is surrounded by a boat-shaped bract, called spathe which is 25–30 cm long. Flowers are white, about 30 cm long and are borne on spadix. Inflorescence is rarely produced by the plant when it is grown indoor.

For pollination, pollens from other plants are usually needed as the stigma of a flower becomes receptive before pollens of the same flowers are produced. Barare and Cheretien (1985) in Canada studied floral anatomy of *M. deliciosa*.

Fruit is a cluster of berries. It develops from spadix which becomes swelled and resembles ears of maize. It is about 25 cm long, 4 cm wide and deep green in colour. When the fruit is ripe, the green colour fades and turns into yellow. At this stage, the fruit is characterized by covering with tile-like, hexagonal scales, which form the outer peels. Fruits require long time to become mature and attain ripeness.

Origin

M. deliciosa is stated to be originated in the rainforest of southern Mexico to south of Columbia. Nativity is believed in the Oaxaca State in Mexico.

Geographical Distribution

From Mexico, the species has at first spread to many countries in the central America, West Indies and northern part of north America. From these countries, it spread to other countries, particularly those having tropical rainforest. In some countries including the Hawaii State in the U.S.A., it is regarded as an invasive weed. In India, the vine is valued for its use as an indoor plant, for creeping along walls and fences and is also planted to construct pergolas and arches in decorative gardens.

Climatic Requirement

The plant grows best in tropical climate but is found to grow fairly well in mild subtropical situation also. Temperature ranging from 20°–30° C with high relative humidity is considered ideal for its growth. Mortensen (1991) in Norway on carrying out pot-culture studies at a temperature range of 15°–33° C observed that 24°–27° C had been best for growth of the plants.

Growth of *M. deliciosa* plants ceases below 10° C and the plants are killed in frost. The plants grow at an altitude of 1.5–2.1 metres. The plant is shade-living but partial shade is found to make better growth of it than in complete shade. Full sunshine damages leaves so as to look them burned, retards growth of the plants and kills them.

Soil Requirement

Soil is not an important factor for growth of the plant except that drainage should be good. For growing in pots, pot-mixture should be made by mixing 30 per cent garden soil with suitable proportions of well-rotten cow-dung, peat, oil-cake, sand, leaf-mould, wood ash *etc.*

Propagation

The plant is propagated from seeds, stem-cuttings, leaf-bud cuttings, ground-layering, and air-layering. Seed propagation is easy and germination takes place within a week. For large-scale propagation, seedlings should be raised in 4–6 cm raised beds made by thorough mixing of well-rotten cow-dung, leaf-mould, sand, oil-cake, wood ash *etc.*

Ground-layering may be done by many methods such as serpentine-layering, tip-layering trench-layering *etc.* In these methods, thick and mature stem should be used and lower part of the leaves should be trimmed. Root formation is high in air-layering. For stem-cuttings, 5–10 cm segments of mature stem having 1–3 nodes should be used and the leaves should be stripped. Application of auxin quickens root formation. Rainy-season is best time to propagate by cuttings and layerings.

Micro-propagation of *M. deliciosa* has been reported by Li yu qi ou and Cuiping (1997) from China and they stated detailed procedure on the most efficient culture-media for *in vitro* culture of the explants.

Culture

M. deliciosa is cultivated to secure fruits only in few countries. However, cultivation of the plant is done in some countries of central America, southern States and Hawaii in the U.S.A., Egypt, Australia *etc.* and also in greenhouses in

England. No special cultural practice is usually adopted to raise the crop in these countries. Weekly application of liquid manure from spring to summer season is found to make better growth of the plants and higher productivity.

The productivity is found to be markedly stepped up by application of fertilizers. Moschini *et al.* (1985) studied the effect of fertilizers by using different nutrient salts in ion-exchange resin. They observed that requirement of nitrogen had been very high. Khattab *et al.* (1987) studied the effect of nitrogen and potassium on growth and productivity of the plant. Darwesh *et al.* (2011) in Cairo University, Egypt studied that plant spacing and nitrogen fertilizers has marked effect on vegetative growth, flowering and productivity of fruits.

Beel (1988) in Belgium studied the effect of three substrates and three irrigation systems of the plant. Growth and tissue composition of the plant were evaluated by Yeh Der Ming (1998) in a sub-irrigation system using three nutrient constituents in tap-water and by the ebb and flow systems and they worked out method that could give rise to plants of high quality.

Pests and Diseases

Insects or animals hardly attack *M. deliciosa* plants or its fruits. But three fungal diseases on the plants have been reported from China (Guandong) by Xiaohui *et al.* (2001). These diseases not only had been observed from *M. deliciosa* but also from some other species of the Araceae family. Takeuchi *et al.* (2011) reported the fungal disease in the plant, caused by *Colletotrichum gloeosporioides* species in Japan.

Harvesting

Flowers in the plants usually appear after three years of planting. When the plants are grown indoor, they rarely produce flowers. Fruits become ready for harvest in rainy-season and this coutinues upto the beginning of winter. Fruits take long time to become ready for harvest after they are set and this may be a year. Determining harvesting time of fruits is very important. It should be done when the base of a fruit is seen to have started to wrinkle and its green colour tends to turn yellow.

Ripening of the fruits is done after they are harvested. Each fruit should be wrapped with thick paper and maintained in warm condition for a few days when they would be seen to have ripened. Fully ripe fruits give off a flavour which resembles a mixture of banana, mango and pineapple.

Yield

Exact yield of fruits of *M. deliciosa* plant is not known. But it is estimated that 15–25 kg of fruits may be obtained from a plant when grown outdoor in a semi-shade situation and when the vines are properly trained to creep on suitable support.

Varieties

No named variety of the plant is known. But a number of types and ecotypes of the plant are found, morphological characters of the leaves of which widely vary. In some types fruits size goes very large. Some types produce variegated leaves which have great demand for use in interior decoration.

Mode of Consumption

Fruits of *M. deliciosa* should not be eaten unless it is fully ripe because, in unripe condition, it contains crystals of calcium oxalate which causes irritation to the throat. Best ripening of the fruit takes place under post-harvest condition after it has attained full maturity.

The entire fruit does not ripen at the same time. Ripening of the fruit starts at the bottom and then ripens gradually upwards. Index of ripening is known when the hexagonal scales of the fruit which look like plates dry out and fall off easily and thereby, the white and soft pulp is exposed. When the fruit becomes fully ripe, its segments come out easily with pressure. The pulp clings to the hard core of the fruit like maize cob but is musky and slimy. The black flecks which adhere to the fruit are floral remnants and these can be eaten with the fruit pulp.

Quality and Nutritive Value

As the name indicates, *M. deliciosa* is a delicious fruit that belongs to the species of the *Monstera* genus. The fruit is considered as one of the best fruits in the world as regard aroma. Besides flavour, the soft texture and taste of the fruit are also greatly favoured.

Nutritive value of the fruit is also high. A report states that 100 gram pulp of the fruit contains 23 gram of protein, 1 gram of fat, 32 gram of carbohydrate, 16 mg of calcium and 10 mg of phosphorus.

Medicinal Value

Leaves of *M. deliciosa* are known to have wound healing properties. Tribal people in the rainforest area wrap wounds with the leaves and then bandage. Infusion made from leaves or roots of the plant is drunk to relieve arthritis in some parts of Mexico. The root of the plant is used as remedy against snake venom in Martinique. Cytotoxic activity of the plant has been reported by Lira *et al.* (2014) in Bangladesh.

Constituents in Developing Fruits

The volatile compounds at three stages of growth of the fruits, *i.e.*, unripe, half ripe and ripe fruits borne by the plants were determined by Spinola *et al.* (2015) in Portugal using suitable procedure, *i.e.*, HS-SPME and GE-QMS. Volatile metabolites numbering 80, some of which were ethyl esters, terpenoids, carbonyl compounds, higher alcohols, lactones had been identified by them.

Undesirable Effect

As stated earlier, unripe or semi-ripe fruits contains crystals of calcium oxalate, *i.e.*, raphides and trichosclereids and these cause irritations, swelling and blistering in throat. In some fruits, these are present in fruits of ripe stage also. Therefore, persons who are sensitive should taste very little amount of the fruit at first to check reaction in the body before eating more. Laxative effect on consumption of much amount of the fruit has been stated by many.

Utilization

M. deliciosa fruits are utilized in many countries in making jam, tangy jelly, canning and drinks. Making wine from the juice of the fruit is also done by tribal people.

Family: Arecaceae

Species: *Borassus flabellifer* Linn.

[Synonyms: *Borassus flabelliformis* Linn.; *B. flabelliformis* Roxb.; *B. tunicatus* Lour.; *B. sundaicus* Becc.; *Pholidocarpus tunicatus* (Lour.) H. Wend.; *Lontarus domestica* Gaertn.; *Thrimax tunicata* (Lour.) Rallis.]

Taxonomic Serial: 42440

The generic name, *Borassus* is derived from the Greek word, Borassos, which refers to spadix of date-palm. The specific epithet, *flabellifer* is derived from the Latin word, flabellatus, which denotes an open fan.

Common Name

In English, the commonly used names of the palm are Toddy-palm, Palmyra-palm, Sugar-palm, Wine-palm. Lontar-palm *etc.* The word, palmyra has derived from the Portuguese word, Palmeira and in Portuguese, it is called Palmeira de palmira.

Palmyra-palm is known in Germany as Palmyra palme, in Franch as Rônier, in Thailand as Tan-yai, in Malaysia as Lontar, in Cambodia as Thnaot, in Vietnam as Loost, in Indonesia as Buah Lontar, Siwalan *etc.* and in Myanmar as Tanbin. The names of the palm in Japan are Ougi yashi, Parumira yashi *etc.* and in China, the names are Shan ye shou tou zong, Shan ye tang zong *etc.*

In Sanskrit, the palm has been stated as Taalah. In Hindi the names are Taad, Trinaraj *etc.*, in Malayalam, Karimpana, Pananungu *etc.*, in Marathi, Tamar, Taal *etc.*, in Konkanese, Eroal, in Gujarati, Taad, in Kannada, Olegari, Taalegari *etc.*, in Tamil, Panai, Panai marom, Karum panei *etc.*, in Telugu, Tatichettu, in Oriya, Talo, Tanlo *etc.*, in Bengali, Taal, in Assamese, Taal and in Urdu as Taad.

Botanical Family

Borassus flabellifer comes under the family, Arecaceae. The family is also known as Palmae and the family name, Palmaceae is not valid now. Arecaceae consists of perennial flowering plants belonging to the Order, Arecales. About 200 or more genera come under this family which consists of 2600–3000 species of shrubs, tree-like, climbers or plants without stem. All these are known as palms. Most of the species grow in tropical climate and some in sub-tropics. The palms are characterized by large, compound and evergreen leaves in most cases which are known as *fronds.*

Palms are mostly mesophytes and some are xerophytes. They may be solitary or clustered. Solitary shoots have crown of leaves at the top. Palms come under monophyletic group of plants and thus, they have a common ancestor. Palms have been classified into 15 major groups in accordance with morphological specialities, which have however, been modified now. N. Uhl and J. Dransfield brought palms under six sub-families.

Roots of palms are fibrous. Prop roots are produced by some palms. Stems are variable and short. They are cylindrical. Leaves are large, evergreen, alternate and are pinnately or palmately compound. They have petioles and are spirally arranged at the upper part and have tubular sheaths at the base.

Inflorescence is most commonly paniculate and subtended by one or more bracts. Flowers are small, white, actinomorphic, mostly unisexual and may be bisexual. Perianth consists of two or three whorls. Sepals are 3, petals are 3, stamens are mostly 6 and filaments may be separate or attached to one another or to the pistil. Fruit is a single seeded drupe. Double coconut (*Lodoicea maldivica*) has largest seed of any fruit in the world. The fruit of this species is not edible.

Genus: Borassus

The genus of the family, Arecaceae comes under the sub-family, Coryophoideae. The species of the sub-family have in most cases palmately lobed leaves. Flowers are solitary. The genus, Borassus comes under the tribe, Borasseae and the sub-tribe, Lataniinae and consists of six species.

Genus, *Borassus* is liliopsid and the palms under the genus are very tall. Leaves are long, fan-like in shape and have spines along the margins of petioles. They are dioecious. Inflorescence is axillary. Male flowers are smaller. Fruits are roughly spherical and yellowish, brownish or blackish in colour. Pulp is fibrous.

Importance

Importance and versatile use of palmyra have been amply mentioned in literature (Anon 1988; Mazumdar, 2004 a, 2014; Chadha, 2012; Gummdi *et al.,* 2016; Jana and Jana, 2017; Jerry, 2018).

In India, palmyra has received much greater importance in Tamil Nadu and West Bengal. In Tamil Nadu, it is regarded to be a celestial tree and is given religious importance (panaiveryamon). It is considered as the official tree in Tamil Nadu. Importance to palmyra is given in other parts of south India also. A district of Kerela, known as Palakkad district is regarded as a land of palmyra trees.

In West Bengal, a number of sweetened products are made with liquid mesocarp of the fruit. The Bengali Hindu families offer these sweetend products to Lord Krishna on the holy day of Janmashtami, *i.e.*, birth day of the Lord. Outside India, palmyra receives high status in Cambodia, Bengladesh, Sri Lanka and other countries. Many trees are found to be grown near the Ankor Vat temple in Cambodia.

It is said that palmyra palm has 800 uses. In India, the palm has a great significance at least on the following six aspects.

(i) The ripe fruit, edible part of which is liquid mesocarp has great fondness particularly to the rural folk.

(ii) Many highly choice food products are made from the liquid mesocarp which are liked by the people in this country. Many of the products are also exported to other countries as they greatly appeal to the people of those countries.

(iii) A number of non-food products are made from various parts of the tree. Accordingly, many cottage- and small-scale industries have grown up in the country centering on them.

(iv) Hardiness, adaptability to adverse adapho-climatic situations and high yield are the special merits for which the palm is preferred for waste land utilization and afforestation.

(v) Robustness of the trunk affords its suitability for use as building materials in huts and foot-bridges to cross streams.

(vi) Tallness, straight growth, symmetry of the crown and graceful appearance of the palm, without losing its structural beauty for hundred years of its growth are the special attributes for which it is sometimes preferred for planting in ornamental gardens and along avenues.

Description of Plant Parts

Plamyra palm (Plate-5) is dioecious. Height of the trunk is 20–25 metres and sometimes more. Girth of the trunk is 1–2 metres. Trunk is greyish, robust and forms rings due to leaf-scars. Top of the trunk which bears leaves and inflorescences is called crown. Growth of the palm is slow in early stage and rapid at later stage.

Leaflets are 60–80 in number and the apex is acuminate. Leaves are simple, palmate, leathery, greyish green and fan-shaped. Each leaf is about 3 metre long and 1–3 metre wide and symmetrically folded along midrib. Petiolar margins have robust black teeth. Inflorescence is axillary. Male flowers are 1 cm long, smaller than female flowers and form semilunar clusters. Calyx of flowers remain persistent all along. Sepals are 3, petals are 3 and stamens are 6. Perianth is fleshy and accrescent Ovary is globose.

Fruit (Plate-6) is spherical to oblate. It measures 10–15 cm in length and 15–20 cm in diameter. Fruits are borne in cluster. They have fibrous pulp embedded with liquid mesocarp which is rich yellow and forms the edible part. A fruit has 1–3 seeds. When 2 seeds are present, it is curved and more curved when 1 seed is present. Each seed is enclosed with a hard, woody endocarp.

Origin

Origin of palmyra palm is controversial. India is believed to be the origin

(Mazumdar, 2004 a) but tropical Africa or south to south-east Asian countries are also considered to be the origin.

Geographical Distribution

Palmyra-palm grows abundantly in all south-east Asian countries, India, Bangladesh, Sri Lanka, and east African countries. It is also much grown in Australia and the Hawaii island in the U.S.A.

In India, the number of trees has been estimated to be 76, 167, 000 of which maximum number is in Tamil Nadu where 44.18 per cent palms are grown (Anon., 1988). Chadha (2012) has stated that total number of palms in India is estimated to be 102 million and of these, 69 million palms are tapable with a production value of Rs. 8,520 million per annum.

Climatic Requirement

Palmyra palm prefers tropical climate but not much sensitive in its climatic requirement. It grows and produces satisfactorily from dry to moist tropics and also in mild sub-tropical situation. The palm requires optimum mean annual temperature around 30° C but is capable to withstand high temperature of upto 45° C. It cannot withstand low winter temperature below 10° C for more than a week.

Annual rainfall ranging from 500–900 mm is favourable for best growth and high productiveness of the palm. However, it grows and produces in low annual rainfall of 250 mm as well as in very high rainfall areas in Assam where rainfall goes to about 5000 mm. In tropical mountain, palmyra palm grows upto an altitude of 800 metres.

Soil Requirement

The palm grows in a wide type of soil from clay to sand. However, sandy loam is most favourable for higher productivity. The palm is drought resistant but it can withstand water-logged soil for a week or more. Soil fertility is an important factor for higher yield performance of the palm (Chadha, 2012). It has been stated that the jaggery which is produced from the sap is more sweet when the palm grows in calcareous soil (Chadha, 2012).

Propagation

Palmyra-palm is exclusively propagated from seed. Adventive embryos are although produced by the seeds but not frequently. Seeds may be sown directly *in situ* or in nursery-beds and the seedlings are transplanted. Sowing should not be done at more than 15 cm depth as in greater depth, it is liable to be rotten. Germination takes place within 45–70 days. Scarification of the endocarp facilitates germination.

Chadha (2012) has elaborated the steps in germination of palmyra seeds. He suggested that seeds should be obtained from fruits which have ripened to 80–90 per cent. The type of nuts that would give rise to palms of low height and female have been indicated by him. Other measures that have been but forward by him are, heaping fruit bunches for 5–6 days to facilitate automatic stripping from bunches, fermentation for easy removal of mesocarp as well as proper method of raising seedlings in the nursery-bed.

Seeds germinate within 20 to 30 days. The plumule comes out initially and utilizes the food matter which is stored in the nut. This part then moves down to the earth like an underground tuber. Roots would be formed when the tubers would mature. On emergence, the roots would take up nutrients from soil, following which leaves would come out by one.

A new method for culturing seedlings has been proposed by Kandiah and Mahendran (1986) from Sri Lanka. They had provided a technique by which, the tubes could be possible to be grown in polythene bags that could be held in hand. Veerasamy *et al.* (1995) had made a study on regeneration of shoots from split shoot apical meristem by exposing the meristem with excision of leaf-sheaths and leaves and after that, vertically incising them into two or four identical segments.

Culture

Culture of palmyra-palm in plantations is not much common in India but plantations are much raised in Tamil Nadu. In other parts, planting is largely done on the bank of pond, along village road, in public compounds, homestead land *etc.* In West Bengal and also in Bangladesh, planting on the banks of ponds is very common and such a ponds is named as *talpukur* (Plate-7).

Chadha (2012) suggested that to set up plantation of palmyra, low-lying land should not be selected. Seeds should be sown at a spacing of 3 metres and in pits of 30 × 30 × 60 cm after filling them with manure and soil. Sowing of a seed should be done at 5 cm below from the soil surface such that the eye of the seed remains downward in the soil. Plant protection chemicals, *e.g.*, Carbofuran and Furadan granules should be poured in the pits at the rate of 50 gram in each pit. According to him, seeds may be sown in any season but August–September would be best time.

Manuring is rarely done in palmyra but application of 10 kg of compost in the first year and increasing the dose every year with the same amount till it receives 70 kg would be of great benefit. Farm the 8th year onwards, 60 to 70 kg of compost should be applied every year in a palm.

Benefit of application of farmyard manure in palmyra has been stated by Velu (1989). According to him, highest sap yield of 301.66 litres from a palm could be obtained by application of 60 kg of farmyard manure Arumugam *et al.* (1994), in Tamil Nadu observed that highest sap yield of 113.82 litres per palm and total soluble solids of 11.02 per cent had been obtained by application of 50 kg of farmyard manure.

Irrigation is of great importance particularly during the first few years of growth but water-logging should be avoided. Chadha (2012) stated that after sowing, irrigation should be given on alternate days for a month. Frequency of irrigation should be twice in a week during second month and at weekly internal afterwords. Ramachandran *et al.* (2004) stated the irrigation systems that are traditionally followed in Kerela. Mulching and inter-cropping may be done in palmyra plantation with benefit. Pruning of 20–30 per cent of leaves in a palmyra-palm desirable and it should be done in phases.

Pests and Diseases

Among insect pests in palmyra palm, rhinoceros beetle (*Oryctes rhinoceros*) is most serious of all. Some other insect pests are red-palm weevil (*Rhyrchophorus ferrugineus*), *Cocotrypes borassi, Chionaspis dilata etc.* The insect pests bore into soft parts of crown, stem, young leaves and also fruits and feed on sap of leaves and flowers.

To control rhinocerus beetle, sanitation should be maintained in the plantation. Cow-dung or other decaying organic material should not be heaped. Beetles may be destructed in holes with hooks. Holes of beetles may be plugged with cotton soaked with kerosene. To control weevil, injection of Sevin at 1 per cent solution may be done through holes. Leaf-eating caterpillars may be controlled by spraying 0.05 per cent Malathion. Grubs which inhabit in soil also cause damages to the palms and to control, soil application of Aldrin–5 per cent or Chlordane dust may be done.

Among diseases, bud-rot caused by fungal species, *Phytophthora palmivora* is very serious. The fungus causes withering of the central shoot of the leaves. Other fungal diseases are stem-bleeding, caused by *Ceratostomella paradoxa*, pin-head spots caused by *Graphiola borassi*, leaf-splitting, caused by *Exosporium palmivorum*, grey-blight, caused by *Pestalotiopsis palmarum etc.* To control fungal diseases, 0.25 per cent copper oxychloride, 0.1 per cent Auriofungin, 0.3 per cent Dithane-Z, 0.25 per cent Blitox, 0.2 per cent DM-45, or 4-5–50 Bordeaux mixture may be sprayed.

Harvesting

(a) Harvesting of Fruits

Palmyra-palm attains maturity to give rise to fruits after many years of sowing seeds and this may be 12 to 15 years or more. (There is an adage in Bengal : A Palmyra palm will bear fruits at the 12th year if it does not touch saliva of cattle, *i.e.*, grazed). The palm produces flowers from the end of February to May. Fruits are ripened and harvested from July to October. Some trees produce off-season fruits in summer and scantily in winter. Fruits are best ripened in the palms but they are harvested slightly before as fully ripened fruits become very soft and such fruits also drop naturally and damaged.

Ripeness of fruits for harvest is judged by smelling them. Ripe fruits emit a sweet and alcoholic smell and they also become soft in texture. Their surface colour also becomes deep brown to black.

For harvesting, it becomes necessary for the harvester to climb up the palm tree with a cloth-bag, strong rope and a sharp, sickle-shaped knife. The harvester climbs up the top of the trunk, employing a technique by the use of a gird of strong rope which he tightly fits to his feet. The practice requires much skill. Nowadays, ladders of bamboo are also used which avoids climbing up the palm.

To harvest, usually a bunch of fruits is cut off and not individual fruit. On cutting the bunch with knife, the harvester puts the bunch or the fruits (as the case may be) into the bag and the bag is brought down by him with the help of the rope. A person stands on the ground who collects the bunch or the fruits if harvesting of fruits is done. Then the harvester pulls the rope so that the bag is received by him. He cuts of another bunch and the same process is repeated till the bunches which are required to be harvested are cut off.

(b) Harvesting of Sweet Sap (= Neera)

Apart from fruits, palmyra-palm produces a highly sweet and excellent juice which is the sap, that is extracted from the palm. In India, this is called "neera." The sap, *i.e.*, the neera may be drunk as such or a number of products are made from it which have been stated later. Interestingly, while the female tree of palmyra produces only fruits, the male tree produces neera, although it does not produce fruit.

Extraction of the juice, *i.e.*, the sap (= neera) is done by the procedure of tapping. Tapping involves technique and only expert persons can do it. The operation is stated below in short.

A palm tree is tapped when it has become 25–30 years old. But tapping may be done even after 15 years of growth when the palm has grown in sandy soil. Tapping may be continued upto 50–60 years of growth of the palm and in some cases, it can be done upto about 75 years of growth.

Both male and female trees can be tapped but sap yield is higher by 30 to 50 per cent when tapped from female than from the male palms.

Tapping is done from spadices, *i.e.*, spikes of flowers. In male trees, flowering shoots, *i.e.*, the finger like projections (= growths) which arise on the main stalk are tapped. In female trees, fruiting bunches in which drupes have developed and have been very small are tapped.

Season of tapping varies and it is usually December to February in male trees and in February–March in female trees.

For tapping, the palm tree needs suitable preparations. At first, all the leaves except a few which are at the top should be cut back but the young bud at the top of the tree should not be cut. Then the racemes are suitably tied, beaten and crushed by placing between wooden battens. This wounding is done to encourage flow of the sap. Beating and crushing are done for three consecutive days and it is done in the morning. After that, in the morning hours for the next consecutive four days, a thin slice is cut out from the upper part of the racemes in order to stimulate sap flow. The sap starts coming out from the eighth day from the punctured, *i.e.*, wounded racemes.

The juice (neera) which comes out is collected in earthen pitchers. The pitchers are suitably hung in the palms by tying with ropes. The pitchers are at first painted with lime in their inner sides to avoid fermentation of the juice. Small amount of sodium benzoate or sulphanilamide may be added to the juice during its collection to avoid fermentation. It should be noted that when the juice should be drunk as neera and not to be made molacess (= gur) from it or not to be made fermented liquor (= wine), it should be clear and alcoholic smell should not come out from it. But such type of juice is difficult to obtain as it is fermented within short time and gives out alcoholic smell.

It is found that if a palm tree is topped for 3 or 4 consecutive years without allowing natural bud bursting, it may die. Hence, it becomes necessary to give break, *i.e.*, resting the tree after three years for a year or two.

Hussain *et al.* (1992) from their studies done in Bangladesh presented results on some factors, *e.g.*, climbing techniques, cutting methods, harvesting and juice collection on tapping and yield of juice, Kandiah and Kokudathasan (1987) had studied on the effect of application of 20 mM solution of EDTA in causing increase of sap flow from the palm.

Yield

Yield of palmyra fruits which are obtained only from female palm trees comes to about 100 to 200 in India in a year while in Sri Lanka, it may be 350 fruits or more in a year. Yield of sap, *i.e.*, juice from a tree is about 100 to 150 litres in a year.

Varieties

Any named variety of palmyra is not grown in India. Some exotic and indigenous varieties have been collected and grown at the Tamil Nadu Agricultural University in Coimbatore.

Mode of Consumption

A number of food products are made from the fruit, seed and juice of palmyra. Liquid mesocarp of the fruits which is a rich yellow and viscous fluid is highly relished either as such or on mixing with thickened milk. Drink is also made with it with addition of sugar, spices and juices of other fruits. Sap or juice of the palm known as neera serves as a highly delicious drink before it is fermented. Jelly-like, translucent seeds obtained from immature fruits are highly appealing materials. Sprouted seedlings are eaten as fresh or cooked. Sweet syrupy fluid is made from the inflorescence.

Quality and Nutritive Value

Ripe fruits of palmyra have been stated to contain 77.2, 0.7, 0.2, 20.7 and 0.7 per cent of moisture, protein, fat, carbohydrate and mineral matter respectively and apart from these, 9 mg of calcium, 33 mg of phosphorus per 100 gram of pulp (Anon., 1988). Using ripe fruits grown in several parts of the western laterite zone tableland in West Bengal, Mandal and Mazumdar (1995) observed that on an average, weight of fruit was 2.66 kg, specific gravity value was 0.88 and stalk-stylar length of 15.1 cm. They observed that in the edible part of mesocarp, the total, reducing and non-reducing sugars and the total titratable acidity (equivalence of citric acid) were 13.14, 6.47, 6.34 and 0.29 per cent respectively while the vitamin C content was 30 mg per 100 gram.

Gopalan *et al.* (2018) from the National Institute of Nutrition (I.C.M.R., Govt. of India) in Hyderabad reported the following nutritive constituents from analysis done at the same Institute with 100 gram of liquid mesocarp of the fruit containing 77.2 per cent moisture.

Protein: 0.7 g; Fat: 0.2 g; Carbohydrates: 20.7 g; Crude fibre: 0.5 g; Minerals: 0.7 g; Calcium: 9 mg; Phosphorus: 33 mg and would provide 87 Kcal (= 364 K Joule) of energy on consumption.

The report had also stated that 100 gram of the tender fruits of palmyra containing 92.3 per cent moisture had the following constituents.

Protein: 0.6 g; Fat: 0.1 g; Carbohydrates: 6.5 g; Crude fibre: 0.3 g; Minerals: 0.2 g; Calcium: 10 mg; Phosphorus: 20 mg; Iron: 0.5 mg and among vitamins, Thiamine: 0.01 mg; Riboflavin: 0.01 mg; Niacin: 0.2 mg and vitamin C: 4 mg. It has been stated that 100 gram of tender fruit could provide 29 K calorie (=121.3 K Joule) of energy on consumption.

(Index: g = gram, mg = milligram)

Seed embryos of palmyra are known to contain 65.6, 12.5, 1.9, 71.5 and 4.3 per cent respectively of moisture, protein, fat, carbohydrates and fibre and 52, 68, 48, 23 and 0.5 mg of sodium, potassium, calcium, magnesium and iron respectively from 100 gram. The seed embryos are known to yield 1,398 kilo-Joules of energy from 100 gram

Medicinal Value

Palmyra is known to have plenty of medicinal properties. Jerry (2018) made a comprehensive review on the medicinal properties of the palm.

All parts of the tree are used in folk medicines in India and in some other countries as well. Fruit is known to have anthelmintic property and in enlarged liver and spleen, it is said to be efficacious. Young sprouted plants are used to relieve biliousness, gonorrhoea and dysentery.

Decoction from young roots has diuretic and anthelmintic properties. Leaves have anti-oxidant property and are helpful in hiccups and gastric troubles. Flowers are laxative and amoebicide. Seed-coat and inflorescence are useful in tumours and cancer. Seed embryo has anti-oxidant property.

Phytochemicals

A number of phytochemical compounds are existent in the fruits and other parts of palmyra palm. Some of these may be mentioned as polyphenols, saponins, carotenoids, glycosides, organo-sulphuric compounds, albuminoids, triterpene, beta-cyanins, quinones, alkaloids, cardiac glycosides, steroidal saponins, flabelliferrins *etc.* Rodia *et al.* (2019) have stated the anti-oxidant properties of mesocarp of palmyra fruits.

Undesirable Effect

Neurotoxic properties had been reported in the flour of palmyra (Arseculeratne *et al.*, 1982). Existence of flabelliferins which are steroidal saponins have also been observed in the fruit (Nikawela *et al.*, 1988), which are toxic compounds. Pollen grains of the flowers of palmyra are known to be serious allergens.

(Many people are reluctant to set any tree of palmyra, coconut or any other tall palm within or near their household compounds for the fear that the pointed spikes of leaves of the palm trees very easily attract electricity in thunderbolt).

Utilization

Quite a number of products are made from the fruits and other parts of palmyra-palm. Some of these have edible value to the human beings and some have other usefulness. Accordingly, utilization of this palm had been discussed under the following two heads.

(a) Food products and

(b) Non-food products

(a) Food Products

The following products which are consumable to the human beings are made from palmyra-palm.

(i) Toddy

Toddy is the fermented product made from the sweet juice, *i.e.*, sap. In many countries and in some parts of India, the unfermented sweet juice is called toddy. But the fermented juice should be called toddy while the unfermented juice should be called as palm-juice or neera. In India, the fermented juice is called by various names, such as tarri, arrack, tar daru *etc.*

The fermented juice, *i.e.*, tarri is a crude form of country-liquor. This raw form (undistilled) of wine is slightly acidic, frothy, pungent and has alcoholic smell. Popularity of this liquor is very high among the rural people in West Bengal, Odisha and other eastern States. In Sri Lanka, fermented toddy is drunk after purification.

Toddy, *i.e.*, fermented juice (= tarri or arrack) is prepared in India by an easy process. Some fermentation of the juice takes place when it oozes out from the tree and this partially fermented juice is fully fermented by the producers. They take the earthen pitchers containing the partially fermented juice to their cottages or to the production centres. In a pitcher, they take half or ¾th of the material and the pitchers so filled are maintained under ordinary atmosphere in shade. Within 8–10 hours, fermentation upto the desired level takes place and the liquor becomes ready for drinking.

As a matter of fact, the juice contains plentiful monosaccharide sugars such as fructose and glucose and the atmosphere contains plentiful micro-organisms which constitute yeasts (*Saccharomyces charomyces, Schizosac charomyces*) and other species of fungi and bacteria.

These micro-organisms take the juice as their substrate, *i.e.*, food and vigorously respire anaerobically inside the juice. Due to such respiration, alcohol (ethyl alcohol) is formed as the end-product and this forms as the fermented juice, *i.e.*, taari (=arrack). The process has many risks. Instead of ethyl alcohol, methyl alcohol may be produced which is deadly poisonous for growing of undesirable micro-organisms in the juice. Insects, spiders, lizards may fall on the juice during fermentation process. Preparation is mostly done in unsanitary condition and this needs improvement.

(ii) Jaggery

The term, jaggery is derived from Portuguese but originally, it has come from Sanskrit. Jaggery is actually a concentrated product which is made by boiling sweet juice of Palmyra (or wild date palm, coconut, sugarcane *etc.*). It is yellowish, brown or dark brown in colour and crystals or molasses or other substances that are present in it are not separated out from it, *i.e.*, they remain present in it.

To make the jaggery, the juice which has been tapped from palmyra-palm is at first heated to about 40° C mixing some slaked lime in it. Then it is cooled to room temperature condition and alkalinity which has occurred in it due to liming is made neutralized. For neutralization, tamarind pulp is used. In improved practice, phosphoric acid is used and the pH is brought down to around 7. Then it is strained through cloth to remove the lime particles and other impurities.

Following that, the strained juice is heated. During the first 15–20 minutes of heating, white scum will appear on surface which is due to impurities inherently present in the juice. The scum should be regularly removed out with a wooden ladle. Heating of the juice should be continued to 120°–130° C, *i.e.*, it should be boiled. Too much frothing may occur during heating and to prevent it, little quantity of milk or coconut water should be added to the boiling juice. Boiling should be stopped when the jaggery becomes ready. Normally, an hour is required to form jaggery but this should be judged by feeling the thickness with hand on taking a little amount and cooling it.

When the product, *i.e.*, jaggery has become ready which would be judged by hand-feeling and good colour, it is cooled to almost room temperature and then transferred to earthen pots. The pots should be moistened by dipping in water before pouring the liquid form of jaggery in them. When poured into the pots, the jaggery will be solidified within an hour. The pots are suitably covered and left for 15–20 days when the jaggery will be crystallized.

On solidification, the jaggery is marketed. Marketing is usually done by cutting the solid form of jaggery into pieces. About one kilogram of jaggery could be prepared from 8–9 litres of palm juice. The jaggery has many medicinal properties. It is also exported to other countries. It is stated that India has a tremendous prospect in jaggery industry (Vengaiah *et al.*, 2013).

(iii) Palm-sugar

It is a brown coloured, crystalline type of sugar. Palm-sugar is produced from jaggery. To make this, the jaggery is placed on a smooth surface and it is left as such for about two months, which is with the idea to evaporate the moisture present in it as far as possible. Then it is taken in a suitable container and poured water in it. After that, it is refined and on refining, brown coloured sugar is obtained.

Nowadays, suitable apparatuses are also used in the factories to make refined sugar. Output of sugar becomes higher by the use of these apparatuses and instruments. Refining is sometimes done repeatedly in the factories by which less brown and whitish sugar is obtained. This sugar has much demand in international markets.

Palm-sugar is used in making various types of sweetened fruit products, toffees *etc.* It has high medicinal properties and also gives high energy on consumption.

(iv) Palm-candy

Palm-candy is a crystalline form of sugary product. In India, it is called tal-michhri, panam kalkandu *etc.*

To make palm-candy, palm-juice is at first boiled as is done is making jaggery. However, so high heating is not done and boiling upto 110° C is considered enough. After boiling for 10–15 minutes, the boiled juice is poured into pre-moistened earthen pots and allowed to cool to room temperature condition.

On cooling the filled-in earthen pots are lightly covered with earthen lids and the lids are tightly tied with the pots. Then the pots with lids are buried under ground to maintain cool atmosphere. The pots are maintained in buried condition for 2–3 months. Then they are removed out from buried condition and by that time, palm-candy will be formed.

Suitable apparatuses are of late, used for crystallization of juice, instead of burying the earthen pots in soil. The apparatuses are insulated and candy could be made within 20–30 days. Earthen pots are not used to fill the juice and plastic or metal containers are employed instead of that.

Palm-candy is rich in sugar, calcium and has many medicinal properties. It is especially recommended to the children when they suffer from cough and cold.

(v) Molasses

Molasses is a by-product which is obtained during crystallization of palm-sugar. The product is rich in sugar and calcium and has high medicinal value. To build up strong bones of children, it is of much help. Molasses is highly beneficial to the cattle also when they are pregnant. It is used in making sweetened food products, *e.g.*, patali, monda etc. and many types of milk products.

In the distilleries, molasses is largely used for clarification and refining to make ethyl alcohol, *i.e.*, ethanol which contains 100 per cent ethyl alcohol, or rectified spirit that contains 95 per cent ethyl alcohol. These are used in pharmaceutics, other industries and in laboratories.

(vi) Cooked Products

Young sprouted plants of palmyra are cooked or roasted and pounded to make a type of meal in Bangladesh.

(vii) Pickling

Small fruits are pickled in vinegar in south-east Asian countries. Tender, succulent, endosperm of immature seeds of palmyra is also like by many (Plate-8).

(b) Non-Food Products

Apart from making products which are consumable to the human beings, a large number of other products which have no edible value are made using different parts of palmyra. Many cottage- and small-scale industries have developed in different parts of India in making these products. Some of these items have been stated in the following but besides these, many other goods are also prepared.

(i) Use of Palmyra Trunk

Truck of palmyra palm is tall, very hard, heavy, smooth, durable, not attacked by insects, animals or microbes and are not easily destroyed in water. Hence, it is used as a material for building hutments in villages. It is also used as a foot-bridge to cross streams and drains. On splitting in the centre, the trunk is used as a raft in ponds.

(ii) Use of Leaves

A palmyra palm produces 35–40 leaves and on drying, they are used as follows:

Thatching of hutments, which is common in south India. Hand-fans, mats, boxes, containers, art-objects *etc.*, are commonly made. On colouring the dried leaves, various types of toys notably, flute, carpet knight (= Sipahi), rattle, chirping bird *etc.*, are made. In West Bengal, a special type of rain-proof is made which has no handle and covers the entire back and head of a person in heavy rain. Midribs taken out from leaves are used in making brooms and brushes.

It may be stated that in ancient time in India when paper was not invented, dried palmyra leaves (= talpatra) were used as writing materials (= punthi) of epics and religious texts.

(iii) Fibres

From the leaves as well as from inner part of bark and pericarp of fruits, tough fibre is obtained which is used to make strong rope, washing utensils and wall painting materials.

(iv) Fruits

Fibrous endocarp on drying is used as an art-object on colouring and drawing figures.

Family: Arecaceae

Species: *Phoenix sylvestrics* (Linn.) (Roxb.)

[Synonym: *Elate sylvestris* Linn.]

Taxonomic Serial: 506744

The generic name, *Phoenix* is Latin for the Greek word which signifies date-palm and the specific epithet, *sylvestris* denotes, of the forest. It needs mentioning in this connection that, *Phoenix sylvestris* is sometimes confused with the allied species, *viz.*, *Phoenix canariensis* but the stem and arrangement of leaflet of the two species and other species have distinctive morphological characters.

Common Name

In English, the species is known as Sugar date palm, Silver date palm, Indian date palm, Sylvester date palm, Indian wine palm *etc.* In Sanskrit, it has been referred to as Madhu Kshira. In Hindi, wild date is known as Khaji, Salma, Sendhu, Thakil *etc.*, in Oriya, Khejuri in Tamil, Itham pinnay, Itchum panny *etc.*, in Telugu, Pedda ita, Ita nara *etc.*, in Kanara, Kullu, Ichal *etc.*, in Malayalam, Katti intha and in Bengali, Desi khejur, Kajar, Jangli khejur *etc.*, are the common names.

Botanical Family

Plants of the family, Arecaceae are tree-like shrubs or climbers, stemless and are perennial. The family comes under the order, Arecales and consists of 180 genera and 2600 species. Most of the species grow in tropical rainforests and deserts and also in the sub-tropics. The palms grow as solitary or clustered.

Leaves of the palm are large, compound and are known as fronds. Leaf forming crown is located at the top of the stem. Vegetative buds are developed by the stem, *i.e.*, trunk which produces new shoots. Leaves of the palms are evergreen and arranged as palmate (fan-leaved) or pinnate (feather-leaved). Inflorescence is a spadix, surrounded by spathes which hardens when the plants become mature. Flowers are white, small and unisexual or bisexual. Sepals and petals are three in number in each and may be jointed in some species. Stamens are usually six. Fruit is mostly a one-seeded drupe (Plate-9) and in some genera, two or more seeds are present in a fruit. The double coconut (*Lodoicea maldivica*) has very large seed which may be 50 cm wide and more than 20 kg in weight. Several sub-families are there within the Arecaceae family.

Genus: Phoenix

The genus is termed *Phoenix* from Greek which may have a historic reference or may be from the sacred bird living in ancient Egypt. The genus comes under the sub-family, Coryphoideae and is known to contain 14 species. They have originated from the Canary island to south-eastern part of Europe and are grown in areas from Turkey to south-east Asia.

The species of the genus are robust, medium or dwarf in size and have pinnately arranged leaves which measure upto 5 metres or more in length. Leaves are sessile or have very short petioles. Plants are dioecious. Flowers are small, 1 cm wide, yellow to brown and are in clusters in panicles. Fruit is a one-seeded berry, elongated, upto 8 cm long which is orange-coloured in young stage and becomes purplish when ripe. Peel is thin, loosely held and the seed is hard and grooved.

Importance

Fruits of *Phoenix sylvestris, i.e.*, wild date have little importance as a commercial fruit in India. This is because, the edible part of the fruit, *i.e.*, its peel is very small in amount and that, the taste of the peel is far inferior than the date of commercial importance, which is *Phoenix dactylifera*. Nevertheless, fruits of wild date are much popular to the tribal and rural folk who are fond of the fruit peel in summer season when the fruits are fully ripe.

However, the palm belonging to this species is of great importance in some parts of India and particularly in West Bengal (Mazumdar, 2004). The reason is, highly sweet sap is extracted from the trunk of this palm in winter season and the sap is processed to make gur (= molasses) of delicious taste. From this gur soft or hard, a number of excellent sweetmeats are made in West Bengal, Tripura some parts of Assam and in Bangladesh. The various types of sweetmeats that are made from the sweet sap are also highly liked by many people in other countries. Centering on preparation of these sweetmeats, plenty of cottage and small-scale industries spring up in the rural sector in different parts of West Bengal, which are usually operated by a section of the rural community during the winter season. The sap in fresh form is also liked by many as a sweet drink.

(Unfortunately, many wild date palms are nowadays cut back in West Bengal and other eastern states to procure land for crop cultivation, widening of roads, construction of buildings *etc.* This brings about lowered production of palm sap. It is advisable that number of wild date trees should be made increased now. This can be easily done by sowing seeds along village paths, tanks, odd places, uneven and uncultivable lands and in forests).

Apart from sap, the pulp, *i.e.*, fruit peel is sun-dried for use in future. Seeds are used in extraction of oil (Kapoor *et al.*, 1990; Bhakhare *et al.*, 1993). In some parts of north-east India, seeds of wild date are used as substitute of arecanut in the use of taking betel leaf with lime.

The trunk of the palm is used to make posts in villages (Mazumdar, 1984; Kamaluddin *et al.*, 1998). Leaves are used in making manually used fans, art objects and as protective barrier along boundaries of land and trees. As fodder for milch cattle, the palm is used which is believed to increase milk fat. The palm is planted along village paths, as wind-break trees and sometimes in landscape gardens to impart naturalistic design. This is an excellent material for utilization of waste land.

Description of Plant Parts

The trunk of the wild date palm is robust, 5–15 metre tall and 30–40 cm wide. It is unbranched and the surface is rough with base of leaf-stalks which are persistent. Leaf-stalks should however, be cut off as the palm grows. Crown of the palm is 8–10 metre wide. Leaflets are 200–250 in number and are arranged in 2 or 3 groups. Lower leaflets are modified into spines. Leaves are 2–3 metre long, recurved on 1 metre long petioles with acanthophylls near base. They are pinnately compound in duplicate, bluish green and persistent.

Inflorescence is 0.8–1.5 metre long, yellowish and erect. Flowers are borne on a spadix covered by a 28–30 cm long spathe and have fragrance. Male flowers are whitish yellow, polyandrous, 6–7 mm long and 4–6 mm wide. Female flowers are 3–5 mm long and whitish. Fruit is 1.8–2.8 cm long, oblong to avoid in shape, peel is orange-yellow coloured when immature and turns dark red to purple when fully ripe. A fruit contains one hard , elongated seed which has grooves.

Origin

Exact origin of the species is not known. However, nativity in India (Pareek and Sharma, 1993) or in area from southern part of Pakistan to India and extending upto Sri Lanka (Singh, 2004) have been suggested.

Geographical Distribution

In India, wild date palm is found to grow in almost all parts and upto an altitude of 1350 metres. About 29 million palms of the species are estimated to be grown in this country (Anon., 1988). Plantations of this palm are not made by man but it grows abundantly here and there, in waste land, scrub forest, along village roads, tanks *etc.* These trees are produced from seeds which are thrown on eating the pulp by human beings and animals. Outside India, wild date palm grows in China, Nepal, Bangladesh, Pakistan, Sri Lanka, south-east Asian countries *etc.*

Climatic Requirement

Wild date plam trees grow well in tropical and sub-tropical climates but size, quality and yield of fruits and yields of trunk sap are found to be higher when grown in sub-tropical situation (Mazumdar, 2004). In tropical mountains, the palm grows upto an elevation of 1400 metres and it survives at low temperature of upto 5° C for a week or so. High light intensity is known to be a pertinent factor in increasing yield of fruits and trunk sap.

Soil Requirement

The palm tree is highly adaptable to inferior soil where other crops do not grow, although yield of fruits and trunk sap are high when grown in alluvial soil.

The palm is extremely tolerant to drought and soil salinity. Yield of fruits and sap of trunk go high when the palm grows in slightly acidic soil (*p*H 6–6.5).

Propagation

Wild date palm trees that are found have all come up from seeds. Seeds germinate easily within a month even in low soil moisture content. However, apart from seeds, off-shoots arising at ground level of the palms may also be cut off and planted to produce new plants (Hayes, 1970). Saini and Dube (1992) studied germination behaviour and storage of pollen grains of the flowers and Bhattacharya *et al.* (1993) had determined phenolic compounds and iso-enzymes of the pollen grains.

Culture

Without any culture, the palm grows well and produces satisfactorily but growth of the palm is greater and also productiveness of fruits and sap yield are high when it is grows in rich soil. Plantation of wild date is not known to be done but if it is so desired, a plant spacing of 1.2–1.5 metre should be maintained. Stagnation of water for more than a couple of days in the soil should be avoided. After the trees have produced fruits in a plantation, excess of male trees, if any should be uprooted.

Pests, Diseases and Physiological Disorders

Any serious pest or disease is not found in wild date palm (Hayes, 1970; Mazumdar, 1984 b) but few insects and grey-blight disease symptoms are often found. Some physiological disorders, *e.g.*, pin-head spots, splitting of leaves, cracking of fruits and premature fall of fruits are sometimes noticed in the palms.

Harvesting

Wild date palms produce flowers after 6–7 years from germination of seeds. Flowers appear during early part to the middle of August and long time is required for the fruits to become ripened.

Harvesting of Fruit Bunches

Bunches of fruits of wild date are harvested and not the individual fruits. Conventional practice is to cut off the bunch before the fruits become fully ripe as the fully ripe fruits are easily damaged during harvesting of bunches. Harvesting of bunches is done at interval of a week or so during summer months and continued upto early part of rainy-season (Mazumdar, 1984). Harvesting of bunches is done manually. However, clipping the bunches with the device as stated under the Chapter on longan would be an efficient method and by this method, harvesting could be done by pulling the rope of the pole, standing on the ground.

After harvest of the bunches of fruits, they are placed on a layer of straw or dried leaves and then covered with straw. Being stored in this way for 3–7 days, the individual fruits are fully ripened, assume purplish red colour, become soft and sweet in taste.

Harvesting of Sap of the Trunk

For the purpose of tapping the sap, the lower leaves along one side of the trunk are at first cut off with sickle so that the inner layer which is soft gets exposed. On doing so, tapping is done after a few days by giving a V-shaped cut with the sickle on the exposed part and thus, a triangular area is scooped out.

The sap, *i.e.*, juice trickles down naturally from the scooped part of the trunk. This is collected in a pitcher which is usually earthen made and to place the pitcher at the proper place, its brim is suitably tied to the trunk with rope. To facilitate accumulation of the sap in the pitcher, a small bamboo slice is attached.

Collection of the sap is done in the evening and the pitcher in which the sap has accumulated by trickling is taken out on loosening the rope with which the pitcher has beem tied. The pitcher containing the accumulated sap is taken out in the early morning of the next day. Then in the next evening, a shallow cut is given on the trunk and again the sap is collected in the same way. After that, the tree is given rest for few days. Rest is given with the idea that more sap would accumulate in the trunk. After giving rest, tapping is again started and in this way, the operation is continued throughout the entire winter season.

The cuts on the trunk are given at alternate sides in a zig-zag manner. That is, the V-shaped cut is at first given at one side of the tree trunk and next time, at the opposite side of the cut and continued in this way.

The sap, *i.e.*, juice of the trunk which is collected is called *neera* in local language in West Bengal. In containing high amount of fructose, this is very sweet in taste. But in ordinary atmosphere, it gets fermented within few hours and turns alcoholic. Fermentation could however, be much retarded by putting some slaked lime in the pitcher during collection of the sap. The neera is preferred by some people as a cooling drink. Major part of the neera is however, suitably heated and processed to make the molasses (= khejur gur) with addition of lime during boiling.

Yield

Yield of fruits of wild date palm is difficult to assess as it differs on a number of factors. In general, a palm produces 8–10 kg of fruits and it may go higher when grown in rich soil, gets water throughout the year and the soil is well-drained. Yield of fruits may be very low if the flowers of female trees are deprived of pollens which may result in the absence of male trees in the neighbourhood of female trees. Yield of trunk sap widely varies and may be 100–300 kg per palm.

Mode of Consumption

Edible matter content, *i.e.*, peel of the fruit is though much less, it is highly sweet, has no sourness and is soft in texture. The sweet sap tapped from the trunk is a cooling drink among the villagers. Many people prefer to drink the sap after it is partially fermented.

Quality and Nutritive Value

Fruits are known to contain 59.2, 1.2, 0.4, 33.8, 3.7, 1.7 and 0.38 per cent, respectively of moisture, protein, fat, carbohydrate, fibre, calcium and phosphorus (Anon., 1988). According to Palmpedia, the fruits contain moisture, sugars (mostly reducing), pectin, protein, minerals, phosphorus, potassium, calcium, magnesium and iron of the order of 66.7, 18.42, 0.51, 1.12, 1.12, 0.042, 0.549, 1.139, 0.006 and 0.007 per cent respectively and 9.42 mg of vitamin C per 100 gram of pulp.

Mandal and Mazumdar (1997) analysed fruits harvested from the trees grown in the southern plain land of West Bengal. They observed average weight of fruit as 2.39 gram, specific gravity as 1.09 and length as 2.47 cm while titratable acidity as equivalence of citric acid as 0.17 per cent and vitamin C content as 20 mg per 100 gram of pulp. Majumdar and Mazumdar (2001) reported appreciable amount of calcium pectate in the pulp. Nutritive value of palm sap has been reported by Barh and Mazumdar (2008).

Fatty acids composition of the seeds had been determined by Kapoor *et al.* (1990) and they stated that the major fatty acids of seeds were palmitic, oleic and linoleic acids. Bhakare *et al.* (1993) reported the composition of seed soil.

Medicinal Value

Fruits of wild date have been stated to have medicinal value in cardiac ailments, anaemia, vomiting, loss of consciousness and has usefulness to provide energy in the body quickly. Meristematic part of the stem which is tender is of help in gonorrhoea. Paste made from the roots relieves nausea.

Utilization

As stated earlier, the sap which is tapped from the trunk of wild date palm tree is of greater significance than the fruit. Molasses, *i.e.*, gur is made from the sap by heating and suitable processing. This is done in wide scale by the villagers in different parts of West Bengal during the winter season when the sap in tapped. Plentiful cottage industry centres have been set up since long past in the State of West Bengal where the rural people prepare the molasses with application of suitable techniques. The operation is mostly done by a section of the rural community and it runs in their families.

The molasses is marketed either in thick form or juicy form, known as *nalen gur*. This is also marketed in solid blocks which is known as *nabat* or *patali* and this is made by suitable drying process employing techniques. The speciality of the wild date molasses is its excellent flavour, for which a number of delicious sweetmeats are made from it which have high demand in the market. However, the sweetmeats are available only in winter season when the flavour is persistent.

Family: Cactacease

Species: *Hylocereus undatus* (Haworth) Britton and Rose

[Synonyms: *Hylocereus trichosatus* Gosselin (Britton and Rose); *Cereus triangularis* major de Candole; *Cactus triangularis asphyllus* Jacquin; *Cereus guatemalensis*]

Taxonomic Serial: 503101

It should be noted that dragon fruit is a general term and it includes fruits of a number of species and also hybrids belonging to a number of species. These species usually come under four genera. Under the genus, *Hylocereus*, there are some species and the genus, *Selenicereus* consists of only one species. Some horticulturists in central America however, opined that about 20 species need also be recognized and named as dragon fruit. Thus, dragon fruit is a species belonging to the family, cactaceae.

Common Name

In different languages, dragon fruit is known as follows.

English: Pitaya, Pitahaya, Night blooming cereus *etc.* Spanish : Pitahaya blanca, Pitahaya roja *etc.* Potuguese: Cato-barse. German: Distilbirne. French: Fruit du dragon. Brazilian: Pitaya vermehla. Vietnamese: Thang long. Korean: Yong-gwa. Chinese: Cardo anana, Cato barse *etc.* Japanese: Pitaya. In Indian languages, there is no specific name of this fruit but some people called it *Daitya phal.*

Botanical Family

The family, Cactaceae comes under the Order, Caryophyllales and consists of about 127 genera and 1760 species. The family is divided into three sub-families. These are, (i) Pereskioideae (= leaf cactus), containing flat leaves and about 20 species, (ii) Opuntioideae containing barbed hairs and about 300 species, (iii) Cactoideae, containing about 1000 species and (iv) Maihuenioideae that contains 2 species.

Species of cactus thrive in drought condition and some species can grow even in extreme drought. However, the species make great economy of water. The fleshy parts present in their body act in conserving water. True leaves are absent in most of the cactus species to protect against animals and also to retard transpirational loss of water from their body. Shape, size and height of the plants differ widely in shape. A species, *Pachycereus pringlei* is known which attains a height of 19 metres or more.

Stems and branches of cactuses have the characteristic feature to contain chlorophyll pigment and hence, they are able to make photosynthesis and thus, able to prepare carbohydrate. Most species have crassulacen acid metabolist system (CAM) which plays part in photosynthesis.

Spines arise from specialized parts known as areoles, which are stunted branches and produce flowers. These are organs by which cactus species are identified. Many of the cactus species are used as decorative plants, many act as protective hedges and some species produce food to the human beings and animals.

Genus: Hylocereus

It has been stated earlier that most of the cultivated species of dragon fruit come under the genus, *Hylocereus*. Whether the genus that comes under the tribe, Hylocereceae is a genus has been known in the recent time. The genus that contains species are epiphytic and are called as nocturnal or night-blooming cactuses. The species are shrubs or vines. They produce long and aerial or hanging roots which measure 10 metres or more.

Importance

Apart from decorative value, the species, *Hylocereus undatus* is now gaining popularity for cultivation in field in some tropical countries to harvest the red-skinned, attractive fruits. The fruits are sweet and pleasant in taste and are of high juice content (Minh *et al.,* 2019). A great advantage in growing this fruit crop is, irrigational cost is almost nil as the species is a cactus and hence, adapts well in deserts or in extreme dryness.

The fruit is rich in many nutritive, medicinal and anti-oxidative properties. A number of preserved products, *e.g.,* jam, jelly, preserves, drinks *etc.* are made from the fruit. However, the species is considered as invasive in some African countries.

Description of Plant Parts

Dragon fruit is an epiphytic, perennial, xerophytic and terrestrial climber which is about 6 metre long and makes quick growth. Roots of the plant are aerial, superficial, sprawling or climbing, very long and are upto 10 metres in length. Aerial roots help the vine to climb upwards and they also absorb moisture from atmosphere.

Branches are 4–7, 5–10 metre long, 10–15 cm thick and with three ribs. Stem segment has three wavy wings which are flat with corneous margins. Leaf is absent. Leaves are converted to spines which are 2–5 cm long, appear as needles and brown to black in colour. Areoles (small area which bear spines or hairs) are 2 mm wide.

Flowers are very large, 25–30 cm long and 15–20 cm wide. They are bisexual with white, scented petals and open at night. Bracteoles are ovate, stamens are 5–10 cm long, declinate and the styles are 5–20 cm long.

Fruit is oval to oblong in shape, 5–10 cm long, 4–5 cm wide and red in colour. It is shredded with green-coloured scales. Pulp is white and embedded with small, black-coloured seeds.

Origin

Exact origin of this cactus species is not known. But it is considered that the nativity of the species might be in the area from Mexico to Pacific Guatemala, Costa Rica, Venezuela, Columbia, Equador, Brazi and Uruguay.

Geographical Distribution

Dragon fruit grows successfully in tropical countries but at present, it is also grown in many subtropical countries where winter is mild although fruit quality is not good when grown in such situation.

Some of the important countries where it is much grown and with desirable quality of fruits are Malaysia, Vietnam, Thailand, Sri Lanka, Florida Caribbean, Mexico, and Hawaii island in the U.S.A. and northern countries of south America. India used to import the fruits from other countries but now it is grown to small extent in Gujarat, Maharashtra, Karnataka, Andhra Pradesh and the total area is considered to be about 400 hectares.

Climatic Requirement

Dragon fruit plant is highly tropical in climatic requirement but it is now found to grow moderately well under mildly sub-tropical climate. Average temperature requirement of the plant is 20°–25° C but it can withstand upto 40° C for some period. Moderate climate prevailing throughout the whole year is considered favourable for the plant.

Frost greatly damages the plant but it recovers from frost injury if it does not continue for prolonged duration. Plants are killed on exposure to 0° C for even one day. Partial shading is favourable for vigorous growth and fruiting of the plant but too much shading is detrimental. Extreme sunshine badly damages the plant and hampers its growth.

Soil Requirement

Soil is not a limiting factor for growth of the plant but drainage is very important. Poor drainage for few days kills the plants in a short time. The plants grow well from clay loam to sandy loam soil but sandy loam with copious organic matter is considered best. Soil pH is an important factor and acidic soil with a pH of 5.3 to 6.5 is most conducive.

Propagation

Dragon fruit plant is easily propagated from seeds. However, stem-cuttings are highly successful and propagation is conventionally done by using stem-cuttings.

For stem-cuttings, lateral branches of the stem segments should be cut off and then, the cuttings should be made, maintaining 20–30 cm length of each cutting. In proper rooting medium, the cuttings regenerate roots when planted in rainy-season or in spring. However, dipping the basal ends of the cuttings in auxin solution, *i.e.*, 50–100 ppm of IBA for few minutes just before planting is very much helpful to induce root formation in shorter time and with production of abundant roots.

Curing of the cuttings before planting is considered to be of much help to induce easy root fermentation in dragon fruit plant and also in other cactuses. To do this, the cuttings should be stored in single layer in a cool and dry place for 4–7 days and then planted for root formation. If treatment with auxin solution is to be done, this should be carried out after curing and then planted directly. Curing of cuttings is however, not essential as root formation takes place even without doing that.

It is advisable that the cuttings which have been cured or not should at first be planted in small pots containing soft and highly rich rooting medium. This is with the idea that roots are produced in them in abundance. After maintaining in these pots for a month or two when profuse roots would come out from the cuttings, each rooted cutting should be carefully uprooted and planted in desired site. Micropropagation of dragon fruit plant has been stated by some workers. Yassen (2002) described the procedure for micropropagation using thidiazuron and NAA.

Culture

To secure human consumable fruits in south-east Asian countries and in northern south America, dragon fruit plant is cropped in organized plantations. For decorative purposes, it is grown in gardens in pots and as hedge. The plant is a vine and hence, suitable arrangement should no doubt, be made to allow the vine to creep it freely without obstruction. This can be done in a simple way by planting low-highted fruit plants such as pomegranate, guava, peach, phalsa, custard apple *etc.* close to the dragon fruit plants. The dragon fruit plants would no doubt, easily creep on these fruit plants.

However, erecting trellises is most effective device for creeping when the vines are cropped for fruits or to set ornamental hedge row. The trellis should be sufficiently strong such that it can endure weight of the vine and the fruits borne by them. To build up a trellis, iron or wooden poles are to be strongly infixed in the ground at desired spacing. Then the nets and tops are to be suitably joined to the poles. As the stems of the vines are soft and jointed, they are easily breakable and to avoid this, any such material should not be used in the trellis system that is liable to cause damages to the plants.

Planting of the rooting cutlings is best done in rainy-season but may also be done in spring after moistening the soil. Spacing of the plants should be 2–3 metres but high density planting may be done at spacing of 1–2 metres. In some countries, high density planting is done for higher production and as many as 1000 to 1400 plants may be grown in a hectare area in this system.

Manurial recommendation is not clearly available but like any cactus plant, application of more and more organic matter would be of great benefit for vigorous growth of the vines and increased production of fruits. Fertilizers are usually not used but liberal application of mixed fertilizers is of benefit. Sulphate form fertilizers would be of advantage which lowers the soil *p*H.

Dragon fruit is a cactus plant and grows well with little or negligible water. Nevertheless, light irrigation at 10–12 day interval is of help particularly during summer and during development of fruits. Any cactus plant with no exception to dragon fruit is tremendously benefitted by covering the soil with suitable mulch that effectively conserves soil moisture, keeps the soil cool and prevents growth of weeds.

Training of the vine at early stage of its growth is an important operation in dragon fruit. It should be done carefully such that the branches arising from all parts of the vine can freely grow on all sides. Vigorous growth that takes place at the top of a vine makes it top-heavy and thereby, causes breakage. Hence, the terminal parts should be occasionally truncated to restrict heavy vegetative growth of the vine. In dragon fruit plantations, broken segments of stem fall on the ground which strike roots and act as weeds. These unwanted plants should be removed whenever observed.

Inter-cropping in a dragon fruit plantation may be done with pineapple or other horticultural or medicinal plants and if cropped, plantation should be adequately manured. Planting of guava, papaya, custard apple, pomegranate, Suriname cherry or other economically important trees or shrubs would be of advantage as the dragon fruit vines prefer semi-shade atmosphere. However, bushy and spreading trees like mango, jackfruit, sapota *etc.*, should not be planted as dense shade of such trees would fall on the vines and hamper their vegetative growth and yield.

Growing dragon fruit plants in pots has the advantage in that, the potted plants can be shifted indoor in severity of cold weather and precipitation which cause damage to them. For pot-culture, large-sized pots which are provided with suitable handles should be used and pots should be very strong.

Pests and Diseases

Animal pests like birds, bats, squirrels *etc.* cause much damage to the fruits. Among the insect pests, aphids and mealy bugs are common which suck sap of the plants. Mites, scale insects and thrips also attack the vines. Sometimes fruit flies pierce ripe fruits and make tunnels.

Some fungal and bacterial species infect dragon fruit plants. The bacterial species, *Xanthomonas campestris* causes rotting of stem. The fungal species, *Dothiorella* makes brown spots on fruit surface. Among other fungal pests, mention may be made of *Colletotrichum gloeosporioides*, *Botryosphaeria dothidea*, *Bipolaris cactivora etc.* Balendres and Bengoa (2019) stated organisms and their control.

Harvesting

Dragon fruit plants produce fruits after 18–24 months of planting. Flowers and fruits appear in 3–4 waves during one season in every year. Ripening of fruits takes place after 40–50 days of flowering. Fruits of the species are non-climacteric and are best ripened in the plants. If harvested before fully ripened, they do not ripen any more under post-harvest condition. Time of harvest of the ripe fruits is best judged by observing colour and texture of the fruits. Harvesting should be done with a secateurs or a suitable pliers with care to avoid damage of the soft, ripe fruits.

Yield

Yield of dragon fruit depends on a number of factors. Average weight of a fruit is 300–500 gram but it may be much higher and goes upto 800–900 gram. Yield of fruits of the vines that are grown on a pole (= support) may be upto 100 fruits per wave. Hand-pollination and illumination at night have been stated to raise yield of this cactus crop.

Varieties

Varietal nomenclature has not been done in dragon fruit but much variations are observed in respect of plant characters and also fruit characters in respect of size, shape, peel thickness *etc.*

Mode of Consumption

The dragon fruits are best relished as fresh however, when they are fully ripe. Small, black seeds are eaten along with pulp without any difficulty. Conventionally, the fruit is cut lengthwise and the pulp is scooped out with a forked spoon. The pulp is also cut into cubes and taken with fruit salad. Some people prefer to make drink using the pulp and drink on cooling.

Quality and Nutritive Value

Ripe dragon fruits satiate all sections of consumers. Fruits improve digestion. They are rich in dietary fibres and anti-oxidants. Immunity is known to be strengthened on taking the fruit. Some mineral nutrients notably, iron and magnesium are rich in the fruit.

Medicinal Value

Dragon fruits are known to have many medicinal properties. They bring under control type-2 diabetes, obesity, undesirable cholesterol level in blood and anaemia. The pigment, lycopene present in the fruit lowers the risk of prostate cancer. For strong bones and teeth, skin health and better eye sight, the fruit is helpful. The chemical, captin present in the fruit alleviates heart disease. Elimination of heavy metals from the body is considered to be a merit of the fruit. Seeds of the fruit which are very small and taken along with the pulp lower the risk of heart diseases.

Sap of the stem has vermifugal effect. The oil contained in the seeds is a mild laxative. Other medicinal properties of the fruits and plant parts are also known.

Phytochemicals

The fruits contain betacyanins which include the chemical, betanin that gives colour to the fruits. Oil of the seeds contains many fatty acids including adequate amount of linoleic and linolenic acids which keep blood cholesterol and triglycerides under control. Stem of the plant contains a C-29 sterol, which is beta-sitosterol.

Undesirable Effect

Consumption of dragon fruit is known to cause allergy in some people. Fruit is rich in fructose sugar which on consumption by a person form glucose and thereby, harmful for the diabetic people.

Utilization

A number of fruit products are made from dragon fruit in different countries. Jam and jelly are most common but apart from these, other products like candy, pudding, yoghurt *etc.* are also made.

Panchal *et al.* (2018) from Ahmednagar in Maharashtra reported preparation of jelly from the fruit and also its storage. According to them, the jelly is not infected by micro-organisms for several months.

Squash made from the juice of the fruit is highly refreshing. Minh *et al.* (2019) from Vietnam investigated with different factors that had influenced the production of dragon fruit nectar. It was observed that best quality of dragon fruit nectar had been obtained by adjusting the *p*H at 4.2, supplementation with 8 per cent sugar and pasteurization at 95° C for one minute. Storage of the nectar had been best in glass-bottle at 3° - 5° C.

Family: Dilleniaceae

Species: *Dillenia indica* Linn.

[Synonyms : *Dillenia specioa* Thunberg.; *D. ovata* Wall. ex Hook.; *D. elongata* Thunberg.]

Taxonomic Serial: 506116

Common Name

Most widely used common name of the species in English is Dillenia. However, other names that are used in English are Elephant apple. Indian catmon, Indian simpoh, Pickle fruit *etc.* In Malaysia, the fruit is known as Simpoh and in Portuguese as Fruita estrela, Flor de abril *etc.*

In different parts of India, the fruit goes by the names as Chalta, Chalita, Karambel, Karmal, Panch phal, Ram phal, Pedakalinga, Thulo tatri, Ovu, Uva, Betta kanigala, Punna, Ulugach, Dabrusi, Jhalpola, Outenga, Ugga kai, Akku, Syalita, Uvaa-thekku *etc.* In Sanskrit, the fruit has been mentioned as Avartaki, Bharija *etc.* (In English, fruit is called Elephant apple, perhaps because it is highly liked by elephants. It needs mentioning in this connection that another fruit, which is *Limonia acidissima L.* also goes by the common name as elephant apple).

Botanical Family

The family belongs to the Super-order, Dillenianae and the order, Dilleniales and consists of ten genera. Trees, shrubs, woody climbers and rarely herbs come to this family. Leaves of the species under the family are simple or pinnately compound, usually spirally arranged or rarely opposite and petiolate. Leaf-blades are leathery with entire or serrated margins. Flowers are bisexual, rarely unisexual, actinomorphic, solitary or in racemes, with imbricate, persistent sepals and white, yellow or red petals. Stamens are numerous, develop in centrifugal manner and staminodes are often present. Carpels are superior, free or hemisyncarpous. Fruits are follicle, polyfollide, berry, capsule or indehiscent and enclosed by sepals. Seeds are with copious endosperm and very small embryos.

Genus: Dillenia

The genus, *Dillenia* is named after the German botanist, Johann Jacob Dillenium and consists of more than 100 species which are distributed in tropical and sub-tropical countries. Trees or rarely shrubs belonging to the genus are evergreen or rarely deciduous with red, grey or brown coloured bark and simple, alternate leaves. Flowers are solitary or in few to many, large and solitary or in terminal racemes with fleshy sepals. Stamens are many and the inner ones are usually longer than the outer ones. Anthers dehisce by apical pores and the carpels are partially joined. Fruit is enclosed by large sepals and dehisces along ventral suture or indehiscent. Seed is usually one or few and arillate or not.

Importance

Dillenia is not considered as a dessert fruit for very high acidity of the tough, fleshy adherent sepals which form the edible part. But for high acidity, it is preferred for cooking, particularly in south India to make sour preparations. For cooking or making pickle, the fruit is much used in West Bengal and Odisha also. The fruit gives off nice flavour when ripened.

The tree has however, great ornamental importance for its graceful appearance, for the large, handsome and fragrant flowers and for uniqueness of the fleshy fruits. Hence, it is planted along avenues.

Timber value is another importance of the tree for which it is considered as a dual purpose tree (Mazumdar, 2004). The timber is very durable and takes a good polish. The sapwood is also of importance although it is attacked by borers. This is used in making posts, rafters, shafts and agricultural implements and is particularly used in textile shuttles. The tree has its importance in making plywood also.

Leaves of the tree are used in wrapping bidis (= country made cigarettes) for their flavour, colour, leathery texture, easiness to use of drying and good burning nature. They are also used as fodder to cattle and goats.

Dillenia tree is also much preferred for planting in forests as fruits are highly liked by many wild animals, *e.g.*, elephants, monkeys, deers *etc.*

In folk medicine, leaves, flowers and fruits of dillenia tree are of great use in India.

Description of Plant Parts

The dillenia tree is usually evergreen but in some places, it remains as semi-deciduos in accordance with season and particularly if the winter is prolonged. The tree reaches a height of 15 metres or more in deep and fertile soil. The crown of the tree is also denser and more spready in rich soil. Bark of the trunk and branches are reddish brown, smooth and thick.

Leaves of the tree are deep green, alternate, mostly terminal and measure 15–36 cm in length and 7–14 cm in width. They are stiff, curved inwards and have conspicuously corrugated surface with impressed veins. Each leaf has 30–40 pairs of side veins. Petioles of leaves are narrowly winged. Flowers are solitary, terminal, bisexual, drooping and 15–29 cm in diameter. Sepals are 5, rounded, 4–6 cm in diameter and thickly fleshy. Petals are white, obovate and measure 7–9 cm. Stamens are many and of yellow colour. They are in two layers and those of the inner layer are taller. Anthers dehisce with two pores. Carpels are 15–20 in number.

Fruit (Plate-10)is greenish yellow and is made up of ripened carpels. It is enclosed by greatly enlarged and thickened imbricating sepals. A fruit measures 5–12 cm in diameter. It is indehiscent with fleshy sepals which are slightly swollen and are persistent. Seeds are 5 or more in a carpel with amorphous mucilage.

Origin

Exact origin of dillenia remains obscure but it is estimated to be in the region form Sri Lanka, eastern India, Bangladesh and Myanmar to Thailand, south-western China, Vietnam, Malaysia and Indochina.

Geographical Distribution

Dillenia trees are grown well in tropical and subtropical climates in many countries. In the south-east Asian countries, *viz.*, Java, Borneo, Sumatra, Malaysia, Vietnam, Thailand *etc.*, the tree is much grown. It is also largely grown in Myanmar, Bangladesh and Sri Lanka. In India, the trees are mostly grown in the peninsular part and in Odisha, West Bengal, Jharkhand and Assam.

Climatic Requirement

Although tropical climate is favoured by dillenia, it grows well in subtropical climate with moderate rainfall. Cold tolerance is fairly good for the tree and it can withstand light frost. Full sunshine is necessary for desirable growth of the tree and its fruiting performance.

Soil Requirement

The tree grows over a wide range of soil conditions but sandy loam and slightly acidic soils are much favourable. Alkaline soil is not suitable and at higher soil *p*H, micro-nutrient deficiency symptoms are usually observed in the leaves. The tree is also moderately salt resistant. Drainage is a very important factor and the trees make stunted growth in poorly drained soil.

Propagation

Dillenia is readily propagated by seed but trees raised from seeds do not retain all characters of the mother tree. On extraction of the seeds from the fruits, they are found to be glued with mucilage on the seed coat and this hampers imbibition of water in them. Seeds germinate within a month without difficulty. However, light exposure and photoperiodic condition are known to have role in germination of the seeds.

Vegetative propagation is also done in dillenia by grafting on rootstocks of the same species. Stem-cuttings are also used but rooting response is found to be higher in air-layering particularly when auxin is applied and the operation is done in rainy season.

El-Kadder and Hammand (2012) in Egypt established an efficient protocol for *in vitro* shoot multiplication in dillenia using nodal explants. Maximum shoot multiplication rate was found to be achieved with MS medium when enriched with 2 mg per litre of benzyl amino purine.

Culture

Dillenia trees are found to grow well without care. However, it is desirable that the soil should be kept moist throughout the year and for this, mulching the soil during summer is a good practice using a suitable and preferably thick, organic mulch. Yield and quality of the fruits are found to be much higher when the trees are grown in well-textured soil and with high organic matter content. Many fruit growers in West Bengal are apt to apply rotten cow-dung on soil around the trees of dillenia in winter season, particularly when they are grown in their homestead land. Size of fruits and the yield performance of the trees are known to be greatly increased by this practice.

In a study aimed at determining nutritional status of leaves, De *et al.* (1996) sampled leaves borne by trees grown in the South 24-Paraganas district in West Bengal. The samples were analyzed for nitrogen, phosphorus and potash contents at monthly interval from the months from December to May. It was observed that all the nutrients had increased and the rate of increase tended to be higher from March to May as regard nitrogen and the phosphorus contents. In Bangladesh, Sajib *et al.* (2014) also determined minerals and heavy metals in dillenia. Bole of the tree, *i.e.*, main trunk becomes very tall and hence, proper training in its young stage is necessary to restrict stem elongation.

Pests and Disease

Pests and disease are not serious problems in dillenia. Occasionally, leaf spot symptoms, fruit borers and bark-eating caterpillars are found in trees grown in southern part of West Bengal. However, in parts in India where there is infestation of monkeys, a great many fruits are eaten and damaged by them. In western lateritic zone of West Bengal, sometimes herbs of elephants intrude into the villages from the adjacent forest in search of dillenia and other fruits and cause serious damage to trees, other standing crops grown and hutments.

Harvesting

Harvesting of dillenia fruits is continued over a long period in southern and eastern parts of India. In eastern India, peak season of harvest is from the end of July and upto September. Fruits are obtained in more than one season in south India.

In Assam, the fruits have been stated to require about 160 days to reach harvest-maturity from the stage of fruit set (Neog and Mohan, 1993). However, in West Bengal and Odisha, 100 to 120 days are usually found to be required by the fruits to harvest.

Maturity of harvest of the fruits is mainly judged by examining their size but their rotund appearance and colour-break, *i.e.*, development of yellow colour on some part of the fruit is also much considered. According to Neog and Mohan

(1993), the fruits are ready for harvest when they assume straw-yellow in colour with a greenish tinge.

Conventionally, each fruit is picked by clipping the stalk on cutting if off with a knife. But a pruner fitted with a spring-attached blade and cloth basket as described in *Dimocarpus longan is* efficient as picking may be done standing on ground, *i.e.*, without climbing on the tree. Yield of dillenia fruits is difficult to assess as it varies on a number of factors, particularly type of fruits, weather conditions and soil fertility.

Varieties

Any named variety of dillenia is not known although many variations are found in fruit and tree characters. Shape of the fruits may be roundish, oblate and sometimes, somewhat obovate fruits are also borne. Size of the fruits from some trees are found to be exceptionally large. Colour of the fruits also varies from green to greenish yellow. Trees may become very tall and the crowns may be highly spready or not. Whether such characters are inherited or controlled by edapho-climatic conditions however, need to be assessed.

Mode of Consumption

Although a fresh and raw dillenia fruit is highly relished by a monkey, elephant or a deer, it is impossible for a human being to do so. In Indian cuisine, the bitter and sour, fleshy sepals of dillenia are used in curries and to make pickles. Adding the sepals in cooking of pulse is very common in south India to make the preparation soury. In West Bengal and Assam, cooking of the sepals is done to make, *ombol* (*omlo*). In parts of eastern India and Bangladesh, the sepals are also cooked with small fishes like *marulla*, small shrimps or prawn.

Quality and Nutritive Value

It has been stated (Gopalan *et al.,* 2018) from the N.I.N. (I.C.M.R.) in Hyderabad that the dillenia fruits contain 82.3, 0.8, 0.2, 0.8, 2.5, 13.4, 0.16 and 0.26 gram respectively of moisture, protein, fat, minerals, crude fibre, carbohydrates, calcium and magnesium per 100 gram and provide 59 K calorie (= 246.85 KJ) of energy on consumption. Mandal and Mazumdar (1995) analyzed dillenia fruits grown in southern part of West Bengal and the results of their study revealed that average weight of a fruit was 592 gram, specific gravity as 0.98, stalk-stylar length as 8.8 cm, total sugars as 0.61 per cent and vitamin C as 20 mg per 100 gram of pulp.

Medicinal Value

Dillenia has medicinal value in its leaves, bark and fruits. Benefit of the fruits in relieving abdominal pain, inflammation, skin diseases, dandruff, cardiac ailments, hepatic disorders and many other ill-health condition has been stated, including anti-leukemic, anti-microbial and anti-diarrhoeal properties. In ayurvedic medicine, the fruit has been given importance for alleviating pitta and cough. Padmavathi *et al.* (2011) have presented a review on morphology, phytochemistry and pharmacological aspects of the tree. Significant anti-leukemic activity of dillenia fruits has been reported by Kumar *et al.* (2010). According to Munmee and Prasad (2013), dillenia fruit controls type-2 diabetes when taken on a regular basis. Juice of dillenia fruits also stimulates hair growth when massaged on scalp (Mazumdar, 2004a).

Phytochemicals

Many phytochemical compounds have been identified and isolated from leaves and bark of dillenia in different parts of the world. In the fruit, presence of tannin, saponin, arabinogalactan, flavonoids, lupeol group of triterpenes such as betulin, betulinic acid and many other compounds have been on record.

Constituents in Developing Fruits

Under Assam condition, physico-chemical changes of dillenia fruits during their development have been stated by Neog and Mohan (1993). Mandal (1997) and Mandal and Mazumdar (1998) under West Bengal condition analyzed dillenia fruits at four equally spaced stages of their development starting from fruit-set and upto attainment of harvest-maturity stage. Size, weight and specific gravity of the fruits, titratable acidity, sugars and the vitamin C content in them were studied with progress of their growth.

Utilization

In India, dillenia fruit is highly preferred in making sweet or sour or hot pickle with addition of spices, molasses, oil *etc.* when long-term preservation in required. For short-term preservation, *chutney* is commonly made. In south-east Asia, Thailand and other Asian countries, jam and jelly are also made with the fleshy sepals. In some countries, beverage with or without fermenting the water-extract of the sepals is also done though to a lesser extent.

Family: Ebenaceae

Species: *Diospyros melanoxylon* Roxb.

[Synonyms: *Diospyros dubia* Wall. ex A. DC.; *D. exsculpta* Bedd.; *D. exsculpta* Dalzell and Gibson; *D. montana* B. Heyne ex A. DC.; *D. roylei* Wall. ex A. DC.; *D. rubiginosa* Roth; *D. tupru* Buch.-Ham.; *D. wightiana* Wall.]

Taxonomic Serial: 26431

The species, *Diospyros melanoxylon* and *D. tomentosa* are closely related. In the former species, leaves are elliptic, narrowed at both ends, tip obtuse or rounded or sub-acute, base usually acute and in the latter species, leaves are ovate and base rounded (Kumar, 2001).

Common Name

In English, the species is known as Ebony, Coromandel ebony (= Coromandel: coast of south-eastern part of India), Black ebony, Ebony persimmon, East Indian ebony *etc.* (Trade name is Ebony).

In India, the species is known by several names and some are as follows.

Sanskrit: Dirgha patraka. Hindi: Kendu, Tendu, Timburni Bengali, Jharkhandi and Oriya: Kend, Kendu, Kand, Tendu. Tamil: Karai, Karundumbi, Tumbi. Telugu: Tuniki, Beedi aaku. Kannada: Abanasi, Balai. Marathi: Tendu, Temru, Timburni. Malayalam: Kari.

Botanical Family

The family, Ebenaceae (also known as Ebony or Persimmon family) is named in accordance with the name of the genus, *Ebenus* by Otto Kuntzee in 1891. Belonging to the Order, Ericales, the family includes more than 750 species of trees and shrubs which are distributed in the tropical, subtropical and also warmer parts in temperate countries. Some species of the family produce ebony, *i.e.*, black wood and some species produce aromatic wood. The species are highly conspicuous in their native ecosystems.

Leaves of the trees and shrubs of the family are usually alternatively arranged but some species have whorled or opposite leaves. Inflorescence is usually a cyme of flowers and sometimes a raceme or a panicle. Most of the species are dioecious. The flower consists of 3–8 petals which are joined at the base. Fruit is a berry and contains tannin for which it is usually not eaten by birds and animals. But ripe fruits of some species serve as food for some wild animals.

Genus: Diospyros

The genus, *Diospyros* has been named from the ancient Greek words, Dios and Pyros, which denote divine fruit or divine wood. More than 700 species of trees, shrubs and small bushes have been included under the genus that grow in tropical and subtropical parts in different countries and some grow in temperate zone also.

Many species of the genus have commercial importance as well. Hayes (1970) has quoted in his book, "Several species of the genus, *Diospyros* bear edible fruit and some also produce ebony, the black tropical hardwood (ablush). Commercially, the value of the wood far exceeds that of the fruit but nevertheless, one species ranks as one of the most important fruits in Japan and China. This is persimmon, kaki, keg fig or date palm, *Diospyros kaki.*" There are some species which apart from importance in producing human consumable fruits are valued as ornamental trees. Many species have plentiful medicinal properties as well.

Leaves of one or two species have tremendous importance in making *beedi*, which is a crude form of cigarette. Fruits of few species have high amount of pectin and are used in good sizing material for textiles.

Importance

Ebony or tendu is a tree, all parts of which are valued barring the root. By far, leaves of the tree are of great worth for their use as wrapping material of tobacco flakes to make *beedi* which is smoked by many Indian people. To make the product, tobacco flakes are wrapped with tendu leaves and then the leaf is folded and tied with string. Although leaves of some other species of trees, *e.g., Butea monosperma, Shorea robusta, Dillenia indica etc.* are also sometimes used as wrapping material for *beedi* to some extent, leaves of tendu offer as the most suitable and enjoyable material and the reason has been stated by Kumar (2001). The leaves of the tree are also exported from India. Some States like Madhya Pradesh, Maharashtra, Uttar Pradesh, Andhra Pradesh, Odisha, West Bengal, Jharkhand, Karnataka *etc.* provide employment to a large number of labourers for collection of tendu leaves and making *beedi*. Plentiful leaves obtained from pruning offer as very good fodder for domestic animals as they contain high amount of protein.

Wood of the tree is hard, tough, durable and hence, is much used for making poles, beams, ploughs, boxes *etc.* The pruned branches are used as good fuelwood materials. All parts of the tree are used in traditional medicine.

Description of Plant Parts

Tendu is a medium sized tree or a shrub which reaches a height of 20–25 metres. It remains evergreen in moist climate but is deciduous in drier region. Bark of the trunk and main branches peels off in rectangular scales and this is the identifying characteristic feature of the tree. Bark is black or dark brown, very thick, hard, corky, deeply fissured vertically and the flakes are rectangular in shape. Heartwood of the tree is black. Leaves are thick, opposite or alternate and coriaceous. They are 30–35 cm long, dark green, elliptic and have petioles. Main source of production of leaves for wrapping are those which arise from root-suckers. Those plucked from upper parts of the trees are stiff, brittle and not flexible and hence, they are not very suitable materials as with those obtained from root-suckers.

Trees are dioecious. Male flowers are 1–1.5 cm long and sessile. Female flowers are mostly extra-capillary or sometimes lateral and are larger than male flowers. Calyx is 4–5 lobed, cordate and the petals are white. Fruit (Plate-11) is ovoid or globose, 3–4 cm wide and is a 1–8 seeded berry. Pulp is yellow, soft and sweet-sour in taste with some astringency. Seeds are compressed, oblong and shiny.

Origin

Most authors are of the opinion that India is the origin of the species. It has also been stated that it is a native of "most woody mountainous countries, *viz.*, India, Ceylon, Malabar, Coromandel" –Roxburgh (Kumar, 2001).

Geographical Distribution

Tendu is grown in the forests of Sri Lanka, India, Pakistan, Nepal and in the south-east Asian countries. In India, it is abundantly grown in Madhya Pradesh, Maharashtra, Odisha and Andhra Pradesh. It is also widely grown in Rajasthan, Uttar Pradesh and Gujarat. The tree is found to be grown in the forests in West Bengal, Jharkhand, Odisha, Andhra Pradesh, Tamil Nadu, Karnataka and Kerala as well, though not to a very large extent.

Climatic Requirement

Tendu trees are sensitive to low temperature and mature trees are more easily damaged. Tolerance to low temperature is however, greater in young trees and they may survive at sub-zero temperature for some period. In the tropical region, the tree grows upto an altitude of 900 metres. The tree prefers a mean annual rainfall of 800–1200 mm. Full sunshine is very necessary for the mature trees but younger trees tolerate partial shade.

Soil Requirement

Soil is not a very important factor for tendu tree provided drainage is good. However, the tree makes best growth in loose and porous soil. The tree produces deep tap-root and in deep soil, it becomes very tall and vigorous. In Florida in the U.S.A., the tree is found to be moderately salt resistant.

Slightly acidic soil with a *p*H of 5.5 to 6.5 is favourable to the tree. According to Hayes (1970), the soil need not be very rich but better crop could be obtained if the nitrogen content in the soil would be high and damage of the tree would not occur even if the soil fertility would be in excess. He mentioned that when the trees were grown in shallow soil, they remained smaller and would give rise to fruits earlier than when grown in deep soil.

Propagation

Tendu is conventionally propagated by seed and any vegetative method of propagation has not been standardized. It is estimated that one kilogram of ripe fruit gives rise to about 200 gram of seeds.

Seeds retain viability for a year in open atmosphere storage. Pre-soaking the seeds for a day is found to improve germination and soaking in warm water gives better result. Seeds may be sown in perforated polythene bag in a medium consisting of equal parts of loamy soil, leaf-mould and sand and when the seedings would become 15–30 cm tall, they may be transplanted. The tap-root which becomes very long in short period should be taken care and not to be broken in transplanting. Direct sowing *in situ* is also a good practice and the seedlings should be thinned out in instalment so as to retain ultimately one seedling to grow which is most vigorous.

Propagation may also be done from the root-suckers. These may be detached from the tree and planted directly or in the nursery and then transplanted. Stump planting of tendu is sometimes done for propagation.

Culture

Tendu tree does not require any manure or culture for its growth but growth of it is vigorous and productiveness is high if it receives inputs. If mass planting is desired, such a site should be chosen for growing the trees which is not shaded by other trees. Deep soil is also preferable as the tree has long tap-root.

As the tree is dioecious, the plantation should have at least 10 per cent male trees to act as pollinizers and these male trees should be well-distributed in the plantation. But it could not be known which one would be female tree until the flowers are produced. Hence, some plants of tendu may be grown in large pots side by side. When flowers will be produced by the trees in the plantation as well as in those grown in pots, the plantation should be reset by transplanting male or female trees which have been raised in the pots. This should be done in such a way that in the plantation, there exists about 90 per cent female and 10 per cent male trees and these should remain in a well-distributed manner all over the planted area.

Planting of tendu may better be done in rows maintaining a spacing of 8–10 metres in rows and 2–3 metres between trees in a row. As the growth of the plants is slow at early stage, nitrogen fertilization is of advantage to boost up their growth. Inter-cropping tendu with other crops may be done successfully. The trees make deep root system and hence, there is a benefit that they would not compete with the inter-crops.

Training the trees by *open-centre, i.e., vase* system at early stage of growth admits adequate sunshine in the interior part of the trees and this would be much helpful for higher production and better quality of fruits as well as leaves. Lopping of branches and young stems is a common practice to encourage new shoots and leaves. Pruning ensures higher production of fruits of larger size and of better quality. The trees bear fruits on old shoots and those which are upto 15 cm girth may be cut back near the ground level. This would encourage sprouting of a large number of coppice shoots. The emerged out shoots would give rise to production of large number of good quality leaves and in the next year, they would be able to produce large number of flowers and fruits. Tendu trees have biennial or irregular bearing habit like mango and pruning the trees would reduce the intensity of the problem of bienniality or irregularity of bearing in them apart from increasing productivity.

Pests and Diseases

Some insect pests and diseases are found to attack tendu trees but they are not serious. The insect, *Trioza obsoleta* attacks leaves which feeds on the sap and forms reddish-yellow galls on them. There are also defoliators, *e.g., Euthalia laudabilis, E. nais, Lamida carbonifera* and *Miresa albipunctata.* Boring between the bark and sapwood is done by the larvae of *Plocaederus ferrugineus*. To control the pests, Chlorpyriphos 20 EC at 0.2 per cent may be sprayed. The fungal species *Stereum lobatum* causes spongy rot in felled timber and leaf spot is caused by *Cercospora kaki*. To control the diseases, Copper oxychloride–0.25 per cent or Dithane M–45 may be sprayed.

Harvesting

Tendu trees produce flowers after 8 to 10 years of planting. Flowers appear in April to June and the fruits ripen in summer to rainy-season, *i.e.*, from March to June.

Yield

Yield of fruits of tendu greatly varies and is estimated to be 100–200 fruits in a tree. (Production of leaves in Madhya Pradesh comes to around 25 lakh standard bags annually and a standard bag of leaves contains about 1000 bundles of 50 leaves).

Mode of Consumption

Pulp of the tendu fruit is yellow coloured, soft in texture, sweet and is taken as fresh. Fruits which are not fully ripe contain tannin and hence, are somewhat astringent. In over-ripe condition, the fruits emit a smell of alcohol.

Medicinal Value

Chintala *et al.* (2012) went on to comment that all parts of the tendu tree are medicinally important and cure a number of ailments. They made detailed pharmacognostic study of leaves, stem, petioles and root in order to establish standards selecting parameters like microscopical studies, proximate analysis, fluorescence analysis and phytochemical screening.

Studies done by various authors revealed that pentacyclic triterpenes that are found in the leaves of tendu have antimicrobial property. The bark of the tree has anti-diabetic, anti-diarrhoeal and anti-plasmodial effects and is also useful in infection by the disease causing fungal species of *Candida*. Flowers are dried and used in folk medicine. It is prescribed in urinary tract infection and in blood purification. Seeds have value in mental disorder and in cardiac disorder. Betulinic acid which is present in the fruit, leaf, stem-bark and wood of the tree has many effects, particularly in intestinal worms, tumours, inflammation, malaria, virus disease and also in cancer of some types.

Utilization

Paste made with the fruit pulp is dried and used by the tribal people. Pellets of 2–3 cm size made by blending fruit pulp of tendu and jackfruit and sun-dried by the tribal people and used when the fruits are not available, i.e., in off-season.

Sahu *et al.* (2012) experimented with preparation of wine from tendu fruits. Sensory evaluation was carried out by them by trained persons and it has been commented upon by them that the wine had been of acceptable standard.

Family: Ebenaceae

Species: *Diospyros peregrina* (Gaertn.) (Gürke)

[Synonyms: *Diospyros embryoptheris* Pers.; *D. malabarica* (Oeser.); *Embryopteris peregrina* Gaertn.]

Taxonomic Serial: 23855

Common Name

In English, the common names of the species are Indian persimmon, Wild mangosteen *etc.*

The names of the species in Indian languages are as follows:

In Sanskrit, it has been referred to as Tinduka or Tinduki. The names in Hindi are Gaab, Kalatendu, Makur kendi *etc.*, in Bengali, Gaab, Desi Gaab, Tendu, Makur kendi *etc.*, in Oriya, Kendu, Dhusaro kendu, Gaub, Gaabo *etc.*, in Tamil, Kattati, Kavikattai, Tumbi, Tumbika, Panicca, Panicika *etc.*, in Telugu, Bandadamara, Gabu, Tinduki *etc.*, in Kannada, Kusharta, Holetupare *etc.*, in Marathi, Temburni, Timburni *etc.* and in Malayalam, Panacci, Pananci, Panancca, Panchi vananji *etc.*

Importance

Fruits of Indian persimmon or Gaab are although not of very high quality and are not much sold in the market but have popularity among the villagers and tribal people in localities where they are grown. The tree has ornamental importance also for which, it is sometimes planted along avenues. The fruits contain high amount of pectin and hence, it is used as a good sizing material for textiles (Biswas, 1944).

The wood of gaab is close grained, moderately hard and heavy and for this quality, it is used in making poles, building materials, making boats *etc.* Besides that, the charcoal made from it is also of good quality (Kumar, 2001).

Description of Plant Parts

The tree is evergreen, dioecious, profusely branched with spready crown and attains a height of 10–12 metres but may be very tall. Bark is greyish black and flaky. Young branches are terete to slightly angular and glabrous.

Leaves of the tree are simple, alternate and distichous. Leaf-blade is 15–20 cm long and 6–7 cm wide, oblong-lanceolate with acute to obtuse apex and glossy. Male flowers are formed in 3–5 flowered cymes in leaf axils. Female flowers are solitary, 4-parted with 4 styles and have 8-celled ovary. Fruit is a berry, globose, 3–5 cm across, reddish velvety, fruiting calyx large and reflexed. Fruit has a leathery rind.

Seeds are 4–8 in a fruit, smooth, embedded in viscid glutinous pulp and turn brown on exposure to air.

Origin

Exact origin of gaab is not known. Singh and Saxena (2008) have indicated India to Malaysia as the nativity of the species.

Geographical Distribution

Gaab grows in Sri Lanka, Bangladesh, Myanmar, Thailand, Malaysia, North and South Vietnam, other countries in south-east Asia, Australia, South Africa, Malagasy, Brazil and other countries in south America. In India, large number of trees are found to grow in Odisha, Jharkhand, West Bengal and along the western ghats.

Climatic Requirement

Gaab is a tropical tree and moist tropical situation is preferred by it. However, it grows in subtropical climate also where winter temperature does not go very low. In the tropics, it grows upto an altitude of 500 metres. For optimum production, a mean annual rainfall of 1500–2500 mm and day temperature of 25°–35° C are required by the tree. Unlike the allied species, *Diospyros melanoxylon*, the trees of which require full sunshine, *D. peregrina* trees grow well in partial shade.

Soil Requirement

The tree succeeds in many types of soil. Depth of the soil is though not very important, in shallow soil, the trees remain smaller and come to bearing maturity earlier than when grown in deep soil. A soil pH of 6–7 is favoured by the trees.

Propagation

Seed propagation is most common for gaab and no vegetative method of propagation is known to be standardized. Seeds have however, short period of viability and hence, should be sown without storing them. Pulp of the fruits should be removed thoroughly and the seeds should be washed repeatedly as the pulp contains chemicals which tend to inhibit germination. Seeds may be sown in nursery beds or in polythene bags and the resulting seedlings should be transplanted. Sowing seeds directly in permanent place is advisable. For this purpose, many seeds should be sown together at one place and the emerged out seedings should be uprooted in instalment, retaining ultimately the most vigorous seedling to grow.

Culture

If mass planting of gaab at a place is desired, similar measures as have been stated in *Diospyros melanoxylon* should be followed. A site with full sunshine and a deep soil are not considered essential for gaab. Maintenance of male trees to act as pollinizers should be done in the same way as described in *D. melanoxylon*.

Pests and Diseases

The same insect pests and diseases that attack *Diospyros melanoxylon* may attack gaab trees also and similar control measures should be adopted if they are attacked.

Harvesting

The trees come to flowering after 7–10 years of growth. Flowers appear in summer to rainy-season and the fruits become ripe for picking after 4–5 months. Yield of gaab is not assessed but a tree may produce 2000 fruits or more in season.

Mode of Consumption

Ripe fruits of gaab are dark coloured measuring 2–5 cm in diameter, have a pleasant smell and the soft pulp is taken as fresh. Fruit is taken only when fully ripe as even slightly unripe fruit is acrid, bitter and oleaginous. It should be remembered that the fruits of neither *Diospyros melanoxylon* nor *D. peregrina* is comparable in taste to the Japanese persimmon fruit, *i.e.*, *D. kaki* which is known as persimmon, kaki, kaki-persimmon, or keg fig and which is one of the most important fruits in Japan and China.

Medicinal Value

All parts of the gaab trees have a number of medicinal properties and in ayurvedic literature, those have been amply mentioned. Leaves of the tree have anti-diabetic property and the bark is used in dysentery, intermittent fever, cholera and bilious fever. Unripe fruits are known to heal sores and wrinkles and have anthelmintic properties. Ripe fruits are of value in dysentery, leprosy, gonorrhoea and oral ulcer. They are also used as antidote for snake venom. The oil derived from the seeds is used in dysentery and diarrhoea. Betulinic acid which is present in the fruit, leaf and bark of gaab is efficacious in many diseases, tumours, malarial fever and also in cancer.

Dewanjee *et al.* (2009) studied the role of aqueous extract made from the fruit in streptozotocin-nicotinamide induced type-2 diabetes in rats. Oral administration of the extract in suitable dose was found to possess significant hypoglycaemic and hypolipidaemic activity.

Phytochemicals

Unripe fruits and bark of gaab are rich sources of tannin. Roots contain glycerides, bark contains myricyle alcohol, saponin and triterpenes, stem contains beta-sitosterol, alpha-leuconanthocyanin, leaf contains triterpenes, fruit pulp contains alkenes, triterpenes and seed contains betulinic acid, beta-amyrin, fatty oil and unsaponified matter.

Family: Elaeocarpaceae

Species: *Elaeocarpus floribundus* Blume

[Synonyms: *Elaeocarpus floribundus* var. *tahensis* (Hend.) Ng; *E. grossus* Wall.; *E. lobbianus* Turcz.; *E. pseudosepicanus* O.C. Schmidt; *E. tahensis* M.R. Hend.]

Taxonomic Serial: 27816

Common Name

In English, the fruit is known as Indian olive.

In different parts of India, it is known by a number of names and some of these are Perinkara, Nallakara, Valiyakara, Jolapari, Jalpai, Desi Jalpai, Avil, Karmava, Chorophon, Bhadrasey *etc.*

Botanical Family

The family, Elaeocarpaceae (Syn.: Aristoteliaceae, Tetrathecaceae, Tremandraceae) is commonly known as Rudrakhsha family, or Blue-marble tree family, or Jamaican cherry family. The family is dilleniid dicot, *i.e.*, a family of more or less advanced dicotyledonous trees, shrubs or herbs and consists of 10–12 genera and more than 600 species which are mostly tropical and subtropical, although some are adapted to temperate climate. In some classification however, these are included in the Santalaceae (Sandal wood) family.

Genus: Elaeocarpus

The name, Elaeocarpus has derived from the Greek words, Elaei, which means wild olive tree and Carpus which means fruit. The genus consists of about 350 species, which are evergreen trees or shrubs and are adapted to tropical and subtropical climates.

The species belonging to the genus are distributed from Malagasy to India, south-east Asian countries, Malaysia, southern China, Japan, Australia, New Zealand and the Pacific islands. The species are however, most popular in New Guinea and Borneo where they abound in to a great extent. Many species of the genus, *Elaeocarpus* growing in different parts of the world have now become extinct. In India, about 25 species are known to be grown now.

Importance

Indian olive and other species of *Elaeocarpus* have great importance as forest trees. The tree is also preferred for planting along avenues. Besides, it offers a very good material for planting on the boundaries of orchards for use as wind-break rows.

The fruits of Indian olive are difficult to consume as fresh for high sourness. But excellent pickle is made from them with or without mixing with other fruits.

The wood of the Indian olive tree is also of great value in making plywood, particle-board, paper-pulp, fire-board and other purposes. It is used in making flexible, light interior materials and art-objects as well.

Mention may be made in this connection that the rudrakhsha tree (*Elaeocarpus ganitrus* Rox.) belonging to the Indian olive genus has great religious importance among the Hindus as its seeds are made into garlands that are worn to worship the lord, Shiva. Such garlands when worn are known to ward off some diseases as well.

Description of Plant Parts

Indian olive tree is evergreen and reaches a height of 15–25 metres. The branches are highly spready and hence, a large area is occupied by the crown. Bark is brown, rough, lenticelled and the inner bark is pale yellow.

Leaves of the tree are simple, alternate, ovate-elliptic in shape, thinly leathery with toothed margins and pointed tips. Petioles are long. Younger leaves are sour in taste. Inflorescence is raceme, 3–12 cm long and is borne in leaf-axil. Flower is 0.5 cm long, cup-shaped and creamy white in colour. Petals are 5, yellowish white in colour, covered by sepals and are divided into many frilly segments. Fruit is drupe, green to dark green in colour, 2–5 cm long, 2–3 cm wide and the surface is smooth. It is bulged at the centre and tapered at both ends. Pulp of the fruit is fleshy and sour. Stone is 3-celled and each cell has a spindle-shaped seed.

Origin

Origin of Indian olive is believed to be in the region from India to Myanmar, *i.e.*, the Indo-Burma region, which has fame to be the origin of mango and many other fruit trees. However, north-eastern Bangladesh, India or Java has also been suggested to be the centre of origin of Indian olive (Pareek and Sharma, 1993).

Geographical Distribution

The tree grows freely as wild and in the forests located in the eastern States, western ghats and the Andaman and Nicobar islands and the Laccadive islands in India. In fact, the tree is found to grow in the whole of the eastern part of the world comprising the countries, *viz.*, Bangladesh, Myanmar, Thailand, Java, Sumatra, Malaysia, Sabah, Borneo, the Philippines, Celebes and other countries.

Climatic Requirement

Indian olive is a tropical tree and thus, hot and humid climate is most favourable. However, the tree grows in mild subtropical climate also but the yield performance and quality of fruits are much hampered in such situation. The tree is capable to grow upto an elevation of 900 metres in tropical mountains and tolerates temperatures upto (-) 5° C for some period. For better fruiting performance, the tree requires open sunny situation.

Soil Requirement

Soil is not an important factor for the tree except that drainage should be good. The performance of the tree is better in sandy loam and forest soils. In rich soil, the trees assume very tall appearance and large, spreading canopy. Soil pH is not very important for the tree but Das *et al.* (2013) stated that mildly acidic soil is favourable for better growth and fruiting.

Propagation

The tree is mostly propagated by seed. Seeds germinate satisfactorily but scarification of the stones is a good practice to facilitate and quicken their germination. Mechanical scarification by rubbing the stones between two sand-papers for sometime is much effective. Das *et al.* (2014) at the Bangladesh Agricultural University compared chemical and hot water scarification of Indian olive stones. It was observed by them that chemical scarification by dipping the stones into sulphuric acid gave highest germination percentage of upto 89.81 and this was followed by scarification on dipping in hot water at 100°C for 12 minutes. The said authors had also examined whether source of procurement of seed had any effect on germination. They collected seeds from four locations but the effect of location of procurement on their germination behaviour could not be observed.

For seed propagation, direct sowing of seeds in permanent place is advisable than transplanting as the latter method exerts shock to the seedlings. The tap-roots of the seedlings are also liable to be broken in uprooting of the seedlings and their transplanting. If however, transplanting has to be done, seedlings should preferably be raised in polythene sleeves having both ends opened, filled with soil-sand mixture, placing the filled in sleeves vertically on sand and regularly watering them in the sleeves.

Propagation by stem-cutting is ordinarily difficult in Indian olive but hardwood and semi-hardwood stem-cuttings may be forced to produce adventitious roots by the use of auxin in suitable concentrations. Response of auxin to regenerate root is greater in air-layering than in stem-cuttings.

Micro-propagation of Indian olive has been reported to give success by application of suitable treatments. Rahman *et al.* (2000) in Bangladesh studied shoot regeneration in internode-derived calli by culturing in modified MS medium with addition of sucrose in various concentrations and maintaining various *p*H levels of the media. It was observed that sucrose at 3 per cent concentration had been most effective and much influence was played by the *p*H of the medium. The same workers (Rahman *et al.*, 2009) had cultured nodal and leaf-extracts and observed that large number of adventitious roots were produced from leaf-derived calli in modified MS medium. Shoot multiplication rate was found to have increased to about two-fold per culture when 15 per cent coconut water was added to the medium.

Culture

Indian olive trees grow satisfactorily and produce fruits without receiving any cultural practice. But when the trees are raised as wind-break or are mass-planted, use of manure once a year in winter is advisable to induce profilic growth of the trees and higher productiveness of fruits.

Training the trees to open-centre or modified central-leader system at early stage of their growth is advisable to prevent them from becoming very tall and thereby, avoid breaking of their branches in manure condition. For improving yield and quality of fruits, pruning the trees in their grown-up stage by heading-back and thinning-out of branches are of great benefit. Fruit size and quality in Indian olive are much increased by ringing of shoots.

Pests of Diseases

Pests and diseases are not serious problems in Indian olive. However, Thiyam and Sharma (2013) from Assam Agricultural University, Silchar reported fungal infection of the fruits, mainly with *Aspergillus* species but also with other fungi.

Mode of Consumption

Indian olive fruits being very sour are difficult to consume as raw and hence, they are suitably processed so as to make them into palatable products. In West Bengal, the fruits are usually taken after making a sour-sweet juicy preparation, which is popularly known as raw jalpai amla (= *omlo*) or ambal (= *ombol*). This is done by crushing the fruits and boiling them in water, adding molasses or sugar, salt and various spices and condiments. This is served with food in lunch or dinner for use of the juice (= jhol) as dessert. In south India, the sourness of the fruits is a likeable feature and is cooked with pulses and curries.

Quality and Nutritive Value

The high sourness of the fruits is due to the presence of a number or organic acids, among which citric acid has been said to be predominant (Utami *et al.,* 2013). Fruit is rich in vitamin C as well.

Mandal, U. and Mazumdar, B.C. (unpublished) analyzed physico-chemical constituents of fully mature and freshly plucked fruits of Indian olive during the end of October, producing samples from trees growing in the South 24-Paraganas district in West Bengal. The results obtained were as follows:

(a) Surface-colour: green to deep-green and in some samples, greyish, (b) Shape: oval to ovoid. (c) Stalk-stylar length: 4–5 cm. (d) Diameter at maximum bulge: 3.1–3.7 cm. (e) Weight per fruit: 3.3–3.9 g (f) Specific gravity of fruits: 1.11–

1.18. (g) Weight per stone: 3.09–3.2 g. (h) Length of stone: 2.3–2.8 cm. (i) Specific gravity of stone: 1.78–1.89. (j) Total soluble solid of pulp juice: 8.6–9.1 per cent.

(Index: g = gram, cm = centimetre).

Medicinal Value

Plentiful medicinal properties of the fruits and leaves of Indian olive are known. Appreciable amount of vitamin C is present in the leaves also.

In ayurvedic system of medicine, fruits of Indian olive are prescribed for various illnesses, notably nervous malfunctioning, epilepsy, Parkinson's disease, dementia, psychic depression and related disorders. In other illnesses also, the Indian olive fruit is advised to consume, such as in hypertension, asthma, arthritis, dysentery, diarrhoea, liver diseases *etc.* Many of the medicinal properties of the fruits and leaves of Indian olive have been mentioned by Das *et al.* (2013) from Tripura.

Phytochemicals

A number of phytochemicals *viz.*, myricetin, myricitrin, mearnsetin, ellagic acid *etc.* have been isolated from the leaves and bark on Indian olive trees. A number of chemicals including phenolic compounds, anti-oxidants *etc.* have also been extracted from leaves and bark of the Indian olive trees that are grown in Malaysia (Utami *et al.,* 2013).

Utilization

Pickle is the most common product that is made with fruits of Indian olive. Either the fruit is used singly or more commonly, it is mixed with other sour fruits such as anola, hog-plum, carambola, dillenia, otaheite gooseberry *etc.* Apart from pickle, other products are also made from the fruit in Tamil Nadu and Kerela.

Family: Flacourtiaceae (= Salicaceae)

Species: (1) *Flacourtia indica* Merr.

[Synonyms: *Flacourtia ramontchi* L'Herit.; *F. sapida* Roxb.; *Gmelina indica* Burn, f.]

and

(2) *Flacourita jangomas* (Lour). Raeusch.

[Synonyms: *F. cataphracta* Roxb. ex Willd.)

Taxonomic Serial: 513616

Common Name

In English, both the above species are usually known as Governor's plum, Ramontchi plum, Batoko plum, Madagascar plum, Mauritious plum, Indian plum *etc.*

The species is recognized by other names in different countries. In Malaysia, it is called Kerk up, in Thailand, Ta-khop-pa, in Laos, Gen, in Vietnam, Bo Quan An, Prunier in Myanmar, Na-yu-wai and in Sri Lanka, Uguressa. The name in French is Prunier de Madagascar, in German, Echte Flacourtie and in Spanish, Ciruela.

In Bengali, Flacourtia in known as Boinchi and according to Chatterjee and Randhawa (1952), the term, Boinchi is used for the species, *Flacourtia indica* and *F. jangomas* also. In Hindi, *F. jangomas* is most commonly referred to as Paniala, although the same name is used for the Mayayan species, *F. inermis*. The species, *F. indica* is called Kantai in some parts of West Bengal, Bihar, Jahrkhand and Odisha. In other Indian languages, the different names of Flacourtia are Bilangra, Kukai, Kakein, Kaker, Sottaikala, Tabspatra, Mullaninchil, Cholkakilai, Kutukali, Kandregu, Saralu, Kanji, Baicha *etc.*

Botanical Family

Many taxonomists have opined that Flacourtiaceae is a 'defunct family' of phanerogams. Its former genera have been placed to various other families and mostly to Achariaceae, Samydaceae and Salicaceae. Hermann Sleumer in 1975 quoted, "Flacourtiaceae as a family is a fiction. Only the tribes are homogeneous".

The classification made by Cronquist indicated that the family included 89 genera and more than 800 species that exist. Among these, many genera including the genus, *Flacourtia* have later been placed in the allied family, Salicaceae in the molecular phylogeny-based classification, known as the APG-II system which has been propounded by the Angiosperm, Phylogeny Group.

Some of the genera which were earlier included in the family, Flacourtiaceae and later transferred to Salicaceae are *Abatia, Aberia, Banara, Calantica, Dissomera, Flacourtia, Homalium, Ludia, Osmela, Ryania, Tisonia etc.*

Genus: Flacourtia

The genus, *Flasourtia* is named to honour Etienne de Flacourt (1607–1660) who was the Governor of Madagascar. The genus is small and contains around 15 species which are shrubs or small trees and they have originated in the tropical regions of Africa and Asia.

The trunk of younger trees is usually guarded by branching thorns. Leaves are pinnately veined and flowers are in small axillary racemes or clusters. Sepals are imbricated and the seed is enclosed in a hard covering which is surrounded by the pulp that forms the edible part.

Importance

Flacourtia indica is of much usefulness. As a fruit, it is though not very delicious but is edible and some better types of fruits in this species are found which are sweet enough in taste. Fruits can be used to make jams and jellies of excellent quality. They can also be dried and stored for later use. The sapwood of the tree is very hard and has good timber value. When closely planted, it serves as an impenetrable barrier for thorny trunk and hence, it is preferred for planting along boundaries. The glistening leaves are very attractive for which it is also planted as an ornamental tree or shrub. Leaves and bark of the tree have also medicinal importance.

Description of Plant Parts

Height of the plant is 3–5 metres when grown as a shrub and may reach upto 15 metres when it forms a tree. The trunk is thorny and the branches are drooping. Leaves are oval to round, pinnately veined and red or pink in colour. Flowers are unisexual or occasionally bisexual. Male flowers are in axillary racemes with slender pedicels. Female flowers are in short racemes or solitary. Fruit is globular, reddish to purplish in colour, 1.5–3 cm across with persistent styles. Pulp is yellowish or whitish. Seeds are 6–10 in layered carpels.

Origin

India and tropical parts of Africa have been conjectured to be the origin of *Flacourtia indica* and *F. jangomas* (Singh and Saxena, 2008).

Geographical Distribution

Flacourtia is found to be grown in Ethiopia, Somalia, Kenya, Tanzania, Uganda, Mozambique, Malawi, Zambia, Botswana, Namibia, South Africa, Madagascar in the African sub-continent and in Asia, it is grown in China, India, Bangladesh, Nepal, Sri Lanka, Thailand, Indonesia, and other south-east Asian countries. In India, the species, *F. indica* is grown almost everywhere in the plain land and much grown in West Bengal, Jharkhand, Odisha, Assam, Tripura, Eastern States, Maharashtra, Goa while *F. jangomas* is largely found to grow at the foothills of the Himalayas.

Climatic Requirement

Both the species of Flacourtia are basically subtropical in their climatic requirement. However, they are grown in tropical and moist subtropical climate also. The trees are drought resistant (Hayes, 1970).

Soil Requirement

The trees grow in wide types of soil but drainage is an important factor as they grow poorly in ill-drained soil.

Propagation

Seed propagation is most common for both the species. Under natural conditions, birds eat the fruits and the seeds are widely dispersed through their excreta. Seeds have low viability period and viability could be maintained over a year in air-dry hermetic storage at 5°C. However, vegetative propagation by inarching, budding, layering and stem-cutting also give good success (Singh and Saxena, 2008).

Propagation by stem-cuttings has been reported by Bora and Das (1998) from Jorhat in Assam and Jayant Kumar and Parmar (2000) from Himachal Pradesh. Both the works have indicated that root formation could be improved by application of IBA in suitable concentrations.

Culture

No culture is usually applied to the trees but irrigation if applied during development of fruits is of much benefit. The tree does very well by trimming of shoots. Fruitfulness is however, a problem for the trees due to insufficient pollination. Hence, several trees should be planted closely together to ensure cross-pollination. Moist environment around the trees favours pollination.

Pests and Diseases

Beetles and larvae of some insects attack the trees which feed on the leaves and damage dead wood. BHC (5 per cent) may be dusted to control the pest.

Harvesting

Tree is usually leafless just before flowering. Flowers appear from December to April and new leaves also appear at that time. Fruits ripen from March to July. Yield of the fruit plant is not ascertained.

Mode of Consumption

Fruits are eaten as raw. Some types are found in which the pulp is fairly sweet. Fruits are stated to be greatly improved in taste by rolling them between the tow palms before eating.

Quality and Nutritive Value

According to the National Institute of Nutrition (I.C.M.R., Govt. of India) in Hyderabad, 100 gram of *F. indica* fruits containing 67.8 per cent moisture has the following proximate constituents. (Gopalan *et al.,* 2018).

Protein: 1.7 g.; Fat; 1.8 g.; Carbohydrate: 22.7 g.; Crude fibre: 4.7 g.; Minerals: 1.3 g.; Calcium: 100 mg. and Phosphorus: 100 mg. It has been stated that 100 gram of fruit provides in body, 114 K calorie (= 477 Kjoule) of energy on consumption.

(Index: g = gram, mg = milligram)

Medicinal Value

Different parts of the tree are used in folk medicine particularly in African countries. They are known to be efficacious particularly in respiratory diseases. Bark of the tree has diuretic effect and is also used in arthritis. Leaf is carminative, astringent and its decoction is used in female diseases. Fruit has its role to cure jaundice and in enlargement of spleen. Romeiras *et al.* (2012) studied the effect of the fruit in neurological disorder in western Africa and Padmaja *et al.* (2013) studied medicinal property of the roots.

Phytochemicals

Bark of the trunk contains tannin which is much used in dyeing of garments. A phenolic glucoside, flacourtin has been isolated by Bhaumik *et al.* (1987) who had also provided structural elucidation of the compound.

Utilization

The fruits are very good material to make jelly and jam. Pickles is also made. Drying of the fruits is done like raisin and date. Fruit juice is also fermented to make wine in some countries.

Family: Guttiferae

Species: *Garcinia cowa* Roxb. ex DC

[Synonyms: *Garcinia umbellifera* Roxb.; ex Wall.; *G. roxburghii* Wight; *G. wallichii* Choicy; *Oxycarpus gangetica* Buch.–Hamil].

Taxonomic Serial: 180103

Common Name

Common name of the species in English is Cowa fruit or Cowa mangosteen.

In India, a number of names of Cowa fruit are used in different parts. In Malayalam, the fruit is called Kowa. In West Bengal, there are several names, *e.g.*, Kau, Kao-gola, Kagphal, Kag-lichu; Kawa *etc.* Tribal names of the fruits is India are Khumi, Kau-gula, Mang-tong, Aangtrisabui *etc.*

Botanical Family

Guttiferae (or Clusiaceae) comprises 50 genera and about 1200 species. The members of the family are evergreen tropical trees or shrubs which usually have milky or coloured sap. Leaves of the species are simple and opposite or whorled, or rarely alternate with no stipules. Flowers are regular, usually bisexual, rarely unisexual and hypogynous. Calyx consists of 2–10, imbricated, often decussate sepals. Petals are 4–12 and stamens are numerous. Gynoecium consists of a single compound pistil of 3–5 or more carpels and an equal number of stigmas. Ovary is superior of 3–5 or more locules and each contains one to many ovules. Fruit is usually a capsule or a berry. The family is divided into two sub-families, which are Clusioideae and Hypericoideae. The family has a characteristic feature in that, besides pollen and nectar, flowers of some species have resin which bees use in construction of their nests and hence, a reward for pollinating bees.

Genus: Garcinia

The number of species belonging to the genus may be upto 300 but this is controversial. The trees of the genus are known by several names such as sap tree, monkey fruit, brindal berry, cambogia, Malabar tamarind *etc.* The trees are small to medium in height and are mostly grown in eastern India and the south-east Asian islands. The fruit of the species is beneficial to reduce obesity, muscular pain, abdominal troubles and in many other ailments.

Importance

Cowa is a minor fruit which grows much in south-east Asian countries and in India, it is grown as wild largely in the eastern States, West Bengal and Odisha. The ripe fruit is sour in taste, gives an uncomfortable feeling in the mouth and evidently, does not stand comparison with ripe mangosteen fruit (*Garcinia mangostana*) which is a delicious fruit belonging to the same genus. Nevertheless cowa fruit is popular among the tribal people in Assam, Mizoram, Tripura and the other eastern States in India.

The cowa tree and fruit have however, other economic importance. The tree gives timber of high quality. Trunk and branches secrete a yellow gum which resembles gamboge and is used to make yellow varnish. The tree parts have many medicinal properties and are used in folk medicine in the eastern States and leaves are also consumed there on cooking.

Description of Plant Parts

The tree is evergreen and reaches a height of 8–12 metres. Bark is smooth, surface is greyish brown and it secretes a yellow sticky exudate. Branches are many borne towards top of the trunk and are horizontal. They are borne towards top of the trunk and are horizontal. Branchlets are quadrangular and drooping. Leaves of the tree are simple, opposite, decussate, exstipulate, glabrous and the margins are entire. Trees are dioecious. Male flowers are 3–8 in number and they are terminal or axillary in umbel. Female flowers are larger in size, usually solitary and axillary. Fruit is a berry, opaquely yellowish brown in colour, ovoid-globose, oblique, 5–6 cm long and 4–5 cm in diameter. Seeds are 4–8, oblong, fusiform and slightly curved.

Origin

Exact origin of cowa tree is controversial but its nativity in India has been indicated (Pareek and Sharma, 1993).

Geographical Distribution

Cowa tree is much grown in south-east Asian islands. However, it is also grown in many parts in Bangladesh, Australia, Polynesia and other countries. In India, the tree grows to a large extent in the tropical rain forests in Assam, Tripura, Mizoram and other States in the eastern part and also in West Bengal and Odisha.

Climatic Requirement

Highly tropical climate with high relative humidity is best suited for cowa tree. The tree however, grows in high altitude in the tropics but growth is less vigorous. Frost even for a short time damages the tree.

Soil Requirement

Soil is not considered to be an important factor for cowa tree but loamy forest soil is conducive. The tree prefers slightly acidic soil. Drainage is an important factor as the trees are found to make stunted growth with high leaf fall where water regularly stagnates in parts of the forests in the eastern States in India.

Propagation

The tree comes off easily from seeds as is found in the tropical forests in eastern India. Experiments done in Vietnam have provided evidence that the tree may be propagated by grafting. Among other grafting methods tested, wedge-grafting was found to give high survival rate and the survival rate was also high when grafting was done in spring than in other seasons.

Culture

Organized plantation of cowa tree is not done and the tree grows in forests and as wild. However, if plantation is desired to be set up, a spacing of 8-10, metres should be maintained between the trees. Application of organic manure and ensuring drainage would no doubt, greatly benefit the trees and step-up productivity of the fruits. Pests and diseases of cowa tree are not significant.

Although there is no named variety of the tree, but many variations are observed in the morphological characters in them, particularly in shape, size and colour of the fruits, floral characters and other features of the tree that grow in the rain forests in the eastern India.

Ripe fruits are obtained from the tree over a long period and to a considerable extent in summer.

Mode of Consumption

Cowa fruit is sour in taste and also gives an unpleasant feeling in the mouth due to sticky juice but it is liked in fresh form by the tribal people of eastern Indian States. In Tripura, fruits are also dried and stored for later use especially to cure dysentery and combat other diseases. Mixing with other fruits and honey to sweeten cowa fruits is also a practice.

Medicinal Value

The fruit and other parts of the cowa tree have been on use in the folk medicines by the tribal people in the eastern States since long time. Fruit is known to be of much help in headache and in dysentery. Leaves are used in spasm and to subside tumors. Bark has been stated to have astringent effect and antipyretic properties. Wahyuni *et al.* (2015) in a study had isolated compounds from the methanolic extract of stem bark of cowa tree and evaluated their cytotoxic effects against breast and lung cancer. The results of their study indicated that stem bark of the tree could be an important source of natural cytotoxic compounds.

Phytochemicals

Fruit pericarp of cowa contains fatty substances and seeds yield a wax-like fat consisting of glycerides of stearic, oleic, palmitic, linoleic and myristic acid. Jena *et al.* (2002) analyzed fresh fruits, leaves and dried rinds of cowa by high performance liquid chromatograph and their findings revealed that the major organic acid present in those parts was hydroxycitric acid while hydroxycitric acid lactone, oxalic acid and citric acid remained present in lesser amount.

Utilization

In the eastern States in India, cowa fruits are very often processed to make sweet, sour or hot pickles which have demand in the market. The fruits are pickled with or without blending with other fruits. *Morabba* is also produced from the fruits but to a lesser extent.

Family: Lythraceae

Species: *Trapa natans* Linn. var. *bispinosa* (Roxb.) Makino (= *Trapa bispinosa* Roxb.)

and

Trapa natans Linn. var. *quadrispinosa* (Roxb.) Makino (= *Trapa quadrispinosa* Roxb.)

Taxonomic Serial: 27170

Common Name

Most commonly used name of the species Water-chestnut. However, it should be noted that water-chestnut is often confused with two other species which are, chestnut and Chinese water-chestnut. Chestnuts constitute several species (8 or 9) belonging to the genus, *Castanea* under the family, Fagaceae, which is known as beech family. They are grown in temperate zone and are deciduous in habit. The species of the genus are grouped as American, European, Chinese and Japanese chestnuts. American chestnut is *Castanea dentata* which is almost extinct due to blight disease. European chestnut (syn. Sweet chestnut, Spanish chestnut, Chestnut) is *C. sativa* the kernel of which is most edible. Chinese chestnut is *C. mollissima* and Japanese (syn. Korean chestnut) chestnut is *C. crenata*. Water-chestnut is a different species *and has no relationship with chestnuts of other species, which have been stated above.*

In English, the common names of water-chestnut are Water-caltrop, Bat-nut, Buffalo-nut, Ling-nut, Devil-pod or Water-chestnut. In French, it is called as Noix aquatique, in German as Wassernuss, in Spanish, Castagna de'aqua, in Swedish as Vattennoet and in Chinese, it is referred to as Lingjiao.

In Hindi, water-chestnut is called Singhara, Pani-patassa, Hingoda, Singhora *etc.*, in Tamil as Singara, Pannimonthan kizangu, in Malayalam as Karimpola, in Telugu as Kubyakam, Pandi gadda, Kubjakam *etc.* in Kannada as Mullu kombu beeja, Singara beeja, Shinghade *etc.* and in Odisha, it is called Pani Singhara, Pani phalo *etc.* In Bengali, the most commonly used name is Paniphal and in some parts, it is called payeenphal.

Botanical Family

Water-chestnut has been placed in the family, Lythraceae under the Order Myrtales. The family is also known as loose-strife family from the genus, *Lythrum*. The family is known to have 31 genera and about 650 species. There are five sub-families in the family and water-chestnut comes under the sub-family, *Trapoideae*.

The family contains mostly herbs but also shrubs and trees. Most of the species grow in tropical countries. Leaves of the plants are simple and usually in pairs, *i.e.*, opposite but in some cases, they are alternate or whorled. They are usually entire and have smooth margins with pinnate venations. Stipules are reduced and exist as a row of minute hairs where present but they may be absent.

Flowers are bisexual, actinomorphic or xygomorphic, purple and often in spikes or clusters. Sepals are 4–8 and may be crumpled. Petals are 4–8 or absent. Stamens are about as many as petals and are arranged in whorls. They are unequal in length. Ovary is superior with one stigma and style. Carpels are 2–6. Fruit is a capsule with 2–6 chambers.

Genus: Trapa

The generic name, *Trapa* is derived from the Latin word, thistle. The genus is magnoliopsid (= with two cotyledons) and has one species.

Importance

Water-chestnut is one of the very ancient food crops in the world. It is known to have been used as food for worship as prayer offerings in ancient China since the Zhou Dynasty. Cropping of water-chestnut had also been a practice in ancient Germany where it was grown as a supplementary food crop. There is evidence that the nut crop which has been referred to as Shrigataka in Sanskrit is grown in India since ancient time.

Water-chestnut is one of the few food crops that grow only in water and hence, it does not face land competition with other agricultural or horticultural crops. Many ditches along railway tracks and highways are made into productive utilization by growing this aquatic crop. Tanks, ponds and some streams and also small rivers are made use of in growing the crop. After harvest of water-chestnut, the ponds and tanks may be used for pisciculture if fertilizers and chemical pesticide are not used in raising the crop.

On drying and grinding the kernels of the nuts, they are used to make many delicious products. Food products made from the water-chestnut flour are offered to gods and many people of the Hindu religion take them on fasting days. One of the attributes of the kernels is, it comprises as a low calorie food and hence, is of help for consumption to combat obesity. Among other medicinal properties, the kernels are useful in cardiovascular ailments and in lowering blood sugar.

Description of Plant Parts

Water-chestnut is an aquatic, floating plant (Plate-12) with submerged stem. Although it is perennial but is cultivated like an annual plant. Roots of the plant are fine, green, long and profuse in number. Stem is submerged and is anchored to the bed of the water body by plentiful roots. Length of the stem is 3–5 metres and sometimes become longer. Stem produces many branches and each branch terminates in a rosette of shoots. It is spongy, buoyant and the submerged part produces stipules and adventitious roots.

Leaves of the plant are of two types. One type comprises submerged leaves which are borne along the length of the stem. Another type of leaves are undivided, float on water surface and have saw-toothed edges. They are ovoid or triangular in shape, leathery, bright, glossy, green at the upper surface and light green on the under-surface with pubescence and have prominent veins. They are 2–4 cm long and 5–7 cm in diameter. New leaves are produced from terminal meristem in the rosette near the water surface.

Flowers are few. They are axillary to the floating leaves, solitary, white in colour and are four-petalled. They have no inner whorl of stamens. Ovary is half-inferior, bilocular with a single, pendulous ovule in each locule. Flowers open above the water surface in the evening hours. Pollination of the flowers takes place by insects. After pollination, the flowers get submerged as the petioles are bent down. One of the two locules and its ovule abort after pollination and on bending of the peduncle into water, a unilocular one-seeded fruit develops. Kadono and Schneider (1986) and Susumu Arima *et al.* (1999) described floral biology of water-chestnut plant.

Fruit (Plate-13) of the plant is a one-seeded nut or drupe and is regarded as a pseudo- drupe. It is ovoid and 2–5 cm long. Fleshy pericarp of the nut is deciduous and covers large horny endocarps (pyrene). Horny endocarps may be 2 or 4 according to species. They in fact, represent sepals. When the fruits mature, they fall and being heavier than water sink down to the bottom of the water and remain there embedded in the mud. They germinate later and some growers uproot the emerged out seedlings for propagation.

Seed of water-chestnut is dicotyledonous, but the two cotyledons are not of same size or shape. Interestingly, one of the two cotyledons is extremely large and the other is extremely small and looks like scale only. Seeds do not retain viability (= germinability) for long time, particularly if their moisture content is much reduced when they are stored.

Origin

Origin of water-chestnut is not confirmed. According to a section of botanists, nativity of the species lies in the region of western Europe while another group suggest Africa to north-east Asia including eastern part of Russia and China. South-east Asian Islands have also been suggested by some to be the origin of the plant.

Geographical Distribution

Water-chestnut has a special merit in that, wherever it has been brought, it is found to grow successfully and gives production except in very cold countries. It is cultivated or is naturally grown from warm temperate parts to all tropical and subtropical countries. For virulent growth in tropical and sub-tropical countries, it has assumed to be a noxious aquatic weed in many parts of the world.

The plant is cropped or is naturally grown to a large extent in southern Europe, Africa, Asia and north America. Important countries are southern part of China, Bangladesh, India, Sri Lanka, African countries, south-east Asia, southern Europe, Caribbean islands and Mexico. In India it is grown from Kashmir valley to Kerala and from Punjab to all the eastern states.

Climatic Requirements

Water-chestnut grows well in tropical and subtropical climates but also grows moderately well in warm temperate climate. In general, subtropical climate is considered better for high yield of the crop. However, varietal difference is considered to be a factor as some types are found to be more productive in subtropics while some are much suited in tropical climate.

For germination of the nuts, a temperature of 12°–15° C becomes necessary and for development of flowers, 20°–30° C is favourable. Full sunshine is a very important requirement for the floating leaves to produce nuts of large size and better quality. Floating leaves are of much importance for photosynthate, *i.e.,* carbohydrate production.

Pond Water Requirement

Depth of water in the pond should not be shallow for growing water-chestnut. Water should not contain much of calcium salts. Slightly acidic water is favourable for a better growth of the plants.

Propagation

Water-chestnut is conventionally propagated by seeds. Nuts are used for propagation and each nut contains one seed which is dicotyledonous. The plants raised from seeds retain almost all the characters of the mother plant. Seeds germinate easily but they are recalcitrant, *i.e.,* do not retain viability for long period, particularly when moisture content of them is much reduced which happens when the nuts are stored. Hence, it is advisable not to use nuts which have been stored for a considerable period.

Selection of nuts is a very important consideration. Only fully grown, healthy, bold, well-developed nuts which are not undersized and misshapen should be used. Germination is low and growth of the seedlings is stunted if nuts of such types are not used. Light has influence on germination. Temperature is also important and maximum germination is found to occur at 15°–25° C (Cozza *et al.,* 1994).

The nuts are allowed to germinatate during November to January. Germination is done in a pitcher or a suitable container in which little amount of water is taken.

Addition of little amount of sand and soil in the container makes quicker germination. Germination takes place within 10–15 days. On germination, the germinated seedlings are taken out. These are carefully sorted out such that only vigorously growing sprouted nuts are taken for broadcasting in the nursery-tank.

Micro-propagation of water-chestnut has been studied by some workers. Agrawal and Ram (1995) achieved success by using Nitsch's basal semi-solid medium. Hoque *et al.* (2006) in Rajshahi, Bangladesh used three species of *Trapa* and embryonal explants were found to be best for callus induction on the medium containing half-strength MS salts with suitable concentrations of 2, 4–D and BA. Explants were found to have best survived by 7- minute exposure to 1 per cent solution of mercuric chloride. Water-chestnut plants may also be multiplied by leaf-rosettes.

Culture

Cultivation procedure of water-chestnut plant which is practised by growers has some difference in accordance with locality, water depth of the pond, agro-climatic situation and other factors. General procedure of cultivation has been described by Mazumdar (1985, 2000).

It is a conventional practice in many parts of India to make the plants partly grown in the nursery tank and then transplanted in the actual pond. For this purpose, the nuts which have been made to sprout are broadcast in the nursery-tank. Nursery-tanks are smaller in size, *i.e.*, one hectare area of a nursery-tank would be enough for raising seedlings to plant four hectare area of pond where cultivation is done.

Water quality of the nursery-tank is of great importance. It should not be polluted with chemicals or other undesirable substances. Weeds, snails, insects *etc.* are to be removed from it before the sprouted nuts are broadcast. Water level of the nursery-tank should be about one metre deep and if the depth of water is less than that, extra water should be poured in it by pumping in from neighbouring tank or other source.

Sprouted nuts are broadcast in the nursery-tank during February–March and sometimes it is done upto April. To do this, the sprouted materials are scattered in the tank. However, a better method would be to press those lightly with feet in the mud at the bottom of the tank and maintaining a planting distance of 0.5–0.75 metre.

Care of the nursery-tank is important. At weekly interval, its water should be vigorously shaken by hand for aeration. During this time, the aquatic weeds, snails and insects are also removed. If the depth of water becomes lower, it should be supplemented with water taking it from other source.

The seedlings should be lifted from the nursery-tank and transplanted in the pond where cropping is to be done during March–April. For late-maturing varieties or types which are usually slow growing, transplanting may be delayed upto May or even upto June. Transplanting of the seedlings should be done, maintaining a spacing of 2–3 metres. During transplanting, depth of water in the pond should be about 1.5 metre and if the depth is less, it should be supplemented with water by pouring in it from surrounding pond.

With progress of growth of the plants, water in the pond should be at least 2 metres in depth. This is particularly important for the vigorously growing varieties or types and this depth should be maintained till the nuts are harvested and if necessary, water should be poured in it.

Enhancing Productivity and Quality

To make increase in yield and quality of the water-chestnut crop, intercultural operations are highly effective. These may be stated as follows:

Water-chestnut crop responds highly to application of manures and fertilizers. Many growers in the central part of West Bengal apply adequate amount of rotten cow-dung and castor cake in the water-chestnut ponds and significant high yield is obtained by them. A section of growers in West Bengal apply 30–40 kg of urea per hectare area of pond after about a month of transplanting and the some dose is applied 20 days later (Mazumdar, 2004 a). It is claimed by the growers that yield performance of the crop is boosted by about four times by such application of fertilizers than when the crop is grown without application of fertilizer. However, this is not confirmed.

Profuse vegetative growth sometimes occurs in many water-chestnut plants. This may be a varietal character of the plants or due to high fertility of the pond water and nitrogen fertilization may be a cause. As excess vegetative growth impedes higher production of the crop, it is justiable that some shoots of these plants should be clipped off with knife or pliers.

Luxuriant growth of weeds sometimes assume a serious problem in the water-chestnut ponds. A number of species of aquatic weeds are found to grow in a pond and among those, *Eicchornia crassites* and *Hydrilla verticillata* are most serious. They cover large area of a pond in a short time and thus, act as impediments for growth of the water-chestnut plants for which yield is tremendously reduced. Eradication of the weeds by pulling up with hands when they are visible in the pond is the only method for getting riddance from this hazard.

Mazumdar (2007 b) in an observational trial noted higher yield and better quality of kernels by pruning of plants, weeding and nitrogen fertilization. Pruning which is done by the growers consisted of trimming of branches by hand and also main shoot of each plant from the onset of rainy-season and continuing upto harvest of the nuts in the middle of November. Nitrogen fertilization comprised of pouring urea as far as possible evenly on the pond surfaces such that 60 kg of it had been applied approximately per hectare basis in two split doses, *i.e.*, first half of 30 kg was applied after a month of transplanting and the rest after a month. Weeding consisted of manual uprooting done by the growers.

Results indicated that all the treatments augmented yield and quality of nuts in respect of carbohydrate and protein content of the kernels and markedly by application of urea.

Pests and Diseases

Among insect pests, Singhara beetle (*Galerucella birmanica*) attacks the plants in most parts of India and throughout the growth of plants. It is carried over to the next crop. To control the pest in the pond and in the nursery-tank, Malathion 50 EC–1 per cent or Carbary 50 WP–0.1 per cent may be sprayed. Dusting of dried and ground leaves and petioles of tabacco is also effective to control.

In the Punjab, Haryana and other parts of north India, grubs and beetles of *Galeucella singhara* cause much damages in rainy-season. Some other insect pests, *e.g.*, *Stenochironomus* species, *Bagous trapae* and *Haltica cyanea* cause damage to the crop. Under south Indian condition, *Rhopalosiphan nympheae* attacks the leaves and cause them to wither. To control the pest, Folidol–E 605 (0.03 per cent) may be sprayed.

The disease which is known as leaf-spot of water-chestnut appears due to infection by *Bipolaris tetramera* (Mck.) Shoemaker cause brown spots with yellowish margins on leaves. The spots merge together and dead areas are visible. The disease is serious under north condition and to control, Dithane-M-45 may be sprayed.

Harvesting

Flowering time of water-chestnut plants varies according to locality. It starts from rainy-season and continues upto the advent of winter. Thus, flowering continues with growth of fruits.

Harvesting of the nuts continues over a long period. It starts at the end of rainy-season and continues upto early winter *i.e.*, in the early part of December. However, peak period of harvest is from September to November. Some growers harvest the nuts prematurely, starting from end of July to catch the early market.

Method of Harvest

Harvesting of the nuts is done in a peculiar manner by floating on the water-chestnut pond. To do this, a raft is locally made by joining two large pitches with pieces of fresh bamboo. Pitchers may be of earthen or plastic made and metallic pitchers are also used, though rarely. The two pitchers which should be of same size and weight are joined together by strongly tying them with two parallel bamboo poles as shown in the (Plate-14). The raft should not be very heavy and should be of such weight that it can be carried with a hand.

For the purpose of plucking the nuts, the raft is brought to the pond. There it is inverted to make sure that no water is left inside any of the two pitchers. Then a seating arrangement is made on the raft before it is made to float the pond. To make this, a strongly built (well-burnt, first class) brick is suitably placed in the middle position of the two parallel bamboo poles which have been tied to the earthen pitchers at both ends.

The person who would harvest should sit on the brick in such a way that his two legs hang on both sides of the bamboo poles, placed parallely. Sitting comfortably on the brick, the harvester would row the raft by paddling with both hands and go on plucking the nuts with his fingers. Besides the raft, the harvester also takes with him one or preferably two large-sized pitchers which are of wide mouth. These large-sized pitchers are made to float along with the raft for collection of the harvested nuts. Little amount of water is poured in these large pitchers. (Plate-15)

Harvesting of the nuts is done by plucking them carefully with fingers. Only nuts are plucked and not any twig. Just after plucking, the harvester puts the nuts inside the large-sized, wide-mouthed pitchers which contain little water and are made to float close to the raft on which the harvester is seated.

The harvester goes on plucking the nuts from the plants and put them in the large-floating pitchers. In this way, when the pitchers are almost filled with nuts, the harvester rows near the bank by hand-paddling and while he rows, he also pushes the large-sized floating pitchers in which the harvested nuts have been collected. Thus, the harvester himself sits on the raft and pushes the pitchers containing the nuts are such that they are ported to the bank of the pond.

At the bank, another person who has been waiting there, carefully takes out the wide-mouthed large pitchers from the pond in which harvested nuts have been collected. The nuts are thoroughly washed by him and then those are transferred to plastic net bags to starain out the adhering water. After that, the nuts are taken in plastic bags and sent to the wholesale or retail markets.

Evidently, the method described above has risk in it and the harvester should be skilled in swimming to save him if he sustains fall from the raft. In fact, no unskilled person should take up the harvesting operation in the above way by rowing in a crude type of raft. In some countries and also in some parts of India, plucking of the nuts is done by a harvester by rowing in small boats in which, risk is less than rowing in a crude raft.

Just after harvest of the nuts, they are made into preliminary grades or classification in accordance with red or green colour of the husk of nuts, 2 or 4 hours that are present in a nut, size and shape of the nuts, freedom from black spots or blemishes on the husks *etc.*

Yield

Yield of water-chestnut is found to vary on a number of factors such as agro-climatic zone where it is cropped, varieties or types, application of agro-inputs, especially manures and fertilizers and on many other factors. It may range from 2000 to 5000 kg nuts per hectare area of pond and may increase to 6000 kg by application of fertilizers, plant protection chemicals, weeding and growing of high-yielding varieties or types. Mazumdar (1982) stated that, in general, average yield of the crop in the central part of West Bengal is 3000 kg per hectare area of pond where no input has been used in its culture.

Varieties

Well-recognized varieties of water-chestnut are not clearly known. However, many types are found to grow in different parts of India which vary according to plant characters, growth of the plants besides size, shape and quality of the nuts. In north India, growers cultivate Kanpuri, Jaunpuri, Basmati, Desi, Nagza, Buriya ke tal *etc.* as varieties.

In Odisha, Haldipada red and green and Balasore red and green are some types which are grown. In West Bengal, Chhoto kanta, Lamba kanta, Char kanta, Hugli, Lal khosa, Subuj khosa *etc.* are names of some of the types that are grown. According to colour of husk, two types of water-chestnut are found, which are red or dark red coloured and deep or deep green coloured. Green-husked nuts have much demand in the market as their kernels are more succulent and have sweetness. Nuts having two or four spines are found to grow together in the same pond.

Mode of Consumption

Kernels of water-chestnut are highly relished in fresh condition in autumn where they are available and as few fruits are found in the market at that time. In north India, many consumers are fond of dried and boiled kernels of the nuts. In West Bengal, dried and ground kernels are in much demand in the market. This is used as a dough to make sweetmeats, confectionaries, hand-made bread *etc.* The products are particularly taken as austere diet by the people belonging to the Hindu religion on the fasting days of religious significance. Overweight people are also advised to take these food products.

Quality and Nutritive Value

Nutritive value, quality and medicinal properties of water-chestnut have been reviewed by Adkar *et al.* (2014).

Mandal and Mazumdar (1995) analysed water-chestnuts grown in West Bengal in respect of weight and size of nuts and dry matter content, sugars, acidity and vitamin C content of the kernels. Subsequently, a comparative study (Mandal *et al.,* 1997) was done on green and red-husked nuts of the *bispinosa* and *quadrispinosa* species (varieties). Results indicated that nut weight, dry matter content of kernels and their total sugars were higher in the red-husked type. Total carbohydrate was found to be lower in the kernels of the *quadrispinosa* nuts.

From the National Institute of Nutrition (I.C.M.R., Govt. of India) in Hyderabad, the nutritive constituents of 100 gram of fresh water-chestnut kernels containing 70 per cent moisture and dry kernels containing 13.8 per cent moisture have been stated by Gopalan *et al.* (2018) as follows.

Protein: fresh - 4.7 g.; dry–13.4 g.; Fat: fresh–0.3 g.; dry 0.8 g.; Carbohydrates: fresh–23.3 g.; dry–68.9 g.; Crude fibre: fresh–0.6 g.; Minerals: fresh–1.1 g.; dry–3.1 g.; Calcium: fresh–20 mg.; dry–70 mg; Phosphorus: fresh–150 mg.; dry–440 mg.; Iron: fresh–1.35 mg.; dry–2.4 mg.; Magnesium: fresh–72 mg.; Copper: fresh–1.31 mg.;

Manganese: fresh–0.85 mg.; Zinc: fresh–1.56 mg.; Chromium: 0.011 mg.; Carotene: fresh–12 mcg.; Thiamine: fresh–0.07 mg.; Niacin: fresh–0.6 mg.; Vitamin C: 9 mg. It has been stated that consumption of 100 gram off fresh kernel would provide 115 Kilocalorie (=481.16 Kilojoule) and 100 gram of dry kernel would provide 330 Kilocalorie (= 1380.72 Kilojoule) of energy in the body.

(Index: g = gram, mg = milligram, mcg = microgram)

Medicinal Value

Water-chestnut kernels are used in preparation of some ayurvedic medicines. Some of the ailments in which water-chestnut kernels have benefit are liver complications, nervous disorders, diarrhoea, dysentery, bronchitis, dental caries, thyroid problems, leucorrhoea, spermatorrhoea, sexual weakness, hypertension in pregnancy *etc.* Dried and ground kernels of water-chestnut are of much demand now around the world for their benefit to avert obesity.

Phytochemicals

A large number of chemical compounds are found to be present in water-chestnut kernels and these are used in many industries. Some of these compounds may be stated as saponins, tannins, flavonoids, phytosterols, phenolic compounds, ursolic acid, cycloeucalenol, tri-hydroxyurs *etc.*

Constituents in Developing Fruits

Changing responses of some physico-chemical components of water-chestnuts at four equally spaced stages of their growth have been stated by Mandal (1997). Later, Mandal and Mazumdar (1998) reported some other components in the kernels of the green-husked nuts of *T. natans* var. *bispinosa* during their growth.

Undesirable Effect

As the water-chestnuts are grown in ponds, it is commented on by many that they may be a source of metallic toxicity and contamination by worms. Babu *et al.* (2013) from their study done at the B.B. Ambedkar University in Lucknow put forward that heavy metals which are toxic to the body come into the water body regularly and make them polluted. These heavy metal pollutants, notably lead, copper and iron enter into the roots of water-chestnut plants and from there, they are translocated to the kernels. They observed that consumption of the kernels cause accumulation of these contaminants in the body and thereby, produce toxic effects.

A section of health scientists is of the opinion that some worms, *e.g.*, flukes during their larval stage swim away to form cysts on the surface of aquatic plants which include water-chestnut also. Thus, the flukes are liable to infect human beings and animals. However, these problems would not exist if the kernels are taken after boiling or drying in heat.

It is stated by many that too much consumption of water-chestnut kernels may produce constipation.

Post-harvest Storage

Water-chestnuts do not keep well for more than 4–5 days under ordinary atmospheric condition. If stored for many days, the kernels becomes softened and give off alcoholic or rotten smell due to hydrolysis of the starch and its fermentation. At low temperature of 6°–8° C and maintaining relative humidity of 85–90 per cent in the storage environment, they may be stored well for 20–25 days.

Placing the nuts on moist sand in cool and ventilated atmosphere, water-chestnuts may be stored for 7–10 days under ordinary environment. Singh *et al.* (2010) studied that water-chestnuts kept at frozen condition exhibited better storage life and the total solids, soluble solids and the total acidity level in the kernels.

Utilization

Many products are known to be made from fresh or dried kernels of water-chestnut. A delicious sweetmeat is made from dried and ground kernels in West Bengal and also in Bangladesh which goes by the name, "paniphaler palo" (Mazumdar, 2020).

The dried and ground kernels are also used in making bread, cake *etc.* and on building with water and sugar, beverage is made. In many Chinese food preparations, kernels are used.

In many countries, bread is made from the dried and ground kernels of water-chestnut. This is considered to be of much help to combat obesity, particularly as the glycaemic index of the kernels is 60.

Many people around the world now consume flour made from the dried and ground kernels of water-chestnut. This is because, flour of this nut, *i.e.*, its dried and ground kernel is free from gluten which is a protein present in the flour derived from wheat, rye, oats and barley and this protein is responsible for the sprue (= caeliac) disease, which is characterized by abdominal distention, weight loss, malabsorption, diarrhoea, bloating, stetorrhoea and other symptoms.

Family: Malvaceae

Species: *Grewia asiatica* Linn.

[Synonyms: *Grewia subinaequalis* DC; *G. hainesiana* Hole; *G. obtecta* Wall.]

Taxonomic Serial: 506089

Common Name

In English, the common name is Phalsa and the same name is used for the fruit in Spanish and German. In Thailand, the fruit is known as Malai Ya-khithut, in Veitnam as Cò-ke-á, in Combodia as Pophilie, in Laos as Nhap and in Pakistan as Dhaman or Phalsa.

In Hindi, Urdu, Marathi, Bengali and Kannada, the fruit is called Phalsa while the name is Unnu in Tamil, Phutiki in Telugu, Chadicha in Malayalam and Shukri in Gujarati.

Botanical Family

The species comes under the sub-family, Grewioideae. The family contains 244 genera and more than 4200 species. The family is composed of nine sub-families, which are however, debated. Phalsa comes under the sub-family, Grewioideae which contains 25 genera and 770 species. Leaves of the species of the family are alternate, often pinnately lobed or compound with palmate venation. Leaf margin is entire or dentate. Stem contains canals of mucus. Flowers are borne in axillary inflorescences. Flowers bear bracts and are unisexual or bisexual. Sepals and petals are 5 and stamens are 5 or more. Pistils are made up of 2 or more connate carpels. Ovary is superior and fruits are mostly loculicidal capsules, schizocarps or nuts.

Genus: Grewia

The genus contains large shrubs or small trees with spreading habit. Leaves of the species are roundish, alternatively arranged along the stems. They are dull green at the upper surface and under surface is paler. Flowers are small, yellow or reddish and with many stamens. They are arranged in clusters. Fruits (Plate-9) are pulpy, small and roundish or ovoid.

Importance

Phalsa has a number of merits. It is a very hardy fruit plant and is capable to withstand adverse situation. Hence, it is highly suitable for productive utilization of inferior land. Drought resistance is a great merit of the shrub. Its juvenile period to produce fruits is short and is a year or two. It gives high yield without receiving care or inputs. Phalsa is more commonly planted as a *filler tree* in the orchard and thus, it makes utilization of vacant space in the orchard. Fruits are highly juicy, sweet sour in taste and are refreshing in summer when they are available. Juice of phalsa is bottled and the product has demand in the market.

Origin

Exact origin of phalsa is not known but origin of the plant in India has been suggested by some horticultural scientists (Sham Singh *et al.,* 1967; Hayes, 1970; Singh and Saxena, 2008; Chadha, 2012). Singh and Saxena (2008) stated that Vadodara region in India might be the origin of the species.

Morton (1987) mentioned in her book that Indian and south-east Asia might be the centre of origin of phalsa. Some taxonomists have however, put forward that Cambodia, Laos and Thailand could be considered as the nativity of the species.

Geographical Distribution

Phalsa is much grown in India. It is also grown in Bangladesh, Sri Lanka, Myanmar, Laos, Cambodia, Thailand and other south-east Asian countries. In India, the shrub grows in Punjab, Haryana, Himalayan foothills in north India, Rajasthan and Madhya Pradesh. It grows to a lesser extent in Gujarat, Andhra Pradesh, Bihar, West Bengal and the eastern States.

Climatic Requirement

Phalsa grows in tropical, subtropical and warm temperate climates but the performance of the fruit crop is best in subtropical environment. Morton (1987) stated that phalsa could be seen to grow in India upto an altitude of 914 metres. The shrub withstands drought to a high extent. It can tolerate temperature upto 45°C in summer and freezing condition in winter (Chadha, 2012).

Unlike in subtropical climate where the shrub is deciduous, in tropical climate it remains evergreen, grows vigorously and produces two crops or sometimes more in a year. Quality of the fruits is however, much better when the fruit crop is grown in sub-tropical climate.

Soil Requirement

Phalsa is grown in almost any type of soil, *e.g.,* in limestone, gravelly or sandy to clayee soil. But productivity is high when grown in sandy loam. Fruit quality is superior when it grows in fertile soil. Drainage of the soil is an important factor as in poorly drained soil, leaves become yellow, *i.e.,* chlorotic and leaf fall occurs to a high extent (However, phalsa is regarded as a waste land utilization fruit plant).

Propagation

Seed propagation is most common in phalsa and the resulting plants retain characters of the mother plants. seeds germinate easily but they are recalcitrant, *i.e.,* lose viability within 90–110 days and hence, stored seeds should not be used.

Stem-cuttings of phalsa are successful by the use of hardwood or semi-hardwood cuttings and with use of auxin. T-budding and ring-budding is sometimes done in Uttar Pradesh and Madhya Pradesh.

Micro-propagation in phalsa has been reported by some workers. Biswas *et al.* (2005) reported *in vitro* propagation and somatic embryogenesis. Wani *et al.* (2016) induced shoot formation from nodal explants with axillary buds on culturing in Woody Plant Medium fortified with sucrose, kinetin and auxin in suitable concentrations.

Culture

In north India, phalsa is usually grown as filler plants in orchards or is interplanted with other crops and is not cultivated as a single crop. Thus, manures, fertilizers and irrigations which are applied to the main crops also benefit the phalsa plants and hence, productiveness is much increased in phalsa.

If phalsa is required to be grown as a single fruit crop, a spacing of 2–3 metres should be maintained between the plants. Sowing seeds *in situ* is done by many growers instead of transplanting seedlings. Double-row system of planting phalsa is preferred by many growers in Uttar Pradesh. Planting of phalsa seedlings is done from February to the early part of rainy-season in north India.

Manuring is not done in phalsa but Hayes (1970) stated that application of 10 kg of farmyard manure every year to a mature bush in the month of February would greatly enhance yield. Experiments done by Saravananan *et al.* (2013) from Allahabad stated that application of 100 gram of nitrogen in combination with 50 gram of phosphorus and 100 gram of potash significantly augmented growth and also yield of fruits by 4.42 kg per bush. Irrigation is avoided in phalsa but one or two irrigations from January to March when growth stars in the plants would be of much help (Hayes, 1970).

In the culture of phalsa, most important oparation is *pruning*. Productive capacity of the fruit crop would be seriously hampered if the shrub is not pruned annually. Pruning is in fact, drastically done by the phalsa growers in north India. (Strangely enough, even putting fire to the bushes and extinguishing by throwing water just before it reaches ground level is also not a myth and is done by some grower the bushes are note destroyed).

Severe pruning is however, unnecessary in phalsa. Pruning of a bush just by coppicing it at 40–50 cm height from the ground level is enough in most cases. Heavy pruning does not produce any good result and would devitalize the plants. Time of pruning differs in localities and is usually done in early winter in most parts.

The horticultural practice known as *bending of branches* which is largely done in guava greatly augments yield of phalsa. By this practice, the vertically growing branches are forced to grow horizontally with rope and as a results, fruit are borne on the whole branches and not at the upper parts only.

Pests and Diseases

Among insect pests, the bark eating caterpillar (*Indarbela* species) are most serious. The insects come in large number and attack many fruit trees. Tunnels are made by the larvae on bark and the affected branches die. To control the insect, as stated in other chapters, whenever, holes would be seen on the branches, they should be plugged with cotton soaked with kerosene or oil emulsion. To destruct the pupae in the soil, soil should be treated with Aldrin or neem cake.

A blackish caterpillar is found to form galls at the apices of growing shoots in phalsa. Some other less serious insect pests in this shrub are mealy bug (*Drosicha mangiferae*), brown beetle (*Anomala* species), aphids (*Aphis craccivora*) *etc.*

Among diseases, leaf spot caused by the fungus, *Cercospora grewiae* is sometimes serious in rainy-season and causes leaf fall. To control the disease, 0.3 per cent Dithane-Z 78 may be sprayed. The rust disease caused by fungus, *Dasturella grewiae* and leaf spot caused by *Phyllosticta grewiae* fungal species are less serious.

Harvesting

Phalsa fruits ripen only in the plants and do not ripen any more if plucked before they are fully ripe. However, there is a drawback in that, ripe fruits are not obtainable from a plant in one or two days. That is, daily pickings become necessary for 10–12 days for complete harvest of fruits from a bush. Thus, harvesting involves high labour coat. Harvest time of fruits, in other words, their full ripening is judged when the fruits change their colour from green to deep red, brown or purple.

Yield

Yield of fruits from a bush of phalsa is difficult to determine as it varies on a number of factors. In general, a bush may produce 4–5 kg of fruits when planted in double-row system. Yield may be high when grown as filler plants in the orchards as they share the manures and irrigations that are applied to the main fruit crop grown in the orchards.

Varieties

Any varietal name of phalsa is not known. These plants which bear juicy fruits are sometimes referred to as Sharbati varieties. Two types of phalsa plants are recognized, which are tall and dwarf in height. Tall type of plants reach a height of 4-5 metres and the dwarf types become 3–4 meter tall. Morphological characters of the tall and dwarf plants have some differences. Yield per plant is higher in plants of tall types and may go upto 5–6 kg, unlike 4–5 kg in plants of dwarf types.

According to Venkataratnam (1963), phalsa plants are also grouped into two types, which are known as 'chhatri-phalsa' and 'bush-phalsa'. It has been stated by the author that the type, chhatri-phalsa is of umbrella shape and the plants of this type are suitable for planting in household compounds. Bush-phalsa on the other hand, is considered for planting inside the orchards as filler shrubs.

Mode of Consumption

Phalsa is a very small fruit (Plate-16) but fully ripe fruits are highly juicy, sweet-sour in taste and refreshing in summer when the fruits are harvested. The juice has thirst-quenching property, for which it has great fondness.

Qualitative and Nutritive Value

Physico-chemical components of phalsa fruits grown under West Bengal condition have been stated by Mandal and Mazumdar (1997) and Mazumder and Mazumdar (2001).

From the National Institute of Nutrition (I.C.M.R., Govt. of India) in Hyderabad, nutritive value of 100 gram of fully ripe phalsa fruits containing 80.8 per cent moisture has been stated as follows.

Protein: 1.3 g.; Fat: 0.9 g.; Carbohydrates: 14.7 g.; Crude fibre: 1.2 g.; Minerals: 1.1 g.; Calcium: 129 mg.; Phosphorus: 39 mg.; Iron: 3.1 mg.; Magnesium: 72 mg.; Sodium 4.4 mg.; Potassium: 351 mg.; Copper: 0.12 mg.; Sulphur: 13 mg.; Chlorine: 86 mg.; Carotene: 419 mcg.; Niacin: 0.3 mg and Vitamin C: 22 mg. Energy obtainable from 100 gram of fruits has been stated to be 72 Kilocalorie (= 301.2 K Joule).

(Index: g = gram, mg = milligram, mcg = microgram).

Medicinal Value

Fruits of phalsa are of benefit in weak heart. They are good sources of anti-oxidants and have anti-cancerous property. Leaves have anti-microbial and anti-cancerous property and stem bark has analgesic and anti-inflammatory role. Haq *et al.* (2013) from the Karachi University in Pakistan have stated a number of medicinal properties of the fruits and other parts of the phalsa tree.

Phytochemicals

A number of chemical compounds are known to be present in phalsa fruits. Among these, polyphenols, anthocyanins, tannins, flavonoids, delphinidine, fatty acids of various types, DPPH (2, 2–diphenyl–1–picryl hydrazyl), ABTS [2,2′–azino-bis (3-ethyl-benzo thiazoline -6- sulphonic acid)], ferric reducing anti-oxidants and other compounds may be cited.

Undesirable Effect

Too much consumption of phalsa fruits cause gastric acidity, belching and other discomforts. From the National Institute of Nutrition, Hyderabad, it has been reported that 100 gram of phalsa fruits contains 200 mg of oxalic acid and 13 mg of phytin P. Oxalates interfere with calcium absorption in the body and phytates bind iron, zinc, calcium and magnesium and inhibit their absorption.

Post-harvest Storage

Shelf-life of phalsa fruits is very low and not more than two days under ordinary atmosphere. In cold storage at 7°–10° C with a relative humidity of 85–90 per cent, fruits may be stored for 7–10 days. Exposure to ultra-violet ray for few hours with application of 5 ppm of PB-2 enhances storage life of the fruits.

Utilizations

Phalsa juice is utilized in making squash particularly in north Indian and it has a demand in the market. In some parts of India, pickles are made from phalsa. Jam is also made from phalsa in some countries.

Family: Moraceae

Species: *Artocarpus chaplasha* Roxb.

[Synonyms: *Artocarpus chama* Buch. Ham.*; A. chama* Buch.-Ham. Ex Wall.; *A. melinoxylus* Gagnep.; *Saccus chaplasha* (Roxb.) Kuntze; *Ficus chrysophthalma* (Miq.) Miq.; *Urostigma chrysophthalmum* Miq.]

Taxonomic Serial: 709040

Common Name

In English, the fruit is referred to as Chaplash, although other names, *viz.*, Thorny jack, Lesser jack *etc.* are also sometimes used.

Botanical Family

The family, Moraceae comes under the Order, Urticales and consists of about 75 genera and more than 1000 species. Plants are trees or shrubs or rarely herbs. They are monoecious or dioecious, evergreen or deciduous and many are thorny. Most of the species secrete milky or watery latex.

Leaves of the plants are alternate, simple, entire, serrate or lobed. Stipules are present and usually caducous.

Inflorescence is axillary, condensed spike, racemose, rarely cymose and capitate. Flowers are small, regular and unisexual. Perianth is mostly 4 in 2 whorls, free or united, valvate or imbricate. Corolla is absent. Stamens in male flowers are equal in number. Filaments are bent inwards in bud or straight. Anthers are 1 or 2 loculed and crescent shaped. In female flowers, calyx lobes are usually 4.

Ovary is inferior or superior, ovule is 1 per locule of 1 or 2, style 1 or 2 and stigma is mostly filiform. Fruit is a drupe and in most cases achenes. Seeds are with fleshy endosperm and curved embryos.

Genus: Artocarpus

The generic names, *Artocarpus* has been derived from two Greek words, which are artos, meaning bread and karpos, which means fruit and the name was proposed by the botanists, J.R. Forster and J.G.A. Forster. The genus contains more than 50 species which are shrubs and trees and grown in south Asian and south-east Asian countries. There are two sub-genera of the genus, which are sub-genus, *Artocarpus*, in which perianth is partly connate and the sub-genus, *Pseudojaca* in which perianth is fully connate.

The species are monoecious and the plants secrete milky latex. Leaves of the plants are small to large and in some species are lobed as in *A. altilis*. Fruit is syncarpous and grows to fairly large size. Few species of the genus produce edible fruits, *viz.*, *A. heterophyllus*, *A. altilis*, *A. lakoocha*, *A. hyperagyreus*, *A. kemando* and *A. hirsutus*.

Importance

The fruits of *A. chaplasha* have edible value and the tree provides good timber. In the south-east Asian countries, the fruits are cooked apart from taking as fresh. Leaves of the tree are good fodder to the pet elephants.

Origin

Nativity of the species is not confirmed but conjectured to be south-east Asian countries.

Geographical Distribution

The chaplasha trees are deciduous with large crown which attain upto about 40 metres in height and are found to be grown largely in China. Myanmar, Thailand, Laos, Malaysia and other south-east Asian countries, India, Bangladesh, Sri Lanka *etc.* In India, the tree grows much in the forests and as wild in Assam, eastern States and in Kerala, Malabar region, Konkan *etc.*

Climatic Requirements

Trees are highly tropical in their climatic requirement. High rainfall and warm, humid atmosphere are highly favourable for growth and productiveness of the trees. Prolonged frost is highly damaging to the trees.

Soil Requirements

Soil is not very important to the tree but drainage should be good. Deep soil is favourable as the tree is deep-rooted. Soil that exist in tropical rain forests is considered ideal for the trees.

Propagation

In the forests, the trees multiply from seeds which germinate easily and the resulting plants grow satisfactorily. A kilogram of seeds contains 1800–2000 seeds.

Rahamat (2014) in Bangladesh studied *in vitro* propagation of *A. chaplasha*. Shoot proliferation was induced from shoot apices and lateral bud explants cultured on MS medium containing different concentrations of BAP, NAA and Kinetin, either alone or in combination. It was observed that 2.5 mg per litre of NAA showed maximum shoot induction when combined with 0.5 mg per litre of BAP.

Culture

Chaplasha tree is sometimes grown in homestead land, public or private compound or along avenues and is found to grow and produce satisfactorily without any culture. However, application of organic manure would seem to be of benefit.

Harvesting

Trees flower in March–April and fruits would be ready for harvest from July to September. Fruits ripen best in the trees but as they become very soft when fully ripened, harvesting should be done slightly earlier.

Medicinal Value

Leaves and fruits of chaplasha are much used in folk medicines in south-east Asian, countries, Bangladesh, Tripura, Nepal and other countries. Chowdhury *et al.* (2013) in Bangladesh studied that leaves of the tree extracted with organic solvents, *viz.*, methanol, ether, chloroform, carbon tetrachloride besides water were observed to have antimicrobial, antioxidant, cytotoxic, thrombolytic and other effects.

Phytochemicals

A number of phytochemical compounds which have industrial importance have been detected in fruits and other parts of the chaplasha trees. Mahato *et al.* (1971) at the Indian Institute of Chemical bilogy in Kolkata identified lupeol acetate, cyclo-artenyl acetate, beta-sitosterol and isocyclo-artenol from the stem bark of the tree.

Family: Moraceae

Species: *Artocarpus lacucha* Buch.-Ham.

[Synonyms: *Artocarpus mollis* Wall.; *A. ficifolius* W.T. Wang]

Taxonomic Serial: 382341

Common Name

In English, the commonly used name of the fruit is Monkey jackfruit but is also called as Red jackfruit. In Malaysia, Tampang and in Thailand, Lokhat is the common name. In commerce, the fruit is known as Lakuch.

Monkey jackfruit has several names in different parts of India. In Hindi, the common names are Deuwa, Barhal, Jeuta *etc.* The name is Tinnipalavi in Tamil and Lekuchamu in Telugu. In Bengali, the common name is Dephal but it is known by other names also, *viz.*, Barhal, Doua *etc.* The fruit is called Badahar in Nepal.

Importance

Although the fruit is not much disposed in urban markets, it is largely consumed by the villagers and tribal people, to whom it is a popular fruit. The tree is hardy, robust and suitable for planting in land of low productive potential.

Planting the tree in the household compounds and in agro-forests is also done. For graceful appearance, it is also planted along avenues. Fruits are much suitable for pickling.

Timber value of the tree is high and is considered by many to have the quality like teak wood.

Description of Plant Parts

Monkey jackfruit trees are deciduous and attain a height of 15–20- metres. The tree secrets milky white latex. Branches of the tree are open and not compact like the allied species, jackfruit. Branchlets are densely covered with pale brown to yellow coloured velvety hairs. Leaves are alternate, 10–25 cm long, elliptical, oblong or oval, pointed, leathery and with entire margin.

Tree is monoecious. Inflorescence is solitary and axillary. Flowers are small and male and female flowers are borne on separate heads. Male flowers are orange-yellow in colour and female flowers are reddish. Fruit (Plate-17) is roundish to irregular in shape and diameter ranges from 5 to 10 cm. Seed is of 10 × 5 mm in size and is surrounded by aril.

Origin

It is believed by many that the species has originated in the sub-Himalayan region in India although this is not confirmed. Hayes (1970) quoted that Bengal might be the origin of monkey jackfruit. However, nativity of the species in south Asia or south-east Asia is also indicated (Singh and Singh, 2008).

Geographical Distribution

The tree grows abundantly in Malaysia, Thailand, Myanmar, Laos, Singapore, India, Bangladesh, Sri Lanka and other tropical countries. In India, the tree is found to be much grown in Assam, Tripura, other eastern States and in the coastal areas of Tamil Nadu, Kerala and Maharashtra (Mazumdar, 2004 a).

Climatic Requirement

Although tropical climate with high rainfall is most favourable for monkey jackfruit, it is fround to grow moderately well in subtropical climate in some parts of north India. However, if winter temperature goes below 10° C for a week, the trees are badly damaged. Frost is highly detrimental particularly for the young trees.

Mean annual temperature for the trees should be 15°–25° C and mean annual rainfall should be 800–2000 mm. Dry atmosphere is desirable for 2–3 months. The tree grows upto an altitude of 1600 metres in the tropics but fruiting performance is poor when the tree grows in high altitude.

Soil Requirements

Deep and porous soil is most favourable for vigorous growth and higher yield. Performance of the tree is poor with high leaf fall in poorly drained soil. High organic matter in the soil is however, of great benefit.

Propagation

Monkey jackfruit is most commonly propagated from seeds. A fruit contains 20–30 seeds which are fleshy. Seeds have thin seed-coats and hence, they should be extracted from fruits without giving pressure so that they are not damaged. Tomar *et al.* (2014) from Allahabad stated that germination of the seeds started from third day onwards after sowing and continued upto 10 days.

Stem-cuttings are ordinarily not successful without application of auxin. Air-layering is found to give good success when done in rainy-season with application of 0.1–0.2 per cent IBA in lanolin paste at the ringed part of stem.

Micro-propagation in monkey jackfruit has been reported by Joshee *et al.* (2002). On removing the seeds from the fruits, they were germinated in 0.8 per cent agar medium containing 0.5 per cent sucrose. They stated the protocol and optimum results for multiple shoot production were obtained by them with MS medium when supplemented with 1.0 to 2.0 mg L^{-1} BA. Rooting was observed in 2–3 weeks in MS medium when supplemented with 2.0 mg L^{-1} of auxin (= IBA).

Culture

Monkey jackfruit trees grow well without receiving culture or agro-inputs. Training and pruning are however, important operations for the trees. The bole of the tree, *i.e.*, the main stem becomes very tall and hence, training by open-centre or modified central leader system is desirable during first few years of growth. Pruning of mature trees should be done annually by heading-back and thinning-out systems. Management of canopy of the trees also needs attention.

In many parts of the south-east Asian countries, monkey jackfruit trees are interplanted with other fruit trees.

Pests and Diseases

Among insect pests, fruit fly causes much damage to the fruits at later stages of their growth. To control, Endosulfan 35 EC–0.035 per cent or Malathion 50 EC–0.05 per cent may be sprayed. Diseases are not serious to the tree parts. Birds cause much damages to the ripe fruits.

Harvesting

Fruits are ripened best in the trees. Ideally, ripe fruits are recognized when their surface colour becomes rich yellow, emit pleasant smell, soft in texture and the pulp colour becomes orange. But fully ripe fruits become so soft that they are easily damaged on plucking. Hence, plucking should be done slightly earlier, *i.e.*, when they have not fully ripened.

Hand-picking of the fruits using a self-supported ladder is a good method of harvest. A harvester of a sickle-shaped knife, spring bamboo pole and net bag as stated in fruits of other species may be used for plucking of fruits.

Yield

Exact yield of monkey jackfruit is not assessed but it is assumed that a tree may produce 70–80 kg of fruits. A fruit weighs 150–200 gram.

Mode of Consumption

The fruits are relished as fresh when they are fully ripe. They are sour-sweet in taste and have a pleasant aroma. The peel can be taken with the pulp. Drying of the pulp is also done for later use.

Quality and Nutritive Value

From the National Institute of Nutritive (I.C.M.R., Govt. of India) in Hyderabad, it has been stated that 100 gram of monkey jackfruit pulp having 82.1 per cent moisture contains the following constituents.

Protein: 0.7 g.; Fat: 1.1 g.; Carbohydrate: 13.3 g.; Crude fibre: 2.0g.; Minerals: 0.8 g.; Calcium: 50mg.; Phosphorus: 20 mg,; Iron: 0.5 mg.; Carotene: 254 mcg.; Thiamine: 0.02 mg.; Riboflavin: 0.15 mg.; Niacin: 0.3 mg. and vitamin C: 135 mg.

On consumption, 100 gram of pulp provides 66 Kilo-calorie (= 276.14 Kilo joule) of energy in the body.

(Index: g = gram, mg = milligram, mcg = microgram)

On analysis of the representative samples of ripe fruits of monkey jackfruit borne by trees grown in southern parts of West Bengal, Sinha Roy and Mazumdar, B.C. (unpublished) obtained the following results.

(i) Weight per fruit: 120–160 gram,; (ii) specific gravity: 0.98–1.04,; (iii) Stalk-stylar length: 7.8–9.8 cm.; (iv) Diameter at maximum bulge: 7.3–8.2 cm.; (v) Number of seeds per fruit: 45–52.; (vi) Total soluble solids of juice: 20°–24° Brix.; (vii) Total titratable acidity of pulp (equivalence of citric acid): 0.72–1.13 per cent.

Medicinal Value

Many medicinal properties of monkey jackfruit have been mentioned in ayurvedic literature. Fruit is known to have remedial effect in liver complication, dysentery, intestinal worms, arthritic swelling, indigestion *etc.* Bark of the tree has anti-diarrhoeal and wound healing effects. Root has purgative action.

Phytochemicals

The sticky and milky latex which is secreted by all parts of the tree contains water-soluble compounds, carutchouc and coagulum which have industrial importance. Heartwood of the tree trunk contains valuable chemicals like artocarpin (ALA–I and ALA–II), resorcinol and oxy-resveratrol. Seeds contain artocarpins, roots contain lupeol and fruits are rich sources of anti-oxidants in containing high amount of carotene.

Utilization

High quality pickle is made from the semi-ripe fruits of monkey jackfruit. Spikes of male flowers of the tree which are sour and astringent are also used in making pickles. In Kerala, pulp of ripe monkey jackfruit is blended with banana and sun-dried after making flat pieces. When hardened, they are eaten with spices or stored for later use. For citrus like flavour, the pulp is made into ice-cream. A stiff jelly is stated to be made from the pulp of monkey jackfruit (Hayes, 1970).

Family: Myrtaceae

Species: *Acca sellowiana* (O. Berg) Burret

[Synonym: *Feijoa sellowiana* (O. Berg)]

Taxonomic Serial: 834974

Common Name

Feijoa is widely used common name for the species, *Acca sellowiana* in English. Other English names of the Acca fruit are Brazilian guava, Pineapple guava, Fig guava *etc.* In Hawaii islands in the U.S.A., the fruit is called as Guavasteen. In Uruguay, the Spanish name, Guayabo del pais is retained as the common name, which is nicknamed as Pineapple guava. Feijoa is very popular fruit particularly in New Zealand and in that country, it is called New Zealand banana. In India, no particular name of Feijoa is known and where it is grown, it goes by the names as Feijoa or Brazilian guava or Anjir.

Botanical Family

Feijoa comes under the family, Myrtaceae, which is also called Myrtle family or guava family or eucalyptus family. The family is very large and consists of more than 5000 species that come under 130–150 genera.

The species belonging to the family are trees and shrubs including a genus of monotypic mangrove, *Osbornia*, which is native to Australia. The species, *Eucalyptus regnans* in this family is known to be one of the tallest trees in the world. The trees and shrubs of the family are halophytic or bifacial in habit and they are usually evergeen. The species are distributed in tropical and sub-tropical climates in the world and some are tolerant to mild temperate zone. The shrubs are usually bisexual, though in some, they are polygamo-monoecious or rarely andro-dioecious. Leaves are alternate to opposite, simple, mostly with entire margin, gland-dotted and aromatic. Inflorescence is terminal or axillary. Flowers are solitary or in aggregated inflorescence. Free hypanthium is present. Petals have a base number of 5 but in some genera, they are minute or almost absent. Existence of plentiful bright coloured, prominent and erect stamens is a characteristic feature of the trees and shrubs in this family. Pistils are single and compound of 2–5 carpels. Fruits are fleshy or not, dehiscent or indehiscent, capsule or berry or drupe.

The family, Myrtaceae has been divided by some taxonomists into two sub-families, which are Myrtoideae and Leptospermoideae. The sub-family, Myrtioideae having 75 genera is recognized in having fleshy fruits and opposite and entire leaves. The sub-family, Leptospermoideae, consisting of 80 genera is recognized in having dry and dehiscent fruits with spirally or alternately arranged leaves, Johnson and Briggs (1984) in contrast, have however, identified 14 tribes or clades in the family Myrtaceae. Molecular Studies done by many scientists have suggested that the baccate, *i.e.*, the flesy fruits had been evolved from the capsular fruits and thus, grouping the family into the said two sub-families does not indicate phylogenetic, *i.e.*, evolutionary history of the family. Hence, many botanists have considered the analysis by Wilson *et al.* (2001). Based on a phylogenetic analysis of plastid DNA, 17 tribes and two sub-families, *viz.*, Myrtoideae and Psiloxyloideae have later been recognized (Wilson *et al.*, 2005).

A number of fruit plants come under the myrtle family, among which, guava is most important. Aromatic oil produced by the leaves of Eucalyptus and some other species is used in many industries. Clove and all spice is this family have importance as spice.

Genus: Feijoa

The genus, Feijoa in the Myrtaceae family had been named so by the German botanist, Ernst Berger, to honour a botanist of San Sebastian in Spain, don da Silva Feijoa. The genus is a kind of magnoliopsid, *i.e.*, a dicotyledonous genus. It has been stated that the specific name of feijoa, *i.e.*, Feijoa (= Acca) sellowiana has been given to honour F. Sellow, "the German, who collected specimens in the province of Rio Grande Sul in southern Brazil".

Importance

Feijoa is given importance for a number of attributes and the pertinent ones are as follows:

(i) The shrub gives fruits which are consumable to human beings.
(ii) For graceful appearance of the shrub, it is planted in ornamental gardens.
(iii) For hardy nature, the shrub is planted in rows to build up a strong bushy and protective hedge.
(iv) For tolerance against adverse edapho-climatic situation, it is grown to utilize land having less suitability to grow other crops.

Fruits of feijoa are liked for their sweetish taste and particularly for their pleasant aroma. They are taken as fresh and are also preferred to make jelly, jam and other products. Even fleshy petals of the flowers have edible value. They are sweetish a minty flavour and are taken as fresh or on using as an ingredient in salad. If petals are carefully removed from the flowers, development and ripining of the fruits are not hampered.

Feijoa shrub is more commonly used as a suitable material for planting as a bushy and ornamental hedge. Leaves are glossy green with silvery grey under surface. Flowers are charming in appearance. Petals are white having purplish tinge on te inner surface. The tuft of stamens are of crimson colour.

For the above merits, feijoa is given importance for planting in gardens in Iran. southern U.S.A. and in other countries.

Description of Plant Parts

Feijoa is an evergreen, multi-stemmed shrub and reaches a height of 3–4 metres or upto 7 metres and even more when grown in favourable soil and climatic conditions. Branches are spready and swollen at the nodes. Bark is thin and grey coloured. Leaves are opposite, bluntly elliptical, thick, leathery, 3–6 cm long and 1.5–2.5 cm wide with short petioles. They are smooth, glossy, green on upper surface and greyish white on the undersurface. Flowers are bisexual, 3–4 cm wide and are borne singly or in clusters. Petals are 4, oval, concave, fleshy, whitish at the outside and purplish at the inside. Stamens are many, erect and purple. Pollination takes place by bees, bumble bees or small birds. Fruit is oblong, ovoid, spherical or pyriform depending on varieties. They are 3–5 cm long, 1–3 cm wide and weigh 15–20 gram. Fruits taste sweetish sour and have an aroma, which resembles a blend of pineapple and strawberry with a tinge of mint. Seeds are small, oblong in shape and may be 100 or more in a fruit.

Origin

The species is considered to have originated in the highlands of southern Brazil, western Paraguay, Uruguay, northern Argentina and Columbia in south America.

Geographical Distribution

Feijoa grows abundantly as wild in south American countries *viz.*, Brazil, Argentina, Paraguay, Uruguay and Chile. It is now grown in the gardens in other countries also, such as in Iran, California in the U.S.A., Jamaica and the Caribbean countries, southern Azerbaijan, France, Italy, Portugal, South Africa, Malaysia and other south-east Asian countries.

Popularity of the fruit is very high in New Zealand where it is called New Zealand banana. Morton (1987) has quoted, "Nowhere has the feijoa received more attention than in New Zealand." Much improvement of the species has been made in that country by selection and the fruits produced there are exported to other countries.

In India, feijoa is grown to a small extent. Hayes (1970) has mentioned of one variety which had been popularly grown in the hills of Tamil Nadu. The fruit had been stated to be red in colour with bronze marketing, or more commonly, dull-green in colour.

In India, feijoa has been introduced much later and is found to grow satisfactorily in the hills of south India and Shimla (Pareek and Sharma, 1993; Singh and Saxena, 2008).

Climatic Requirement

Feijoa is a typically subtropical shrub and its performance is best where the winter is cool but not freezing and the summer is moderate with low relative humidity. It is estimated that 100–200 chilling hours below 7° C may be necessary in winter for feijoa to produce optimum fruits and of better quality fruits. The shrub, however, withstands as low as (-) 11.1° to (-) 9.4° C (Sham Singh *et al.*, 1967) but prolonged exposure to such low temperature kills the shrub. Annual rainfall of 750–1000 mm is considered optimum and full sunlight is necessary for production of better quality fruits. High wind in summer is harmful and causes premature drop of fruits.

Soil Requirement

Soil is not an important factor for feijoa except that it should be well-drained. Loamy soil with a *p*H around 6 is preferred by the shrub. In England, it has been found to do well in soil having relatively high content of chalk (Morton, 1987). Feijoa has been found to be moderately salt resistant in Florida, U.S.A.

Propagation

Seed propagation is very common in feijoa and the characters of the mother plants are much retained in the resulting seedlings. Germination takes place in 3–6 weeks at 15°C. If the seedlings are raised in nursery beds, the fungal disease, damping-off causes great harm to them. To avoid this, sterilization of the beds before sowing is advisable as is done in papaya (Mazumdar and Ali, 1998). The shrubs raised from seedlings take 4–6 years to produce optimum fruits.

Stem-cuttings give fair degree of success in feijoa with the use of auxin. Zhang Meng *et al.* (2009) studied on the combined effect of strata (media) and auxin in softwood-cuttings of feijoa and observed that perlite with humus-leaves-soil and soaking in 1000 mg L-1 IBA gave best results. It was observed by them (Zhang Meng *et al.*, 2009) that semi-hardwood stem-cuttings gave better results than hardwood-cuttings.

Ground-layering in feijoa is done in New Zealand and France but it takes time to produce well-developed roots. Success by air-layering is found to be much higher than stem-cuttings and ground-layering.

Grafting is done in feijoa by adopting methods, *e.g.*, whip, whip and tongue and veneer. Success by grafting is not reported to be high but the grafted plants produce fruits in much less time than in other methods. Souza (2013) stated a grafting technique using young shoots as scion and equally young rootstock stems. The technique was found to give high success of upto 92 per cent.

Experiments on micro-propagation have been done by many workers, who have described the procedures (Oltramari *et al.*, 2000; Wang Dan *et al.*, 2007; Zhou Li Juan *et al.*, 2008; Guerra *et al.*, 2013).

Culture

Feijoa does not need much care but during the first few years of growth, liberal use of manures, fertilizers and irrigations are much helpful to boost up growth of the shrubs as they are slow to grow at early age. However, any type of deep soil work and hoeing should be avoided as the shrub being shallow-rooted, many absorbing roots are liable to be cut off.

Where organized planting is desired, plants may be set at a spacing of 2–3 metres. Space should be less in order to build up hedge-rows. Close planting is advisable to keep moist environment around plants in low rainfall area to prevent fruit drop. Some varieties of feijoa are self-unfruitful and for them, two or more plants of the varieties *viz.*, Choiciana, Superba *etc.* should be interplanted to supply pollens so that they may become fruitful.

At early stage of growth, training the shrub to a single-trunk or espalier system is desirable. At bearing maturity, annual pruning has a great influence in feijoa for fruiting performance. The fruits are borne on new shoots and in order to encourage growth of such shoots, pruning of the shoots which have borne fruits should be done after harvest. Pruning should be done by heading-back and thinning-out of shoots. Water-suckers and under-growth, should always be removed.

Pests and Diseases

Pests and diseases are usually not serious problems in feijoa. In southern part of Europe and in California and Florida in the U.S.A., black-scale and wax-scale insects and leaf-rolling caterpillars are known to cause some damage. However, fruit-fly may sometimes be a serious problem in feijoa fruits. In Victoria and New South Wales in Australia, feijoa growing is not encouraged as the fruit is feared to be a prime host of fruit-fly insect which damages other principal fruit crops also. Among diseases, some fungal infections cause damages and among these, leaf-spotting fungus, thread-blight, mushroom root rot *etc.* are more common. Crown rot and root rot diseases in feijoa have been stated by Vitale *et al.* (2008).

Harvesting

Feijoa shrubs raised from seeds come to flowering after 4–6 years of growth. Those from stem-cuttings and air-layers flower earlier and bearing maturity is still less in grafted plants. Actual flowering time of the shrub varies much in countries, *e.g.*, in Uruguay, it is in November, in northern Florida and in southern California in the U.S.A., in April–May and the flowering time in England is July. Fruits are ready for harvest after six months of fruit-set.

Timely harvest of the fruits is an important aspect in feijoa. The fruits are best ripened in the shrubs (Schotsmans *et al.,* 2011). Sugar content increases much during the last part of development of the fruits and if harvested earlier, they do not become much sweet in taste. Pulp of the fruit is white and opaque in immature fruits and its appearance turns clear and jelly-like when it becomes ripe. However, determining optimum stage of harvesting is difficult as the colour of the peel which is waxy-grey, green or bluish-green in immature fruits does not change much in fully ripe fruits.

Texture of the fruits may be an index for harvest and when the fruits are felt somewhat soft by hand pressure, they should be harvested. Instead of harvest from the plants directly, many growers prefer to collect the fruits which have naturally fallen on the ground. Shaking the shrubs to cause dropping of the fruits is also done but fruits which are fallen naturally or are made to drop by shaking the shrubs are liable to be damaged.

Yield

Production of 1000 fruits from a feijoa shrub is considered optimum and 2000 fruits is regarded to be a high yield. In India, feijoa gives low yield where it is grown and it is usually 100–200 fruits from a shrub. The possible reasons may be insufficient pollination, lack of proper pruning and damage by birds. It is observed that fruit size and yield are greatly increased when fruits are set from cross-pollinated flowers, even if the flowers are self-fruitful.

Varieties

Feijoa has quite a large number of varieties that are grown in different parts of the world. Morton (1987) described the characteristics features of a number of varieties that are grown in New Zealand, Australia, south American countries and California in the U.S.A. These are, Andre, Besson, Coolidge, Choiceana, Superba, Triumph, Mammoth, Apollo, Gemini, David, Roundjon, Magnifica and Hirschvogel. Ducroquet *et al.* (2007) described the features of the Brazilian cultivars of feijoa in respect of their origin, plant and fruit characters and disease resistance capacity.

Mode of Consumption

On removal of the peel of feijoa fruits, the pulp quickly turns brown on exposure to air. To avoid this, just after removal of the peel, the fruit should be dipped in a weak solution of common salt or sugar or in water to which, juice of sour lime fruits has been added.

Pulp of ripe fruits satiate much, particularly for its aroma which resembles a blend of pineapple, strawberry and with a tinge of flavour like that of mint.

The pulp has two parts, which are a clear jelly-like part, *i.e.*, the seed-pulp in the inside and at its outer side towards the skin is the opaque and the relatively firm part having a slightly granular texture.

Apart from the fruit, the fleshy and purplish petals of feijoa flowers are also relished for their sweet taste and minty flavour. Development and ripening of the fruits are not hampered if the petals are plucked off.

Quality and Nutritive Value

The various components in 100 gram of pulp of feijoa fruits have been stated to be as follows by Morton, (1987).

Moisture: 84 g,; Protein: 0.9 g.; Fat: 0.2 g.; Carbohydrate: 10 g.; Ash: 0.5 g.; Potassium: 166 mg.; Sodium: 5 mg.; Calcium: 4 mg.; Magnesium: 8 mg.; Phosphorus: 10 mg.; Iron: 0.05 mg. and Vitamin C: 28–35 mg. The fruit has been stated to be rich in water-soluble iodine compounds which ranged from 1.65–3.9 mg per kg of fresh fruit.

(Index: g = gram, mg = milligram).

Quality and nutritive value of feijoa fruits belonging to 11 genotypes had been assessed by Beyhan *et al.* (2011) in Turkey. They reported that weight, length and width of the fruits and soluble soilds and acidity of the fruit pulp had been respectively ranged as 18.57 to 40.0 gram, 27.74 to 59.95 and 23.42 to 39.43 millimetre, 8.3 to 14.2 and 3.33 to 4.50 per cent. Results on the analysis done with dried fruits of these genotype in respect of nitrogen, phosphorus, potassium, calcium, magnesium, iron, manganese, zinc and copper have been presented by them.

Medicinal Value

Feijoa fruits are highly valued for their medicinal properties. Karami *et al.* (2008, 2013) studied liver toning activity and Monforte *et al.* (2014) studied gastroprotective effect of the fruits. Anti-inflammatory and anti-bacterial effects and other medicinal value of feijoa fruits have also been studied by other workers.

Post-harvest Storage

Feijoa fruits do not keep well for more than a few days under ordinary temperature and in such environment, the pulp turns brown due to warmth, although colour of the peel does not change much. In low temperature and high relative humidity, fruits may be stored for 1–3 weeks. Under frozen condition, quality of the fruits is not lost for several months.

Effect of application of calcium chloride on post-harvest storage of feijoa fruits has been studied by Ramirez *et al.* (2005) and the chemical was found to have prolonged the storage life. Weight loss of the fruits was less by application of higher concentration of calcium chloride at each storage temperature, besides other beneficial effects that were observed.

Utilization

In containing high amount of pectin of superior jelly-grade, excellent jelly and jam could be prepared from the fruits. Many other products are also made either with the fruit alone or mixing it with other substances. Pudding, pies, fruit-sponge cake, pastry fillings, dumplings, fritters yoghurt, ice-cream and beverages are some examples. Fruit extract is known to be utilized in making wine also.

Family: Myrtaceae

Species: *Psidium cattleianum* Sabine

(Synonyms: *Psidium chinensis* Hort.; *P. littorale* Raddi; *P. humile* Vell.; *P. coriaceum* Mart. ex O. Berg; *P. variabile* O. Berg)

Taxonomic Serial: 27240

Two botanical varieties of the species are recognized. These are, (i) *Psidium cattleianum* var. *lucidum*, fruits of which are yellowish in colour and roundish (= spherical) in shape and (ii) *P. cattleianum* var, *acre* fruits of which are reddish to purplish in colour and oval to elongated in shape.

Common Name

Most commonly used name of the fruit in English is Cattley guava. The name is ascribed to honour the British horticulturist of the 19th century, Sir William Cattley. There are also other English names of the fruit, which are, Peruvian guava, Chinese guava, Japanese guava, Cherry guava, Pineapple guava, Strawberry guava, Hill guava, Small guava, Purple guava *etc. P. cattleianum* var. *lucidum* is known by the common names as Lemon guava, Guayabo *etc.* The common names of cattley guava are different in different languages. In India, the names such as Lal amrud, Lal anjir, Beguni peyara *etc.* are used by the people in places where the cattley guava is planted.

Genus: Psidium

Psidium as a genus was described by Linnaeus in 1753. The genus comprises 100 or more species, which are evergreen trees and shrubs having nativity to the warmer parts in the western hesmispheres in Mexico, central and south America, West Indies, Galapagos islands and the southern part of the U.S.A.

The genus is magnoliopsid in that, the seed has two cotyledons and the embryonic leaves appear after germination. Leaves are simple and opposite. Flowers are produced in clusters. Petals are 5 in number. Plants contain guavin–B and quercitin glycosides.

Many species of the genus bear edible fruits among which most important is guava. The trees and shrubs of the genus grow in tropical and sub-tropical climates and one species is known to be fairly cold tolerant, which is grown in the coastal areas in south-east Asian countries.

Importance

Cattley guava is a minor fruit and is considered as one of the important relatives of the common guava, *i.e., Psidium guajava*. However, in India, it is usually regarded to be next in importance to the common guava.

Although not so important as a fruit crop, importance of cattley guava plant is much more recognized for its beauty (Morton, 1987). Its leaves are dark green in colour, glossy and the new leaves are reddish. Bark of the tree is cinnamon brown in colour. Flowers are white and have a sweet smell. Fruits are shiny, reddish, purplish or yellowish. For these attributes, the tree is usually preferred for planting in parks and ornamental gardens, where, it is planted as a solitary tree in open space to draw attention of the visitors, or in rows as a decorative hedge, or in combination with bougainvilleas to climb on them for support.

Cattley guava has a number of merits. It is a very hardy shrub and grows well in adverse soil where farming potential is usually low. Adaptability to any change in the environment is a special feature of cattley guava. Its productiveness is very high and much higher than the common guava. Pectin content of the fruits is very high and is of high jelly-grade, for which excellent jelly and jam can be prepared from them. Wood of the tree has also economic use.

Nevertheless, in many countries and notably in south-eastern U.S.A., many countries in the south and central America, Hawaii and other Pacific islands, Tanzania, South Africa, Mauritius, Australia, New Zealand *etc.*, cattley guava is regarded to be a serious invasive weed. The reason is, the plant multiple easily and quickly and covers large areas of agricultural, pasture and forest lands and make them unsuitable for growing any crop. The fruits contain abundant seeds, which are dispersed by birds, other animals and especially by the feral pigs. The plants are hardy and when uprooted, they grow again in rain. They are difficult to be controlled even by chemical weedicides. For these reasons, the plant rampantly grows and covers more and more area. It is even branded as one of the hundred most noxious weeds in farm land.

In India however, cattley guava is not known to be grown so virulently in any part of the country such that it may act as a weed. The shrub is grown now in parts of the southern and eastern regions of the country as a minor fruit crop and also in large ornamental gardens as decorative and bushy hedge.

It may be opined that in India, cattley guava needs much attention now for growing in agro-forests, social-forests, as hedges and wind-breaks along orchards, in large decorative gardens, along avenues or where mass planing is desired to utilize inferior land. The reason is, in order to make guava jelly, the common guava fruits are largely used in this country. Considerable proportion of guavas produced in the country being utilized in making jelly, people are deprived to relish this nice, dessert fruit. Evidently, cattley guava fruit have much less importance than the common guavas, *i.e., Psidium guajava* for use as fresh fruit but it may be mentioned that in containing high amount of pectin of superior jelly-grade, excellent jelly can be prepared from them and the jelly is better than when made from common guava fruits. Hence, Mazumdar (2018) has stated that by giving emphasis on this minor fruit in making guava jelly, the pressure that is exerted now on the common guava fruits to make jelly could be lessened. Besides jellying potential of the fruits, the shrub needs attention for growing in land of low fertility status also.

Description of Plant Parts

Cattley guava is an evergreen shrub or a small, compact tree. The height goes normally to 4–6 metres but may go upto 8 metres in favourable soil. Leaves are smooth, entire, opposite, elliptic to oblong, shiny with leathery texture and may be upto 13 cm long. They are dark green and the new leaves are reddish in colour. Branches are roundish and the bark is cinnamon-coloured, which peels off from wood. Flower is bisexual, white, 2.5 cm wide, borne singly in axil and has 5 aromatic petals and many stamens. Fruit is globose, hard, 4–6 cm in diameter and has attractive colour. Seeds is 3–5 mm long and numbers 70–100 or more in a fruit.

Morphological differences of young shoots, flowers and fruits of *P. cattleianum*, *P. guineense* and *P. guajava* (= common guava) have been given in the Chapter on *Psidium guineense* in the book.

Origin

The species is native to south America and is considered to be in the region from the low lands of south-eastern Brazil to north-eastern Uruguay.

Geographical Distribution

Cattley guava is grown in many tropical countries and also in humid zones in countries of subtropical climate. Some countries that may be mentioned are Singapore, Malaysia and other south-east Asian countries, Australia, New Zealand, Tanzania, Mauritius, South Africa, West Indies, southern and south-eastern parts of the U.S.A., central and southern American countries, Hawaii, Pacific islands *etc.* In India, the plant is now grown as a minor fruit crop or is planted in ornamental gardens in the southern and the eastern parts of the country.

Climatic Requirement

Cattley guava is able to grow in a wide range of climatic condition within the tropics but also grows satisfactorily in the moist subtropical climate. The shrub is more hardy than the common guava but the yellow-fruited botanical variety (*Psidium cattleianum* var. *lucidum*) is less hardy than the red-fruited species (*P. cattleianum* var. *acre*). Yield performance of cattley guava is very high when it is grown upto an altitude of 800 metres but it gives fruit upto an altitude 1500 metres in the tropics. It survives low temperature of upto (-) 5.5°C for some period. Full sunshine during fruit development of the fruits gives rise to fruits of better quality and colour.

Soil Requirement

Soil is not an important factor for growing cattley guava. The tree grows and gives fruits even in very poor and adverse soil condition that would barely enable other fruit plants to grow. Salt tolerance capacity of the fruit tree being high, it may be grown in coastal area. The tree tolerates soil alkalinity to a considerable extent. Tolerance to drought and flood are special attributes of cattley guava and the red fruited species is much more tolerant than the yellow fruited botanical variety. However, the plant is shallow-rooted and low moisture content of the surface soil is harmful as it causes heavy leaf fall, drying of twigs and low productivity. Good drainage is necessary to the shrub for vigour and higher productivity.

Propagation

Cattley guava can be easily propagated by seeds as germination has no problem (Mazumdar, 2018). Fairly true-to-type plants as with the mother plant are obtained when they are raised from seeds (Hayes, 1970). Propagation by cutting-off water-suckers from the basal parts of the plants is also done. Stem-cuttings are difficult to form roots in cattley guava but success is obtained by treatment with auxin. Voltolini and Fachinello (1997) reported root formation in hardwood-cuttings by application of suitable concentrations of IBA and particularly when the stock plants were grown in 70 per cent shady situation. Dutra *et al.* (2000) studied on the effect of shading and application of IBA and Phloroglucinol on leafy stem-cuttings in which terminal buds were cut off. They observed that Phloroglucinol decreased root number per cutting while root numbers increased by treatment with 2000 ppm of IBA and on maintaining in 70 per cent shade.

Stool-layering gives high success and in this method, there is advantage to obtain many rooted layers from a single stock. Propagation by stool-layering has been studied in cattley guava by Mishra *et al.* (2007) and Saroj (2012) under north Indian condition. Air-layering might give high success particularly by the use of auxin and when it is done in rainy-season.

Budding is not done in cattley guava as it is difficult to perform this operation in this plant, whose bark is thin and hence, lifting the bud from the rootstock and fixing that on the scion are almost impossible. However, a budding method known as chip-budding, in which a chip of wood which remains attached to the bark is retained in the bud-piece may be tried in this plant and also in other plants whose bark is thin and difficult to be separated out from the underlying wood. Detached method of grafting, *e.g.*, whip, whip and tongue, side- and veneer-grafting may also be done to get success. The species itself is also a strong rootstock for the common guava. Experiments have been done with success on micro-propagation of some *Psidium* species which are called guava relatives and including cattley guava

Culture

Normally, relatives of guava are grown in India without artificial care and culture. But nowadays, mass planting of these species is done in some places for waste land utilization and industrial use of their produce. Accordingly, where organized planting of these minor fruit crops is desired, the plants should be set at a spacing of 4–5 metres on both ways. For planting to build up hedge-rows, 2–3 metre spacing may be maintained between the plants in a row and row spacing may be given 5–6 metres.

The tree responds very well to application of manures, fertilizers and summer irrigation. But productiveness in terms of quantity and quality are tremendously stepped up by judiciously done training and pruning operations. Hayes (1970) went on to comment, in improved guava culture, "Where other cultures are neglected in guava, it is training and pruning that cannot be disregarded." The same is true with the relatives of guava as well. Foliar application of liquid manure to increase production has been stated by Mazumdar (2018).

During the first few years, training of cattley guava plants should be done by *vase* (= open-centre) system such that they remain dwarf and highly spready which would enable to admit adequate light and air to reach in the inside of the plants. Pruning of the mature trees should be done both by heading-back and thinning-out of branches to avoid sparse growth as well as over-crowding of shoots as the case may be. Bending of shoots as is done in guava is of great advantage in case of guava relatives also for higher productivity. Like the common guava, *feeder roots* of the guava relatives are superficial and hence, they often suffer from moisture stress. To avoid this, liberal irrigation in summer is of great advantage. Mulching of the soil is a very good practice to conserve soil moisture and control of weeds. Water-sprouts that arise at the base of the shrub should be regularly trimmed as they do not produce fruits but compete with the mother plant in respect of nutrients and moisture.

Pests and Diseases

All guava relatives are hardy and pests and diseases are not serious problem to them. Some insect pests that attack cattley guava are thrips, scale insects, fruit-fly, bark eating caterpillar, mealy bug *etc.* but the extent of damage by their attack is not significant.

Thrips cause damage to flower parts and Chlopyriphos (0.02 per cent) may be sprayed before flowering to control them. Scale insects suck plant sap and to control, Rogor 30 EC or Phosphamidon (0.03 per cent) may be sprayed. To control bark eating caterpillars, Carbaryl (0.1 per cent) may be applied on trunk and main branches. Mealy bugs suck sap from leaves and young shoots and to control, soil treatment with Aldrin or Thimet or Malathion and banding base of the plant with polythene are effective. Fruit-fly insects lay eggs under the skin of fruits and the maggots enter into the fruits which rot. Reddy and Vasugi (2008) stated destructive effect of this pest. To control, Malathion 50 EC (0.05 per cent) may be sprayed. Hanging of methyl euzenol bottle traps near the plant is also effective. Among diseases, brown spots on leaves and fruits due to anthracnose disease is common. To control, Copper oxychloride (0.3 per cent) or Difolatan (0.3 per cent) may be sprayed.

Harvesting

Seed-borne cattley guava shrubs flower after 2–3 years of transplanting. Flowers appear in November–December and the ripe fruits are obtained from spring to autumn. In the southern part of India, two crops are obtained, which are in July–August and January–February. In highly tropical and moist climate, some fruits may be available almost all throughout the year.

Fruits are best ripened in the trees and do not ripen any more after harvest. Hence, only ripe fruits should be harvested. However, Morton (1987) suggested that as the ripe fruits are highly perishable, those which had not attained full ripening, *i.e.*, those which are semi-ripe fruits should only be plucked. This may have necessity if storage of the fruits for some period is required.

Harvesting should be done in instalment and as described in the Chapter on *Dimocarpus longan*, a tree-pruner fitted with a sickle-shaped blade, basket and pulling rope is efficient for the purpose. Shaking the shrub to cause dropping of the ripe fruits and then collecting them from the ground is although done by many but fruits are damaged by this practice due to falling on ground.

Yield

In rich soil of high moisture content, yield of cattley guava may be 15 tonnes from a hectare area of plantation in two crops in a year.

Varieties

Although a great many variations in plant and fruit characters of cattley guava are found, named varieties of the plant are not much known. John Riley is a variety of large round fruit having dark red skin and cherry red pulp. The fruit is of excellent flavour and its ripening continues over a relativity long period. Dinesh and Reddy (2001) have stated accessions of different *Psidium* spcies for their fruit characters from Bangalore.

Mode of Consumption

Cattley guava fruits have sweet-sour taste, pleasant flavour and lack of muskiness which is obtained in the common guava (Subramanyam and Iyer, 1993). Flavour of the fruit resembles strawberry which however, is less pronounced in those of the botanical variety, *P. cattleianum* var. *lucidum*. Fruits are best relished when they are fully ripe.

Fruits are taken as fresh and there is also a practice to make a paste of the pulp on blending with ginger, sugar and common salt and then to take.

Quality and Nutritive Value

The nutritive constituents in 100 gram of cattley guava fruits containing 85.3 per cent moisture has been stated by the National Institute of Nutrition, under the I.C.M.R. (Govt. of India) in Hyderabad are as follows (Gopalan *et al.,* 2018).

Protein: 0.1 g.; Fat: 0.2 g.; Carbohydrate: 9.0 g.; Crude fibre: 4.8 g.; Mineral matter: 0.6 g.; Calcium: 50 g.; Phosphorus: 20 mg.; Iron: 1.2 mg.; Thiamine: 0.03 mg.; Riboflavin: 0.02 mg.: Niacin: 0.3 mg and Vitamin C: 15 mg. Consumption of 100 gram of pulp gives rise to 38 Kilocalorie (= 158.97 K Joule) of energy in the body.

(Index: g = gram, mg = milligram)

Morton (1987) has stated that analysis of ripe fruits of cattley guava done in different countries had shown the following constituents.

In red fruits: Seed, water, ash, crude fibre, protein, fat and total sugars respectively as 6, 81.73–84.9, 0.74–1.5, 6.14, 0.75–1.03, 0.55 and 4.42–4.46 per cent respectively. In yellow fruits: The same constituents had been in the order of 10.3, 84.2, 0.63–0.75, 3.87, 0.8, 0.42 and 4.32–10.01 per cent respectively. In both red and yellow fruits, the vitamin C content was reported to be 22–50 milligram per 100 gram of pulp. The energy value for 1 kilogram of pulp in both two types of fruits had been stated to be 268 Kilocalorie.

Medicinal Value

Cattley guava has therapeutic effect against many ailments. Acaricidal effect of an active compounds had been isolated from the fruits by Oh Minseok *et al.* (2014). They also put forward antimicrobial and antiproliferative effects on human cancer cells from extracts made from the fruits.

Phytochemicals

The essential oils that are responsible for the flavour of the fruits had been extracted by Chalannvar *et al.* (2012) in South Africa. They identified 53 chemical compound which accounted for 61 per cent of essential oils and the major compounds had been stated to be caryophyllene oxide. The fruit is also a rich source of pectin of high jelly-grade and hence, excellent jelly and jam could be prepared from it. Vriesmann *et al.* (2009) have reported types of pectic (= polygalacturonate) substances in the fruits and their characteristics.

Constituents in Developing Fruits

Fruit developmental study undertaken by Gatho *et al.* (2000) in Brazil revealed that the fruits required four months to complete their development. It had been observed by them that dry matter content in the fruits continued to increase until the end of their development.

Lozano Veronica and Luis Fernando (2013) studied the evolution of some components of the non-enzymatic defence system with analysis done during growth and ripening of the fruits. It was observed by them that the antioxidant activity including total phenols and non-phenols such as vitamin C was highest during the growing period of fruits. The antioxidant activity had decreased and other compounds such as anthocyanins appeared during the ripening period.

Post-harvest Storage

Morton (1987) stated that ripe fruits of cattley guava being highly perishable could not be kept for more than 3–4 days at room temperature condition. She commented upon that if shipping of the fruits was desired, they should be harvested before the stage of full ripening. In regard to cold storage, she cited that Hawaiian-grown cattley guava fruits when slightly under-ripe were stored at 0°–2.22° C for a month but those were found to be shrivelled and decomposed. Hence, still higher temperature should be maintained during storage.

Utilization

Cattley guava fruits are rich source of pectin of high jelly-grade. Hence, excellent quality of jelly and jam could be made from the fruits. Fruit cheese and drink can also be made from the fruits.

Family: Myrtaceae

Species: *Psidium friedrichsthalianum* (O. Berg) Niedenzu

[Synonym: *Calyptropsidium friedrichsthalianum* O. Berg]

Taxonomic Serial: 506171

Common Name

The indigenous name of the species in Costa Rica is Cas. In English, the common names are Costa Rican or Costa Rica guava and also Chinese guava.

In Guatemala, the common name is Guayaba acida, in Columbia, Guayaba agria, in Panama, Guayaba de agua and in Ecuador, Guayaba del Choco. The common name in Mexico is Guayaba motes and in France, Goyavier de Costa Rica or Goyavier du Costa Rica. There in no particular name of the fruit in India and where it is grown in this country, the name, Costa Rican guava is used. In some parts of India, it is called Chinese guava.

Importance

Morton (1987) has stated, "Perhaps the most noteworthy of the lesser species of *Psidium* is *P. friedrichsthalianum*." The fruit is very sour and hence, has not been much used as an edible fruit. However, in containing high amount of pectin of high jelly-grade, it is much preferred in making jelly.

This small, evergreen tree which has a *cylindrical bole* and good branching habit from near the ground has however, an *attractive* appearance. Hence, it is considered as an ornamental tree in the central American countries and in those countries, Costa Rican guava is much preferred for planting in decorative gardens. The species also serves as a highly suitable and hardy *rootstock* for the common guava.

Description of Plant Parts

Costa Rican guava is a small evergreen tree and is usually with a straight and cylindrical bole. It branches near the ground and the peeling bark is brown. The growth habit is similar to that of common guava. The tree is evergreen.

The tree reaches a height of 6 to 12 metres. The trunk is slender and the bark is reddish brown with greyish patches. Young branchlets are wiry, quadrangular and dark reddish in colour. They are minutely hairy. Leaves are 5–12 cm long and 2.5–5 cm wide. They are elliptic, oblong-elliptic or oval in shape, pointed at the tips, gland-dotted and thin. Upper surface of the leaves is dark and smooth and lower surface is pale. Flowers are usually borne singly. Calyx is closed but splits into irregular segments when the flower expands. Petals are white, 2.5 cm wide and fragrant. Stamens are many, about 300 in number and are 1.25 cm long. Fruit is round or oval, 3–6 cm long and the peel is yellowish. Pulp is white and highly sour. Seeds are few, flattened and 0.5 cm long.

Origin

Costa Rican guava is stated to have been originated somewhere in central America.

Geographical Distribution

The tree is widely grown in Costa Rica, especially in its temperate zone highlands. It grows naturally in the Cauca and Magdalena valleys in Columbia, throughout central America, in Oaxaca in southern Mexico and in the south-east Asian countries. The species has been introduced in some parts of India and the fruiting performance is known to be satisfactory.

Climatic Requirement

Costa Rican guava grows in tropical and subtropical climates and upto an altitude of 1000 metres. The tree grows well in areas having annual rainfall of 1200–600 mm. Open sunny situation is highly favourable for better growth. The tree is highly sensitive to low winter temperature and is injured by frost.

Soil Requirement

The tree succeeds in a wide range of soil condition. Slightly alkaline soil with a pH of 7.1–7.2 is however, preferred by the trees. Drainage is very important as the trees are badly suffered in poorly drained soil.

Propagation

Costa Rican guava is mostly propagated by seeds. Seeds germinate easily and pre-soaking the seeds in warm water for 24 hours facilitates germination. Germination is delayed when sown in cool soil and quicker when sown in warm soil.

Propagation can also be done by hardwood or medium hardwood stem-cuttings and with application of auxin. Like guava, mound-layering could be done in Costa Rican guava. Mishra *et al.* (2007) have reported 8.88 per cent rooting after two months of mounding. Saroj (2012) in Meerut in Uttar Pradesh had obtained 9.53 per cent rooting in mound-layering but the rooted plants failed to establish when planted. Air-layering gives success in Costa Rican guava by application of auxin. Micro-propagation in Costa Rican guava has been successfully done by Kumar *et al.* (2003) using shoot tip explants and maximum sprouting was observed by them when the materials were collected in spring season and with 2 cm long explants.

However, Costa Rican guava in itself is a compatible and highly suitable rootstock for the common guava. The rootstock is resistant to wilt disease and nematodes and hence, it is used in many guava growing countries.

Culture

Costa Rican guava grows and fruits well without receiving any care but application of manure greatly increases productivity and quaility. Size and yield of fruits however, markedly enhance by proper training and pruning practices. Training the trees to open-centre or *vase* system is considered best. Flowers are borne on new shoots and to encourage production of such shoots, pruning is of great advantage by trimming the old shoots.

Pests of Diseases

Attack by pests and diseases is insignificant in Costa Rican guava. However, attack by fruit-fly (*Bactrocera dorsalis*), tea mosquito bug (*Helopeltis antonii*) and spider mite are sometimes found although to a lesser extent.

Harvesting

Costa Rican guava is a slow growing tree and the trees coming from seeds start fruiting after 4 or 5 years of growth. Fruits ripen best in the trees and are of attractive colour when fully ripened. However, like the common guava and the other relatives of it, the fruits become much softened in ripe condition due to hydrolysis of pectic substances and hence, plucking of the fruits before full ripening is advisable.

Varieties

True varieties of Costa Rican guava are not known but many variations in physico-chemical characteristics of fruits and tree characters are met with. Vasugi and Dinesh (2007) have reported genetic variability of the species from Bangalore.

Mode of Consumption

Fruits of Costa Rican guava being very sour are difficult to take as raw. Morton (1987) quoted, "Early Spaniards complained that eating the raw fruits set the teeth on edge." Most commonly, jelly and other products are made from the fruits. Fresh fruits are also tasted on blending with sugar, salt and spices.

Quality and Nutritive Value

Morton (1972) has stated that analysis of the fruits in Guatemala showed, moisture: 83.15, Protein: 0.78–0.88, Carbohydrate: 5.75–6.75, Fat: 0.39–0.52, fibre: 7.90 and ash: 0.80 per cent respectively. It has also been stated that pectin content of the ripe fruits had been very high. Fruits are known to contain high vitamin C.

Medicinal Value

The medicinal value of the extracts made from Costa Rican guava fruits against chronic obstructive pulmonary disease has been reported by Flores *et al.* (2013). According to them, extract made with ethyl acetate had highest antioxidant activity and this fraction had anti-inflammatory action.

Phytochemicals

Pino *et al.* (2002) in Cuba isolated volatile compounds from Costa Rican guava fruits by steam distillation process. From the extract of aroma, 173 components were identified by them. According to them, amount of aliphatic esters and terpene compounds were possible factors that determined unique flavour of the fruits. Flores *et al.* (2013) isolated a number of phenolic compounds in the fruits.

Utilization

As stated earlier, the fruits of Costa Rican guava are very sour and hence, they are difficult to consume as raw. However, the fruit is a rich source of pectin of high jelly-grade for which it is much valued in making jelly. Apart from making firm jelly, fruits are also utilized in making preserve, pie and fruit cheese.

The common guava of commerce which is botanically known as *Psidium guajava* has reputation in making fruit jelly of high standard. Mazumdar and Mukhopadhya (2011) and Mazumdar (2020) have stated in their books that if jelly is made from the common guava along with combination of fruit pieces obtained from some other relatives of guava, *i.e.*, closely related species of *Psidium guajava*, the jelly that could be made would be of much better quality, consistency, standard and also of high nutritive value. For this purpose, selection of fruits of different species of guava should be made intelligently. The proportion of fruit samples that should be added in the mixture should also be made judiciously, besides that sugar and acid that are to be added should be in optimum proportion.

Family: Myrtaceae

Species: *Psidium guineense* Swartz (R)

(Synonyms: *Psidium guyanense* Pers.; *P. molle* Bertol.; *P. araca* Raddi; *P. schie-deanum* Berg.; *P. costaricense*; *P. laurifolium; P. schippii; P. ovideum; P. rotundifolium*).

Taxonomic Serial: 506172

Morton (1987) stated that this species of the guava relative had been the subject of much confusion. According to her, "For a long time it was considered distinct from the guisaro, *Psidium molle* Bertol (Syn. *P. schiedeanum* Berg.) but now these names as well as *P. araca* Raddi treated as synonyms of *P. guineense* and all the corresponding colloquial names should be applied to this one confirmed species".

Common Names

In English, the names that are commonly used for the species *Psidium guineenese* are Guinea guava, Brazilian guava, Wild guava, West Indies guava, Castilian guava *etc.* In other languages and countries, the fruit is known by other names and some of these are as follows:

Brazilian: Guayaba brava. Peru Sacha guayaba. Bolivian: Chobo diondan. Afrikaans: Brasiliaanse koejawel. Costa Rican: Cas extranjero. Portuguese: Araca-azedo. Spanish: Guayaba agria. French: Goyavier du Bresil. German: Stachelbeerguave.

There is no particular name of Guinea guava in Indian languages and where it is grown, it is called Jangli amrud.

Importance

Guinea guava is a minor fruit in India and is grown in some parts of India now. Evidently, the fruit is not comparable to the common guava of commerce in respect of quality. Nevertheless, these small fruits are not disliked by those who have tasted them. Fruits have a strawberry-like flavour and are highly suitable to make jelly. They are also preferred for baking and preserving. A merit of this fruit is, unlike common guava, it remains firm when ripe and hence, it is able to withstand transport under ordinary temperature. Hybrids have also been produced by crossing this species with common guava. Wood of the tree is strong and has good timber value. In some parts of India, the shrubs is preferred for growing in land of low farming potential. However, in many countries, this is regarded to be an invasive weed.

Description of Plant Parts

Guinea guava is a relatively slow-growing, evergreen shrub of 1–3 metres in height but in rich soil, it becomes very tall so as to attain a height of 6–8 metres and look like a tree. Leaves are greyish and measure 3.5–14 cm in length and 2.5–8 cm in width. They are oval-shaped and the leaf-blade is stiff and sometimes have toothed edges. Undersurface of the leaves is glandular and covered with pale reddish hairs. Young shoots are hairy and cylindrical or slightly flattened. Bark is greyish in colour.

Flowers are borne in leaf-axils, either singly or in a cluster of upto 3. Petals are white and fragrant. Stamens are highly prominent and 150–200 in number. Fruits are round or pear shaped (= bell-shaped), firm even in ripe condition and are 1.5–5 cm wide. Peel is yellow-coloured. Pulp is white in the inner side and yellowish at the outer side. Fruits have a flavour like that of strawberry. Seeds are many and embedded in pulp. They are hard in ripe fruits.

Morphological differences between *Psidium guineense*, *P. cattleianum* and *P. guajava* could be stated as follows.

(a) Young stem–In *P. guineense*: these are hairy (= pubescent) and almost rounded in cross-section (= sub-cylindrical). In *P. cattleianum*: these hairless (= glabrous) and truly rounded in cross-section (= cylindrical). In *P. guajava*: these are hairy (= pubescent) and are 4-angled in cross-section (= quadrangular).

(b) Flowers–In *P. guineense*: usually 3 flowers (occasionally 1) are borne in leaf-axils. The hairy (= pubescent), dull green leaves have 6–10 pairs of lateral veins. In *P. cattleianum*: flowers are borne singly in leaf-axils. The hairless (= glabrous), glossy green leaves have 6 or 7 pairs of lateral veins. In *P. guajava*: flowers are usually borne singly (occasionally 3) in leaf-axils. Somewhat hairy (= pubescent) dull green leaves have 10–20 pairs of prominent lateral veins,

(c) Fruits–In *P. guineense*: fruits are very small (1–2.5 cm long), round or pear-shaped (= pyriform). Peel is yellow coloured. In *P. cattleianum*: fruits are relatively large (= 2–4 long) and round in shape. Peel is purplish–red or yellow–coloured. In *P. guajava*: fruits are much larger (3–11 cm long), round or pear-shaped or ovate, depending on varieties. Peel colour is greenish, cream–coloured, yellowish or light orange. In some varieties, peel is dotted with red or brown spots, *e.g.*, in the cv. Chittidar.

Origin

The origin of the species is tropical America, in the region from Mexico to Argentina, including parts of the Caribbean countries.

Geographical Distribution

Guinea guava is much grown as wild in Mexico, West Indies, Guatemala, Panama, Guyana, Suriname, Venezuela, South American countries, South Africa,

Tanzania, Kenya and other tropical countries. In some countries, it is considered as an invasive weed. It is not much grown in India.

Climatic Requirement

Although guinea guava grows best in tropical climate, it thrives successfully in subtropical situation and arid-zone. The shrub prefers open sunny situation for better performance. It grows well upto an altitude of 1500 metres in the tropics. Very low temperature is tolerable to the plant but for a short period only.

Soil Requirement

Guinea guava grows very well in coastal area in Brazil. The shrubs grows in a wide range of oil condition. However, it does not develop satisfactorily on light sandy soil (Morton, 1987).

Propagation

Where guinea guava is grown in the world, it is exclusively propagated by seed. Seeds germinate well but very slowly and may take several weeks to a few months to germinate. Seeds are very hard and hence, mechanical or chemical scarification loosens the seed-coat, for which germination is facilitated. Low temperature delays germination of the seeds and for this, the temperature of the soil or the medium should be maintained at around 25° C. Propagation by ground–layering (= stool layering), air-layering and stem-cuttings are also successful and auxin is of help to regenerate adventitious roots.

Grafting is not done in guinea guava but the species itself having resistance against soil-borne fungi (that cause wilt disease) is employed as a rootstock for the common guava. Experiments have been done with success on micro-propagation of some species of *Psidium*, including *P. guineense*.

Culture

Guinea guava is not grown with any cultural practice. However, manuring and summer irrigation greatly increase yield and quality of the fruits. If mass planting of the fruit crop is desired to be done, plant spacing of 6–9 metres should be maintained. The plant responds markedly to training and pruning practices like guava. Methods of training and pruning as stated in *P. cattleianum* may be adopted for the purpose.

Pests and Diseases

Pests and diseases are not serious problem in guinea guava. Some of the pests as stated under *P. cattleianum* attack this fruit crop also and to control them, measures that have been stated there may be adopted. As the fruit is hard even in ripe condition, attack by fruit fly is not common. An evergreen parasitic plant, *viz.*, mistletoe (*Psittacanthus angustifolius*) is sometimes found which sucks the plant sap when grows over it. Manual removal should be done to eradicate this phanerogamic parasite.

Harvesting

Plants raised from seeds attain bearing maturity within 3–4 years. They flower from late spring and early summer. Fruits are harvested before winter and from September to November. Like other guava relatives, fruits are best ripened in the plant. Although ripe fruits are not easily spoiled, harvesting just before full ripeness is advisable, particularly if the fruits are needed to be stored for some period. Harvesting should be done in instalment and with suitable implement. For harvest, shaking the plants to cause dropping of the fruits is not desirable as the fruits may be damaged.

Yield

Yield of guinea guava may go upto 12 tonnes when the shrubs are grown in rich soil.

Varieties

Named varieties of the guinea guava are not known. The species has been crossed with common guava (*P. guajava*) to produce hybrids (Landrum *et al.*, 1995). Morton (1987) stated that the hybrids which were produced by crossing guinea guava and the common guava had been dwarf, hardy and heavy bearing.

Mode of Consumption

Guinea guava fruits are sweet-sour in taste with a strawberry-like flavour and make good appeal. Seeds are small and taken with fruits. Although taken as raw, fruits are also processed as described in cattley guava.

Medicinal Value

Fruits of the species are helpful in reducing blood sugar. Decoction of bark and root is used to treat diarrhoea in Brazil. In Costa Rica, leaf extract is used to treat common cold, varicose vein, ulcer and other diseases.

Phytochemicals

Fruits are known to contain flavonoids, tannins and essential oils. Peralta-Bohorquezo *et al.* (2010) isolated volatile compounds from the fruits by headspace solid-phase micro-extraction and simultaneous distillation solvent extraction and identified by GC-FID and GC/MS.

Utilization

Excellent jelly can be prepared from the fruits of guinea guava and according to Morton (1987), this jelly is superior to that made from common guava. To make jelly, fruits which have not fully ripened should be used as their pectin content is much higher. Fruits are also suitable for baking and preservation (Morton, 1987). Fruit–cheese can also be made from the fruits as in done in common guava.

Family: Myrtaceae

Species: *Syzygium cumini* (Linn.) Skeels

[Synonyms: *Calyptranthes jambolana* Willd.; *Eugenia cumini* (Linn.) Druce; *Eugenia jambolana* Lam.; *Myrtus cumini* L.; *Syzygium jambolana* D.C.].

Taxonomic Serial: 505419

A wild species, which gives small fruit of inferior quality is *Syzygium fruiticosum* D.C. that grows in the foothills of Himalayas (Singh and Saxena, 2008). Pareek and Sharma (1993) stated eight species of wild jamun growing in the Nilgiris and western ghats.

Common Name

In English, the most widely used common name of the species, *Syzygium cumini* is Black plum. Other names that are used in English are Java plum, Malabar plum and Jambolan. The Indian names, Jamun and Jaman are also nowadays used in English to refer to Black plum. In English, Black plum is sometimes called Blackberry or Indian blackberry. But Blackberry is a misnomer as there is another fruit which goes by the name as Blackberry, the botanical family and species of which are altogether different from the Black plum, *i.e.*, Jamun.

In other languages, the fruit is known by other names. The names that are used in some other languages or countries are as follows.

Brazil: Jalao, jambo, Jambolao, jamelao, Jambola *etc.*, Fiji: Kavika India. French: Faux Pistachier, Jambolon, Jambolanier, Jamelongue *etc.* German: Jamblanapflaumenbaum, Wachsjaambuse *etc.* Spanish: Yambolana, Ciruelo de Java, Guayaboo, Pesgooa *etc.* Portuguese: Jamelao. Suriname: Jameon, Druift *etc.* Indonesia: Jamblang, Jewet *etc.* Java: Duwet, Djoowet *etc.* Malaysia: Jambul, Jamelong, Jaman, Jambulana, Jiwat *etc.* Thailand: Wa, Ma-ha, Hakhiphae *etc.* Myanmar: Thabyay–hpyoo. Nepal. Ban Jamum, Jambu, Phanir, Kainyu *etc.* Tanznia: Mzambarau. Kenya: Lushanaku.

In Sanskrit literature, the fruit has been referred to as Jambu. The tree is grown in all tropical and subtropical parts of India and the local names of the fruit that are used in different parts of the country are as follows.

Jaam, Jambool, Jambhul, Jambu, Jaman, Jamudo, Jamukoli, Jambuvu, Kalojam, Mahadan, Naaval, Nagai, Nawar, Nerudu, Neralehannu, Phalani, Phalinda, Ra Jamun, *etc.* and besides these, there are also many other names that are in vogue in different parts in the country.

Genus: Syzygium

The genus, *Syzygium* is largest in the family, Myrtaceae and comprises about 1200 species. Formerly, the genus was taxonomically confused with the genus, *Eugenia* and there are many species which were classed as *Eugenia* have now been placed in the genus, *Syzygium*. The genus, *Syzygium* is however, poorly known and hence, needs to be critically analyzed. Most of the species in this genus are evergreen trees or shrubs and grow in tropical and sub-tropical climates, although some thrive in temperate climate.

Native range of the species extends from Malagasy to Hawaian islands and from India and China to south-east Asia, southern Australia and New Zealand. At least 70 species are known to occur in Australia.

Syzygium genus is magnoliopsid, *i.e.*, the seed contains two cotyledons which give rise to produce two embryonic leaves. Leaves of the tree are simple, usually opposite and aromatic. Many species have economic importance and are grown as ornamental trees for their attractive foliage, some produce fruits which are edible to the human beings and some have importance as spices. Nevertheless, some species grow as invasive weeds in some countries where they are even considered as ecological threat.

Importance

Black plum, *i.e.*, Jamun has a special significance in India. The fruit tree is usually not much cropped in organized orchards in this country but plentiful trees make their existence in natural condition and in forests in all tropical and sub-tropical parts in this country. The trees come up easily from seeds which are thrown here and there by the people on consuming pulp of the fruits, by birds and animals and the seeds of fruits which are naturally fallen from the trees. Two or three adjoining villages in any tropical and subtropical parts in India could hardly be seen that has not at least one fruit tree of mango and jamun. It is also true that even very poor boys and girls in tropical and subtropical India could hardly be met with who have not tasted jamun fruits during season of their production.

In India, not only that the jamun tree is grown naturally here and there in abundance but people also raise the trees by sowing seeds in household compounds and places to secure fruits and leaves from them, fuelwood from their thin branches and the timber from the trunk and thick branches of the tree. The tree is highly suitable for afforestation to build up large forests and also for planting in social and agro-forests. Jamun trees are highly preferred for planting along orchards boundaries to serve as wind-break trees and along roads to serve as shade trees.

The enormous medicinal value of the fruit and other parts of the jamun tree were known to the Indian people since ancient time and has been amply mentioned in the Sanskrit and Ayurvedic literature where it has been termed as Jambu. It has been mentioned in the epic, Ramayana that Lord Ramachandra and his associates used to take jamun fruits during their exile in the forest. In the Hindu and Buddhistic cosmology, Jambudweep which is regarded as one of the Great Seven Continents is said to be named from Jambu, *i.e.*, Jamun. The Hindus in Gujarat regard jamun to which they call jamboon as the secred fruit of the gods.

All parts of the jamun tree have their usefulness to man. Fruits are delicious in taste and are highly relished especially to quench thirst in summer. Drink, other products and even fermented liquor are made from the juice of the fruits. Seeds being rich in protein, carbohydrate and calcium serve as highly nutritive cattle feed (Hayes, 1970). Both fruits and seeds have tremendous medicinal value and are particularly efficacious in diabetes of type–2. Leaf juice is a remedy in painful gums. Wood of the tree is hard, not attacked by worms, resistant to water for which the trunk and the thick branches are used to make doors, windows, cart-wheels, railway sleepers *etc.*

The tree is very hardy, highly adaptable to adverse soil situation and hence, serves as an effective as well as economically important tree to make use of land having low farming potential. It can also withstand water–logged situation for a considerable period. The highly robust nature of the tree is very much effective to resist high wind and serves as protective and productive wind–break along orchard boundaries. For dense foliage and highly spreading branches, jamun is an ideal shade tree and for graceful appearance, it is planted along avenues. The tree has also great suitability for afforestation in tropical and subtropical climates.

Productive capacity of the jamun tree is high and the fruit quality and yield of it greatly increase with much less use of agro-inputs. New varieties having relatively large fruits and of better taste are available now. The tree gives high economic return and the risk factors are insignificant. For these worth, a section of fruit growers in some parts of India have now started to raise plantations of jamun. If the area of the fruit crop goes on increasing in the country and cropping is done in organized orchards, jamun might be regarded to be a major fruit crop in the country in future, rather than as a minor fruit crop.

Description of Plant Parts

The jamun tree is evergreen and lives long which may be 100 years or even more under favourable condition. It attains a height of about 30 metres and may reach greater height when grown in deep, well-textured and fertile soil. Trunk of the tree is usually straight. Wood is reddish grey, rough, moderately hard and dark in the centre with no distinct heartwood. Bark is rough and brown or dark grey with light green upper part.

Dense foliage is a characteristic feature of the tree. Leaves are simple, entire, opposite, elliptic to broadly oblong, 5–25 cm long, 2.5–10 cm wide, smooth, glossy, somewhat leathery, dark green on upper surface and light green on undersurface. Younger leaves are of pink colour. Leaves are characterized by prominent, yellow–coloured midribs which are conspicuous in mature leaves. Turpentine smell is emitted by the leaves.

Flowers are borne in branched clusters on old twigs at the back of the leaves which are 5–7 cm long. Each flower is 6–8 mm long, slightly fragrant, almost stalkless with cup-shaped, light green coloured conical base. Calyx is cup-shaped and petals are 4, white coloured, roundish in shape, concave and united into a cup. Stamens are numerous and each stamen is 5 mm long. Pistils are with inferior ovary. Ovules are numerous, tiny and stout. Mishra and Bajpai (1975) described the floral biology and anthesis of the jamun flowers.

Fruits (Plate-18)of jamun are ovoid to oblong or elliptical in shape depending on varieties or types and they are 1.3–5 cm long. They are 1-seeded berries which appear in cluster and are almost stalkless. The fruit colour starts from pink to crimson and turns purplish black on maturity. A variant of the tree is known to produce white coloured fruits. Seed is 1–2 cm long, strongly astringent and is slightly bitter in taste. Sometimes irregularly shaped, 2–5 seeds which are angular remain compressed together into a mass so as to resemble a single seed. Cotyledons are pale green in colour.

Origin

Jamun was formerly considered to be a native of India, Myanmar and Sri Lanka (Morton, 1964). In the Indian books, nativity has been broadly referred to the region in the south-east Asia or south Asia and the countries *viz.*, Malaysia, Myanmar, Bangladesh, Sri Lanka and India (Singh and Saxena, 2008; Hayes, 1970; Pareek and Sharma, 1993; Anon, 2021). It appears that the exact origin of jamun remains controversial and needs to be ascertained.

Geographical Distribution

Jamun is grown naturally in almost all tropical countries and to a considerable extent in the countries or regions of mild subtropical climate. Some of the countries that may be named are eastern part of Australia, southern part of China, the Philippines, Sarawak, Borneo, Java, Sumatra, Malaysia, North and South Vietnam, Laos, Cambodia, Thailand, Myanmar, India, Bangladesh, Sri Lanka, Malagasy, Tanzania, Kenya, Suriname, Trinidad *etc.* In Brazil, the fruit tree seems to have been introduced from India by the Portuguese sailors. In Hawaii islands in the U.S.A., the tree is regarded to be an invasive species and although naturally grown, it is seldom cropped in organized plantations.

In India, jamun trees abundantly grow from the north to south and in all States from the east to the west and north to the south. It thrives upto 1300 metres in the Kumaon hills and produce satisfactory fruits upto an altitude of 800–900 metres in the Himalayas and the Nilgiri hills.

Climatic Requirement

Jamun grows satisfactorily in a wide range of tropical climate and also in mild subtropical situation. Adaptability to climate is however, found to depend on varieties, tree height, soil moisture and other factors.

The tree grows upto an altitude of 1300 metres in the tropics but optimum fruiting is hampered above 800 metres. Mean annual temperature of 5°–45°C is considered optimum. The jamun tree never becomes leafless in the year but prolonged dry and hot weather may cause large number of leaves to fall. Frost in winter for long period damages the young plants which may even be killed. Mean annual rainfall required by the tree is 900–1200 mm.

An important climatic factor that has a great influence on jamun tree is early rain in monsoon. The size, colour and taste of the fruits are markedly improved when the tree receives early rains (Sham Singh *et al.*, 1967; Bajpai and Chaturvedi, 1985; Mazumdar, 2004; Anon, 2012). Pre-monsoon showers in West Bengal (popularly known as *kalbaishakhee*) and adjoining States in Bihar, Jharkhand, Odisha and Assam are found to have great impact on fruiting performance in jamun.

Soil Requirement

Jamun is not very exacting in its soil requirement. The tree is known to grow on a wide range of soils, which may be calcareous, saline, sodic soils and marshy areas (Anon, 2012). However, very heavy and light soils are not good for the tree (Mishra and Bajpai, 1975). In regard to soil *p*H, Gurbachan Singh *et al.* (1998) in Haryana studied the response of jamun trees to soil *p*H of 8.1, 8.4, 8.7, 9.4 and 10.0. It was noted that after 90 days of planting, the trees showed 60 per cent survival at 9.4 pH. For vigorous growth of the trees and high yield, deep, loamy and retentive soil with high organic matter are considered most suitable.

Propagation

Jamun tree is largely propagated by seed but vegetative propagation methods are also much followed at the present time.

Seed propagation is easy and is followed to a large extent in India. Seeds have no dormancy and they germinate within 15–20 days. Sowing in vertically placed, both ends opened polythene sleeves filled with mixture of garden soil, leaf–mould, sand and wood–ash is ideal for efficient water–management. If this is not done, seeds may be sown in 15–20 cm raised, well-prepared nursery beds at a depth of 4–8 cm and maintaining a spacing of 20–25 cm. Seedlings become ready for transplanting in permanent places in the following spring or in the rainy-season. To avoid transplanting shock, seeds may be sown directly in permanent places. To do so, several seeds should be sown at one place. When the seedlings grow to few centimetres in height, those which are less vigorous should be uprooted. This should be continued at interval of several days, such that ultimately only one seedling which is most vigorous is left to grow in the permanent place.

Jamun seeds have a drawback in that, they are recalcitrant, *i.e.*, lose viability within two weeks or so when stored at ordinary room temperature after extraction from the fruits. Hence, fresh seeds as may be obtained from fully ripe fruits in summer to rainy-season should only be used. Viability of the seeds may be restored for some period, if they are stored at low temperature and reducing their moisture content by partial drying in shade or by fanning. Anandalakshmi *et al.* (2005) conducted trial in Coimbatore, aiming at prolonging viability of the seeds when they were stored. By manipulating the moisture content of the seeds and regulating temperature of storage at different levels, it had been observed by them that storing seeds having 10 per cent moisture content in them and at 20°C prolonged viability upto 180 days with 92.5 per cent germination. In regard to various containers, storing seeds in polybags was found to give encouraging results.

Like mango and many other fruit trees, jamun is a highly cross–pollinated tree and hence, highly heterozygous in its chromosomal make up. Accordingly, plants true to the mother plant are not obtained when they come up from seeds. However, a large proportion of the seeds are polyembryonic and in the polyembryonic seeds, the seedings(s) emerged out from the nucellar embryos retain all characters of the mother tree. If the seedlings from such embroys are carefully excised and planted at rudimentary stage of growth, they will be true-to-type to the mother tree. Mazumdar (1990) observed under West Bengal condition that out of 200 seeds sown, 31.9 per cent were polyembryonic and average of 3-4 seedlings were protruded per seed.

Stem–cuttings are ordinarily not successful in jamun but roots regenerate by the use of auxin and particularly when leafy stem–cuttings are used and maintained in mist–chambers. Success is obtained in air–layering also by the use of auxin and when the operation is carried out in rainy-season. Chovatia and Singh (2006) observed success in air–layering by ringing the shoots and application of auxin. Propagation by budding is usually not practised in India. In Indonesia, modified Forkert budding is employed with success. Among grafting methods, veneer grafting is done in other countries and to some extent in many nurseries in India.

Micro-propagation in jamun has been studied by many workers with success and they have put forward the protocol (Jain and Babbar, 2000; Remashree *et al.,* 2003; Rudriamampionona *et al.,* 2008) Mishra *et al.,* 2005 made a review of the propagation methods.

Culture

Where a new plantation of jamun is desired to be set up by planting many trees at one place, a spacing of 10–12 metres between the trees and rows is considered optimum. For wind-break purpose, spacing should be much less and the distance between trees should be 5–8 metres in the row.

When seedlings are raised in nursery–beds or polythene sleeves, these may be transplanted in the following spring or summer. Care should be taken in transplanting that seedlings remain straight, stem parts are not buried and tap–roots are not broken or twisted. Planting in pits which are pre-filled with soil–manure mixture as described by Mazumdar (2004 c) is of great value to bring about vigorous growth of the plants at their young stage. On doing so, the trees reach bearing maturity earlier and productiveness of the trees is also increased.

Sowing the seeds *in situ* in permanent place avoids transplanting shock and thereby, mortality of the seedlings after transplanting could be averted.

Application of manure is although less common in jamun trees but organic manure is of great help. Yield and quality of the fruits are also greatly improved by application of potassium fertilizer (Mazumdar, 2004 a). Beneficial effect of iron, zinc and magnesium application in jamun trees under Madurai condition has been put forward by Balakrishnan *et al.* (2000). Jamun trees are usually grown without irrigation but irrigation in summer is highly beneficial for the trees. In plantations, irrigation by furrow, border, basin and ring methods (Mazumdar, 2004) are better than flooding and check method, if surface irrigation is needed to be given.

At young stage of growth of the plants, training by *open-centre* or *modified central leader* system will be much helpful to make the trees dwarf, spready, bring higher about productiveness and to secure fruits of better quality. Regular pruning of jamun trees is not necessary but dead, dried and criss–cross branches should be removed throughout the year. Flower and fruit drop are much common in jamun, in which only 12–15 per cent flowers give rise to fruits. The drop take place in three distinct stages. Application of 60 ppm of GA_3 solution at full bloom and subsequently after 15 days of initial fruit set is stated to reduce drop (Bajpai and Chaturvedi, 1985).

For converting jamun trees of inferior types to improved varieties, top- working the trees may be done as described in the Chapter on *ber*.

Pests, Diseases and Physiological Disorder

Pests and diseases are usually nor very serious in jamun but in more humid condition, they may become serious. Many growers do not take much care to control pests and diseases for the difficulty in spraying the trees due to their tallness and consider to be uneconomical also.

Some of the serious pests and diseases in jamun may be stated as follows.

Among insect pests, white fly, thrips and leaf-eating caterpillar are most common. White fly (*Dialeurodes eugenia*) damages jamun tree in all parts of India. The surface of the fruits gets wormy appearance due to their attack. The insect is difficult to control by application of insecticides. The affected fruits should be plucked and destroyed. Soil around the tree trunk should be dug up to destroy the maggots in the affected fruits and the pupae which hibernate in the soil. Sanitary condition should also be maintained in the soil around the tree.

Thrips (*Rhipiporothrips cruentatus*) suck sap from undersurface of jamun leaves. It has been observed by Dahiya and Lakra (2001) in Haryana that infestation by the pest started from February and continued upto June and the peak activity was noted to be in the month of April.

Leaf-eating caterpillar (*Carea subtilis*) has been reported long ago from Coimbatore. The insect feeds on the leaves and may completely defoliate them. Among insecticides, Rogor and Malathion proved effectiveness.

Animal pests are perhaps more serious in jamun than insects in some parts. Squirrels, birds and crows, bats *etc.* cause extensive damage of fruits. Scare–crows, catapult *etc.* may be used to scare them. A device which produces loud retorts like gun shot after every few minutes (Sham Singh *et al.*, 1967) had been on use decades back to drive away the birds. Monkeys also cause great loss of fruits.

Among diseases, leaf spot and fruit rot due to fungal infection (*Glomerella cingulata*) which cause light to reddish brown spots on leaves are commonly found. When fruits are infected, they shrivel and rot and are very sweet due to disease–stress condition. Diathne Z–78 (0.2 per cent), Copper oxychloride (0.3 per cent) or Bordeaux mixture (4:4:50) or any other copper fungicide may be applied to control the disease.

Harvesting

Jamun trees come to bearing maturity at the age of 9–10 years when raised from seeds and 5–6 years in case of budded plants (Anon., 2012). However, in many places, the bearing maturity in trees that have grown from seeds is found to be still earlier, *i.e.*, after 5 or 6 years from transplanting.

Flowering time of jamun ranges from the end of February to early part of May and the actual time differs according to places. Under north Indian condition, it has been stated that flowering in jamun starts during March and continues upto April (Anon., 2012). In the eastern India, flowering time is almost the same or is slightly delayed in many places. Bajpai *et al.* (2012) from the Central Institute for Sub-tropical Horticulture in Lucknow stated that under subtropical condition, flowering period ranged from the third week of February and continued upto the second week of March.

Fruits of jamun become ready for harvest, starting from the month of May and continue upto early August, in accordance with localities but in most places, it starts from early June and continues upto the end of July. Under north Indian condition, fruits ripen and are ready during this period, *i.e.*, June–July or with the onset of rains (Anon., 2012). Late ripening of jamun fruits takes place in some areas. In areas where there is a northeast monsoon of the east–coast, fruits ripen from the middle of August and continues upto the middle of September. Under south Indian conditions, fruits are harvested twice in a year.

At what ripeness, the fruits are to be plucked from the trees is a very important consideration. Jamun is a non–climacteric fruit and the fruits ripen best in the tree. That is, the taste of the fruits is best and most sweet only when they are fully ripened and this occurs when they remain on the tree. Mandal and Mazumdar (unpublished) observed that sugars content in the jamun fruits borne by trees grown in the Purulia district in West Bengal increased by 7–8 per cent during the last 10–15 days of development till the fruits were fully ripe and become very soft. When the fruits were harvested earlier, they did not ripen any more under post-harvest condition. Thus, jamun fruits should be plucked only when they are fully ripe and not before that for palatability.

However, it is a known fact that fully ripe fruits of jamun become very soft and at this stage of ripeness, their harvesting or handling are difficult. Damage by birds is also very high in such fruits. Hence, from the practical point of view, many fruit growers harvest the fruits when they have not attained full ripeness, *i.e.*, few days before they are fully ripe. All fruits do not attain such ripeness at the same time and hence, harvesting should be done in instalment and this should be continued for about a month.

Method of harvest is conventionally done by plucking the individual fruits. To do this, the pickers climb the trees with a cloth bag and the plucked fruits are collected in the bag. When the bag is almost filled with fruits, it is sent downwards on the ground with the help of a rope. On the ground, another person takes out the fruits from the bag and then with the rope, he sends the empty bag to the picker who is seated on the tree. The process is continued. For large–scale harvest, more than one picker climbs on the same tree at different sides and picks fruits from the respective side. However, an efficient method may be to use a pruner as described under the Chapter on *Dimocarpus longan*.

Yield

Yield of jamun varies widely on many factors. On an average, a tree grown from seed may produce 70 to 90 kg of fruits. Yield is lesser in case of budded or grafted trees and may be 50 to 70 kg. Yield is higher in trees grown under south Indian conditions. Application of manure, fertilizers and cultural practices markedly step up yield of jamun fruits.

Varieties

It has been stated (Pareek and Sharma, 1993) that 8 species of *Syzygium* excluding *S. cuminii* are found to grow wild in the Nilgiris and western ghats and some of these yield edible fruits.

Although jamun is a delicious fruit and grows in many parts in India, unfortunately any standard, well-recognized and well-defined variety of this fruit in this country has not been clearly established.

A survey done in Pune and Ahmednagar districts of Maharashtra showed evidence of a wide variation of fruit weight ranging from 3.5 to 16.5 gram, pulp content from 54.29 to 85.71 per cent, total soluble solids from 4.5 to 17 per cent and acidity from 0.160 to 0.554 per cent and from these, promising types of the Nos. 4, 13, 14 and 15 were identified (Keskar *et al.,* 1989).

A superior type of jamun is grown in Gujarat, which goes by the name as Paras. At the Narendra Deva Agricultural University at Faijabad, Uttar Pradesh, among other types or strains, Narendra jamun–6 is known to be the desirable quality (Pareek and Sharma, 1993).

Ra jamun (also called in some localities as Ram jamun) is stated to be a variety, or perhaps a type or a group rather than a variety, which is highly acclaimed and grown in Uttar Pradesh and other parts in northern India. The fruit is fairly large and measures 2.5–3.5 cm in length and 1.5–2.5 cm in width at the maximum bulge. Shape of the fruit is oblong to oval, surface colour is deep purple or bluish black and pulp colour is purplish pink or sometimes greyish. Fruit is highly juicy and sweeter than fruits of other types and its stone is small. The fruit ripens in June–July. Chadha (2012) stated that this is the most common type of jamun that grows in north India.

A variety or type of jamun is heard which goes by the name as Katha. This is an inferior type having small fruits which are sour and of poor quality (Gopalan and Mohanram, 2000).

On a survey undertaken by Patel *et al.* (2005) with jamun trees grown in Uttar Pradesh and Jharkhand, 32 genotypes of desirable fruit characters had been brought to light. Field studies on jamun have also been done by Rakesh and Shivanna (2015) in Karnataka as regard fruit characters and the areas having superior seed sources have been stated by them.

A number of types of jamun grown in the southern part of West Bengal have been studied by Mazumdar (Mazumdar, B.C., unpublished). Fruits of a type which are claimed by the local people to have high anti–diabetic potential were analyzed for their physico-chemical quality (Mazumdar, 2016). The fruits are found to be small in size having a length of 0.9–1.4 cm, diameter of 0.8–1.2 cm and weight of 1.3–1.5 gram and of inferior in quality as compared to the other types grown in the locality. However, high shelf-life of the fruits, absence of belching on their consumption and dwarf nature of the trees are attributes apart from medicinal value of the fruits. The fruit of this type have demand in the locality for medicinal value, particularly to cure gastric ulcer, and diabetes–type 2, although they are of poor taste.

Mode of Consumption

Jamun fruits make a good appeal for their taste, soft texture and pleasant flavour. Fully ripe fruits melt in the mouth and are used as dessert. Conventional way of taking jamun fruits is to take them in a container, sprinkle table salt to that and shaking the container thoroughly covering with lid so that the salt is uniformly mixed with the fruits. The salted fruits are then taken.

Fruits which are not fully ripe are sometimes soaked for new hours in salt solution and then taken. Some people prefer to add sugar to the fruits which have not been fully ripened and then take those. Making sherbet from the jamun pulp is also a preference to many. Mixing jamun sherbet with water extract made with bitter gourd is advised to people afflicted with type-2 diabetes. Seeds of jamun are also taken for their medicinal value.

Quality and Nutritive Value

The different nutritive constituents determined from 100 gram of jamun fruits containing 83.7 per cent moisture have been presented at the National Institute of Nutrition, (I.C.M.R., Govt. of India) in Hyderabad as follows (Gopalan *et al.,* 2018).

Protein: 0.7 g.; Fat: 0.3 g.; Carbohydrate: 14.0 g.; Crude fibre: 0.9 g.; Minerals: 0.4 g.; Calcium: 15 mg.: Iron: 0.43 mg.: Magnesium: 39 mg.; Potassium: 55 mg.; Sodium: 26.2 mg.; Sulphur: 13 mg.; Copper: 0.07 mg.; Zinc: 0.16 mg.; Manganese: 0.15 mg.; Chlorine: 8 mg.; Chromium: 0.005 mg.; Carotene: 48 mcg.; Thiamine: 0.03 mg.; Riboflavin: 0.01 mg.; Niacin: 0.2 mg.; Vitamin C: 18 mg. and Choline: 7 mg. Consumption of 100 gram of fruit provided 62 KCalorie (= 259.4 K Joule) of energy.

(Index : g = gram, mg = milligram, mcg = microgram).

Gomes *et al.* (1986) have stated a number of sugars and galacturonic acid in jamun fruits. Quality of jamun fruits borne by trees grown in the South 24–Paraganas district in West Bengal had been determined by Mandal and Mazumdar (1997).

Medicinal Value

If a list of fruits is made having benefit in diabetes of type–2, jamun will undoubtedly come to the top of the list. Plentiful studies have corroborated this fact which have however, been brought forward here.

Significant sugar reducing effect in blood by the use of jamun fruit has been stated by Gupta and Saxena (2011) in their experiments done with rats in which diabetes was at first induced with administration of the chemical, streptozotocin.

Not only the fruit but also the seeds have their therapeutic value in diabetes and hence, are recommended in ayurvedic and unani systems of medicine. Seeds contain the alkaloid, jambosine and the glycoside, jambolin, which are known to prevent diastolic conversion from starch to sugar. Seeds may be taken as fresh but in traditional medicine, it is also advised to take them after drying. The dried and ground seeds are easy to consume and can be stored for long time under ordinary atmosphere and hence, the diabetic people can take those throughout the year.

The fruits, seeds, leaves and the bark of jamun tree have remedial properties in other ailments as well. Some of these may be stated as indigestion, diarrhoea, anaemia, dysentery, hepatitis, spleen disorders, urinary tract infection, gout, blood purification, slow heartbeat, sunburn, bed wetting of children *etc.* They are also mild astringent, carminative and diuretic agent. In containing leutin, the fruits are known to slow down ageing process. Chemopreventive, radioprotective and neoplastic role of jaumn fruits have also been studied. Swami *et al.* (2012) and Ayyanar and Subash–Babu (2012) in their reviews have brought to light the various therapeutic properties of jamun.

Phytochemicals

Analyses made with fruit, seed, leaf and bark of jamun tree have shown the presence of many bioactive chemicals and many of them have therapeutic potential. The most common compounds having existence in fruit pulp and seed are jambosine, jambolin or antimellin, maleic acid, oxalic acid, gallic acid, betulinic acid, ellagic acid, tannin, cynidine glycoside, oleanolic acid, flavonoids, essential oils, friedelin, anthocyanin, glucoside, isoquercetin, kaemferol, myrecetin and also many other compounds (Sah and Verma, 2011; Ayyanar and Subash Babu, 2012).

Constituents in Developing Fruits

Bajpai and Chaturvedi (1985) quoted the works of Shukla, which stated that fruit growth of jamun had been characterized by three distinct phases. In the first phase that occurred from 15 to 52 days after fruit set, growth had been slow, in the second phase that lasted from 52 to 58 days after set, growth had been rapid and in the third phase that occurred from 58 to 60 days, growth had been slow. The changing pattern of photosynthetic pigments at four stages of fruit growth had been determined by Rao and Charyuhi (1989) and they observed various increases and decreases. Various constituents in jamun fruits during their growth had been analysed by other workers as well (Venkitakrishnan *et al.,* 1997; Garnde *et al.,* 1998).

Undesirable Effect

Consumption of fruits and seeds of jamun does not apparently produce any harmful effect in the body if moderate amount is taken. However, too much consumption of fruits may give rise to symptoms of belching and flatulence in many people.

Some anti–nutritional factors in jamun fruits, *e.g.*, 89 mg of oxalate and 2 mg of phytin in 100 gram of pulp have been reported from the National Institute of Nutrition in Hyderabad (Gopalan *et al.,* 2004). Oxalates are known to interfere with calcium absorption and phytate causes poor absorption of iron from the food and hence, these constitute are anti-nutrients. Tannin present in the jamun seed is also an anti-nutrient in that, it interferes absorption of iron. The possible danger of anti-nutrients does not usually arise unless high amount of them go inside the body of a person.

Post-harvest Storage

Jamun fruits have very thin peel and the pulp contains about 85 per cent moisture. Hence, they are highly perishable and cannot be stored well for more than 3–4 days under ordinary room temperature condition.

Storage life of the fruits could however, be increased at low temperature condition. Maintaining a temperature of 8°–10° C and a relative humidity of 85–90 per cent in the storage atmosphere, fruits may be stored well for about 3 weeks. Lakshmi (2013) studied post-harvest storage and value addition of jamun fruits and observed that fresh fruits are successfully stored upto 42 days at 8°–10° C, 28 days at 15°–16° C and 7 days in room temperature condition.

Pre-cooling of fruits and storing them in perforated polythene bags is an effective method of storing them well for some more days than when stored them without pre-cooling (Anon., 2012).

The effect of treatments with 50 and 100 ppm of gibberellic acid, 1.0 and 1.5 per cent of calcium chloride and paraffin wax on storage of jamun fruits, kept in with or without perforated polythene bags had been studied by Ayar *et al.* (2011) at the Anand Agricultural University in Gujarat. It was observed by them that treatment with 1.5 per cent calcium chloride and storing in perforated polythene bags proved to be the best treatment. According to them, the treatments effectively reduced the physiological loss in weight and also loss due to spoilage and thus, had been useful in maintaining a balance between ascorbic acid and sugars content in the fruits during storage.

In an observational trial, the author (Mazumdar, B. C. unpublished) used fully ripe fruits after approximately 12 hours of harvest. Fruits were dipped in week emulsion of lime in water for 4–5 minutes, partially dried after which throughly ground fenugreek seeds were mixed with the fruits at the rate of approximately 2 gram per 100 gram of fruits and then stored under ordinary room temperature condition, keeping in perforated polythene bags. When the bags were opened after 10 days, most of the fruits were found to keep well and had not been much softened while most of the untreated fruits, identically stored in perforated polythene bags were found to have much softened and broken when examined after 4 days.

Utilization

Although fully ripe jamun fruits are highly relished in their fresh condition, they are also processed to make products. Sherbet and squash are very commonly done from ripe jamun fruits and are highly liked by the Indian people (Mazumdar, 2004). Jam is made with the fruits and in making so, other fruits that are rich in pectin are added so that the product in formed with a desirable gel. Vinegar is also made from the fruit and for this, semi–ripe or unripe fruits are used than those which are fully ripe.

Processing of jamun fruits to make dairy–based products, *e.g.*, shrikhand, yoghurt and lassi has been experimented with by Lakshmi, (2013). Wine is made from the juice of jamun fruits in Goa (Gopalan and Mohanram, 2000).

For antioxidant property of the fruits, they are added with other fruits also and taken. The fruit pulp is sometimes mixed with wheat flour in making dough for preparation of hand–made bread. Kapor *et al.* (2015) assessed for antioxidant activity and quality of chapattis, *i.e.*, roti for supplementing whole wheat flour with jamun pulp at 5, 10 and 15 per cent. Anthocyanins, total phenols and antioxidant activity of the chapattis could be observed to have highly increased on supplementing with 15 per cent jamun pulp. From their study, they concluded that wheat flour mixed with jamun pulp had improved nutritional and antioxidant status of the chapattis.

Apart from fruit pulp, dried and ground seeds of jamun are also mixed with dough of wheat flour by many to make hand–made bread, which lowers the blood sugar level.

Chadha (2012) stated that fruit nectar could be prepared from jamun juice. For this, 1 kg of juice (°Brix–10, acidity–1.5 per cent) should be blended with 0.61 kg of sugar, 10 gram of critic acid and 2.38 litre of water.

Family: Myrtaceae

Species: *Syzygium jambos* (Linn.) Alston

[Synonyms: *Syzygium jambos* Alston; *Eugenia jambos* Linn.; *Jambosa jambos* Millsp.; *Jambosa vulgaris* D.C.; *Caryophyllus jambos* Stokes].

Taxonomic Serial: 505420

Common Name

Most widely held English name of the fruit is Rose–apple. It has been quoted by Morton (1987) that, "the term, rose–apple is so widely employed that the species has few alternate names". Nevertheless, other English names are also used for Rose–apple, which are Jambos, Jambu, Malabar plum, Wax jambu, Jamrosa, Mountain apple, Plum rose, Cloud apple *etc.* For religious significance, Rose–apple tree is called Oudh–tree.

In different languages in India, Rose–apple is termed as Gulab jamun, Golap jam, Boga jamuk, Panneeralai hannu, Jambakha, Jambhool, Chhota pijuli, Pannirkoya, Paninir champa, Jambu neredu *etc.*

Importance

Rose–apple is an excellent fruit that appeals all sections of consumers for its rosy fragrance, sweetish taste with practically no sourness and the spongy and cushion–like texture which is neither hard nor very soft (Mazumdar, 2004 a). the tree has great ornamental importance as well for its graceful appearance, long leaves, long showy stamens and the pale yellow fruits.

There are also the following economic benefits of the tree parts of rose–apple. Branches of the tree are used to make hoops for large sugar casks in Puerto Rico, wood is used to make furniture, spokes for wheels, frames for musical instruments like guitar and violin and also makes very good charcoal for industrial use. Leaves yield essential oils that are used in perfumery industry. Flowers of rose–apple are rich sources of nectar of high quality and the honey derived from that is of good colour and of high acceptance in the market. In having high pectin, fruits have a good jellying potential. Fruits, flower and leaves of the tree have many medicinal properties as well.

Rose–apple tree is very hardy and thrives even on lime and sandstone having little organic matter. The tree is resistant to wind and water–logging. It is planted in the Peruvian part of the Amazon to protect against the frequent floods.

Description of Plant Parts

Rose–apple is a small to medium tree rather than a shrub. It is evergreen and reaches a height of 7.5 to 12 metres and may even be more in retentive and fertile soil. Crown is dense, slender with spreading and low branches. Overall width of the crown usually exceeds the height. Leaves are opposite, oblong–lanceolate or narrow, elliptic or tapering to a point. They are thick, glossy, shiny, long and measure 10–22 cm in length and 2.5–6.25 is width. The newly emerged leaves are pinkish in colour and later, they turn green. Flowers (Plate-19) are 4–5, together in terminal cluster, creamy white or greenish white in colour. Each flower is 5–10 cm wide and consists of about 300 conspicuous stamens.

Fruit (Plate-19) is 3–5 cm long at the stalk–stylar end and 5.0–5.8 cm wide at its maximum bulge. Large forests are produced by trees. Roy (1953) indicated tetraploidy in the species. Fruit is greenish–yellow to dull yellow in colour. Shape, size and colour of the rose–apple fruit resemble some less common species of guava and hence, is often confused. Texture of the fruit is soft and is somewhat crispy and crunchy. The centre is hollow and in the hollow centre of each fruit, 1–4 brown, rough coated medium–hard, more or less rounded seeds are present, which are 1–1.6 cm thick. Seeds loosen from the inner wall of the fruit and thus, when shaken, they give rattling sound.

Origin

It has been stated by different authors that south-east Asia is the origin of rose–apple (Hayes, 1970; Pareek and Sharma, 1997; Mazumdar, 2004 a; Singh and Saxena, 2008).

Geographical Distribution

Rose–apple is grown in many tropical countries in the world. Some may be named as all the south-east Asian countries, India, Bangladesh, Sri Lanka, South Africa, Jamaica, Bermuda, the Bahamas, Cuba, southern part of the U.S.A., Mexico to Peru, Puerto Rico, Guatemala, Virgin islands, Honduras, Panama, the Galapagos, Hawaii in the U.S.A. *etc.*

In India, the fruit is largely grown in West Bengal, Assam, the eastern States and Odisha. It is also grown in south India where it thrives upto an elevation of 1350 metres in the tropical hill (Hayes, 1970).

Climatic Requirement

The tree is basically tropical and its growth and fruiting performance are ideal when grown in the warm and moist tropics. However, it adapts well in mild subtropical climate also. It is found to thrive well at relatively high altitude in the tropical countries, *e.g.*, upto an altitude of 1200 metres in Jamaica and 2300 metres in Equador but the fruiting performance is affected when grown at higher altitude (Morton, 1987). This is true in south India also where the tree grows upto an altitude of 1350 meters but the productiveness and fruit quality appear to be less satisfactory than that in the plains. The heat tolerance of rose-apple tree has been evaluated by Yamada *et al.* (1996) in Japan by determining chlorophyll fluorescence. The tree is much hardy and can withstand high wind to a considerable extent.

Soil Requirement

Soil is not an important factor for rose-apple and it grows well in inferior soil of low depth. It also grows in sandy and limestone soil, having very little organic matter. However, productivity of the tree goes much higher and the quality of fruits is greatly increased when it grows in deep, loamy and well-drained soil. The performance of the tree is better when the *p*H of the soil is slightly acidic.

Propagation

Rose-apple is conventionally propagated by using seeds but all characters of the mother tree are not retained in the trees raised from seedlings. However, there is advantage in that, the seeds are polyembryonic and the vegetative, *i.e.*, apogamic embryos that are present in the seeds are nucellar (Roy, 1953). The plants that come out from these nucellar embryos retain all characters of the mother tree. Hence, if the plants that arise from these adventive embryos are carefully detached and separated out from the other seedlings that come out from the same seed at rudimentary stages of their growth and then suitably grown, they will give rise to trees that will maintain all characters of the mother tree.

Root regeneration has been observed in rose-apple with stem-cuttings by application of auxin under Indian condition and success has been achieved when semi-hardwood rather than hardwood cuttings were used (Morton, 1987). Sharma *et al.* (1989) obtained success in stem-cuttings and it was observed by them that rooting performance was best and was recorded upto 96.8 per cent when the cuttings were treated with 2500 ppm of NAA, in combination with Ethephon at 1000 ppm. Root formation in cuttings as affected by growth-regulators and seasonal influence had been studied by Hore and Sen (1994) under West Bengal condition.

Air-layering done in spring season on treatment with 1,000 ppm of the auxin, NAA has been found to give 60 percent success under Indian condition (Morton, 1987). Martins and Autunes (2000) in Brazil studied air-layering of rose-apple, using various concentrations of IBA and it was observed that higher concentrations of the auxin were effective.

Budding is usually not successful in rose-apple but some grafting methods bring about good results.

In vitro responses of adventitious embryos has been studied by Litz (1984), culturing on modified MS medium.

Culture

No cultural practice is in fact, given to the rose-apple trees grown in India. But, some culture however, renders much benefit to invigorate the trees, step up yield and production of better quality fruits.

At young stages of growth, mortality of the seedlings is usually high. Hence, in hot-weather, they should be protected against scorching heat by provision of suitable thatch as has been described in setting up of orchards (Mazumdar, 2004 a). Where there is danger of frost, mulching of the soil is also necessary. Training the trees by *open-centre, i.e., vase* system (Mazumdar, 2004) is of great advantage to make the trees dwarf, spready and thereby, for higher yield performance and production of better quality of fruits. Lopping of branches may be done to encourage new shoots.

No manure or fertilizer is given in rose-apple trees. But where the soil is less fertile, a semi-annual feeding of a balanced fertilizer, such as 6-6-6 NPK has been recommended. Laminar nitrogen, phosphorus and potash contents of rose-apple trees growing in the central parts of West Bengal has been determined by De *et al.* (1996). From the analyses done by them at monthly interval from December to May, it has been observed that all the nutrients had initially declined during the first few months but tended to rise with advancement of summer season. The results were indicative of application of manures during or before winter as the case may be. El-sidding *et al.* (1998) had studied the combined effect of nitrogen and water on growth pattern and gas exchange characteristics of rose-apple. Their findings among other things had indicated that addition of nitrogen had a compensating effect on the reduction of dry matter accumulation in presence of water stress.

Rose-apple trees are hardy and capable to tolerate semi-arid situation. However, severe moisture stress in soil hampers vigorous growth of the trees and results in small-sized fruits besides causing heavy fall of fruits. Hence, liberal irrigation during summer season and during growth of fruits is highly beneficial.

Pruning is although not done, heading-back of shoots which bore fruits ensure higher productivity in the following year. In some countries where rose-apple trees are planted in coffee plantations, drastic pruning is done to encourage highly vigorous growth of the rose-apple trees.

Production of seedless fruits along with thick pulp has been stated to have been produced by treating opened flowers with auxins, *viz.*, naphthoxy acetic acid (NOXA) and naphthalene acetic acid (NAA) or chloride-substituted phenoxy acetic acid, *i.e.*, 2, 4, 5-Trichlorophenoxy acetic acid (2, 4, 5-T) (Morton, 1987).

With the objective to make improvement in the quality of rose-apple fruits, Mandal and Mazumdar (2000) sprayed the trees with 1 percent solution of potassium chloride or zinc sulphate, or 50 ppm solution of IBA or GA_3 after 12-15 days of fruit-set. Treated fruits were plucked after 44 days of fruit-set when they had attained harvest-maturity stage and the proximate constituents were determined. It was observed that weight, length and diameter of the fruits as well as dry matter, reducing sugars and acidity of the pulp had increased over the untreated control but no significant increase could be observed in case of specific gravity of the fruits and non-reducing sugars and vitamin C content of the pulp. In general, zinc, sulphate solution had been the most effective treatment in causing improvement of fruit quality.

Pests and Diseases

Pests and diseases are not serious problems in rose–apple. But some fungal diseases are observed in the trees. In humid atmosphere, the leaves are subject to a sooty mould from aphid secretions. Leaf spot disease caused by the fungi, *Cercospora* species, *Gloeosporium* species and *Phyllosticta eugeniae*, algal leaf spot caused by *Cephaleures virescens*, black leaf spot by *Asterinella puiggarii*, root rot caused by *Fusarium* species, mushroom root rot caused by *Armillariella* species, anthracnose caused by the fungus, *Glomerella cingulata etc.* are known to infect the tree parts (Morton, 1987).

Harvesting

Rose–apple fruits ripen during summer to early rainy-season but the actual time differs according to environmental condition that prevails in a locality. Under the conditions of central part in West Bengal, ripe fruits are harvested from the middle of February and continued upto the middle July. The peak season is however, April to May. In the South 24–Paraganas district in the central part of West Bengal, there are some trees which produce sporadically and bear at least few fruits almost all throughout the year, except in winter season. In south India, peak season of harvest is March to early part of May.

It has been stated that in Jamaica and Puerto Rico, rose–apple trees bloom and fruit sporadically almost all the year though somewhat less in summer. In Florida in the U.S.A. and Bahamas, the main season of plucking is May to July (Morton, 1987).

Hand–picking is the conventional method of plucking of the fruits in India, climbing on the trees with cloth bags as is done in jamun.

Yield

The central part of a rose–apple fruit is hollow and hence, the fruits are lighter in weight with specific gravity that may range from 0.6 to 0.7. Hence, the number of fruits that would be borne in a tree although might be large in number, their yield in terms of weight would be low as compared to other fruits. Under Indian condition in general, where no cultural inputs are given to the rose–apple trees, a tree may produce 2–2.5 kilogram of fruits in a year.

Variety

The rose–apple that grows in India is of plenteous types. It is found to differ widely in size, shape, colour, texture and flavour of the fruits as well as on growth characteristics of the trees. But no varietal assortment is done although this has a great necessity. In other countries, *e.g.*, Thailand, Malaysia *etc.*, varietal classification is usually done considering the colour of the fruit and its consistency.

Mode of Consumption

It will perhaps be difficult to find out a person who disrelish rose–apple fruits. The author made a survey (Mazumdar, B.C., unpublished) of consumers in Kolkata, in regard to their rating of a number of fruits that are obtained in summer and rainy-season. Adults and adolescents numbering 5 from each group and gender were sampled replicating three times. It was revealed that rose–apple was organoleptically ranked as one of the three best fruits considering the traits like taste, flavour, texture, size, freedom from sourness, presence of undesirable features *etc.* Lack of adequate sweetness had though been considered a shortcoming but this could not stand in the way when other qualities were considered together and especially the nice, rosy fragrance that surpassed everything.

Rose–apple fruits satiate consumers when they are taken as fresh and the fruit is conventionally taken in this way. In West Bengal, some people have however, the preference to make the pulp into pieces, mix those in sweetened curd and take. Taking syrupy water containing small pieces of the pulp is also preferred by many.

Quality and Nutritive Value

Food value of rose–apple fruits per 100 gram of pulp has been presented by Morton (1987) as follows.

Protein: 0.5–0.7 g.; Fat: 0.2–0.3 g,; Carbohydrates: 14.2 g.; Fibre:1.1–1.9 g.; Calcium: 29.0–45.2 g.; Magnesium: 4 mg.; Phosphorus: 11.7–30 mg.; Iron: 0.45–1.2 mg.; Sodium: 34.1 mg.; Potassium: 50 mg.; Copper: 0.01 mg.; Sulphur: 13 mg.; Chlorine: 4 mg.; Carotene: 123–235 I.U.; Thiamine: 0.01–0.19 mg.; Riboflavin: 0.028–0.05 mg.; Niacin: 0.521–0.8 mg.; Vitamin C: 3–37 mg. and Calories: 56.

(Index: g = gram, mg = milligram I.U. = International Units).

Physico-chemical constituents of rose–apple fruit grown in West Bengal have been studied by Sinha Roy (1996), Mandal (1997) and Mandal and Mazumdar (1997).

Nutrient status of the rose–apple fruits as presented by the National Institute of Nutrition (I.C.M.R., Govt. of India) in Hyderabad (Gopalan *et al.,* 2018) from their analytical studies have been presented in the following.

The results represent the amount of the nutrients per 100 gram of fruit pulp that contains 89.1 per cent moisture.

Protein: 0.7 g.; Fat: 0.2 g.; Carbohydrates: 8.5 g.; Crude fibre: 1.2 g.; Mineral: 0.3 g.; Calcium: 10 mg.; Phosphorus: 30 mg.; Iron: 0.5 mg.; Magnesium: 4 mg.; Sodium: 34.1 mg.; Potassium: 50 mg.; Copper: 0.01 mg.; Sulphur: 13 mg.; Chlorine: 4 mg.; Carotene: 141 mcg.; Thiamine: 0.01 mg.; Riboflavin: 0.05 mg.; Niacin: 0.4 mg. and Vitamin C: 3 mg. Consumption of fruit provided 39 Kilocalorie (= 163.17 Kilojoule) of energy on the body.

(Index: g = gram, mg = milligram, mcg = microgram)

Medicinal Value

The fruit of rose–apple is known to be helpful for brain and liver. Seeds are used in diarrhoea and dysentery. Bengum *et al.* (2015) have stated antioxidant and antimicrobial properties of the stem bark of rose–apple tree.

Constituents in Developing Fruits

Changing pattern of nitrogen, phosphorus, potassium and iron contents in rose–apple fruits from the bud–stage to their harvest–maturity condition had been analysed by Chattopadhyay and Ghosh (1996). They observed that the nitrogen content reached to the maximum level at 20 days after fruit–set, phosphorus and potassium were higher at the higher maturity periods and the iron content was maximum in the fruits whn the y were 30-day old.

Mandal and Mazumdar (1998) had determined physico-chemical constituents of rose-apple fruits at four equally spaced stage of their development till attainment of harvest-maturity condition. It was observed by them that the gain in the fruit weight had been maximum during advanced stages of fruit development, while specific gravity, fruit size, dry matter content, at acidity and the vitamin C had maximum accumulation at early stage of development. Gain of sugars in the fruits was lesser at the intermediate stage of fruit development.

Undesirable Effect

The seeds of rose-apple fruits are said to be poisonous. Roots, stems and leaves of the tree are also poisonous in containing hydrocyanic acid. An alkaloid jambosine has its existence in the bark of the tree and roots (Morton, 1987)

Utilization

Rose-apple fruit is rich source of pectin of superior jelly-grade. Hence, the fruit is a good material in making jelly and jam. Canning of the fruit is also done in syrupy water. Besides, the fruits are utilized in making candy, pickle *etc.* Sometimes they are cooked with rice to add flavor. It has been stated (Morton, 1997) that long ago, fruits were distilled to make rose water in Bengal in India. However, this is not done now and rose petals are used to make rose water and rose essence.

Family: Myrtaceae

Syzygium samarangense (Blume) Merrill and Perry.

[Synonyms: *Syzygium javanicum* Miq.; *Eugenia javanica Lam.; Eugenia samarangensis* O. Berg.; *Eugenia alba; Eugenia formosa; Myrtus samarangense*]

Taxonomic Serial: 506170

Common Name

The species *Syzygium samarangense* goes by a number of common names in English and some of these are Water apple, Wax apple, Star apple, Malay apple, Java apple, Jamaican apple, Royal apple, Cloud apple, Mountain apple, Love apple, Bell apple, Wax jambuetc.

In other languages and countries, the fruit is known as follows.

Samoa: Nonuvao. Indonesia: Jambu air, Djambo semarang *etc.* Malaysia: Jambu ayer rhio. Taiwan: Lian-woo, Bell-fruit *etc.* Phillippines: Makopa. Thailand: Chom pu kao. Vietnam: Roi, Man *etc.* Sir Lanka: Pini jambu, Jambu *etc.* Bangladesh: Jamrool. France: Jamalac. Caribbean : Corazon. Cuba: Cajuilito, Jamaica: Jamaican apple or Otaheti apple, Indonesia: Jambu air, Malaysia: Jambu ayer.

In India, the *Syzygium samarangense fruits* have high fondness in north-eastern States. In West Bengal, Assam, Tripura, it is known as Jamrool. In northern India, the common names are Safed Jamun, Amrool *etc.*, in Kerala, the name is Painir champaka and in Andhra Pradesh, the fruit is called Gulabijaa-michetu.

Importance

Wax apple is not very much sweet or appealing in taste. But the fruit, tree and its parts have many distinctive features and some of these may be stated as follows.

(i) A very good attribute of this fruit is its excellent thirst quenching property. Even a very thirsty person gets relief with a bite of a wax apple. For this quality, in games, sports and swimming, athletes are given wax apple or pear to get relief from exertion. (ii) Although the fruits are not very sweet, they are not sour and do not have any disagreeable trait. In fact, the soft, juicy, crispy, spongy texture of the fruit pulp and its pleasant flavour are liked by many. (iii) The leaves, bell-shaped (= pyriform) fruits which are of various colours and the shape of the tree are attractive in appearance and hence, the tree is sometimes preferred for planting along avenues and parks. Wang *et al.* (2010) states that wax apple trees are planted along avenues and around homes in Taiwan for their handsome dark, evergreen foliage and the production of bright red fruits during the summer. (iv) Tree is hardy and requires very little attention. (v) Dwarf varieties are suitable for growing in pots. (vi) Fruits have relatively high keeping quality under ordinary atmosphere. (vii) Processing and preservation of the fruits could be done to render those in highly palatable forms. (viii) Fruits are recognized as low-calorie food-stuff. (ix) Many medicinal properties of the fruits, flowers and leaves of wax apple trees are also known.

Description of Plant Parts

The tree is evergreen and reaches a height of 5 to 15 metres. Crown of the tree is wide spreading. Trunk is short, 25 to 30 cm thick with flaking bark. Leaves (Plate-16) are opposite, nearly sessile, elliptic to oblong and rounded or slightly cordate at the base. They are 10-25 cm long, 5- 12 cm wide, yellowish to dark bluish green in colour and aromatic. Flowers are borne in drooping panicles of 3 to 30 at the branch tips. They arise from axils and almost on any part pt on the trunk and branch. They are 2 - 5 cm in diameter and yellowish white in colour. Petals are four in number and stamens are numerous and long. Fruit (Plate-17) is berry and pear-shaped. Shape of the fruit is however, not same for all varieties or types. According to varieties, colour of the fruit may be white, pale green, green, pink, red, crimson, purple or deep purple. Either the whole fruit or the base may be coloured. Fruit peel is very thin and glossy. Pulp is white, spongy, dry to juicy, crispy, almost crunchy and is mildly fragrant. Seeds are 1 to 2 in number or nil. They are rounded and 0.50.8 cm wide.

Origin

The nativity of wax apple has been stated to be in the area that includes the Greater Sunda islands, Malaysia, Indonesia and the Andaman and Nicobar islands in India.

Geographical Distribution (834)

Wax apple is largely grown in south-east Asian countries and the important countries are Taiwan, Thailand, Cambodia, Laos, Vietnam *etc.* It is also much grown in northern Australia, Antilles, Suriname and in the African countries in Zinzibar, Pempa *etc.* and in Bangladesh. In India, the tree grows abundantly in West Bengal and Andaman and Nicobar islands, East Godavari in Andhra Pradesh, Tamil Nadu, Puducherry, Laccadive islands, Goa and throughout Kerala.

Climatic Requirement

Wax apple is highly tropical in its climatic requirement. It grows best in moist tropical climate having a fairly long dry season. Full sunshine is required for better development of fruits and for proper development of the colour of fruits in case of varieties having coloured fruits. In India, it grows upto an altitude of 1220 metres in the tropics.

Soil Requirement

Rich soil is conducive to production of large-sized fruits. Trees make better growth in relatively heavy soil than where the soil is deep. Tolerance to drought and also salinity of the soil is low for the tree.

Propagation

Wax apple is propagated by a number of methods. Seed propagation is very common but as the seeds are recalcitrant and thereby, have short viable life, old seeds should not be used for raising trees. However, all character of the mother tree are not transmitted to the in trees raised from seeds.

Stem-cutting is successful in wax apple particularly by application with auxin. Sarkar *et al.* (1984) have observed that application of auxin in combination with phenolic compounds had been much helpful in root formation of hardwood stem-cutting. The effects of three common auxins, *i.e.,* IAA, IBA and NAA and 6-BA on the success of stem-cuttings in wax apple had been studied by FuPing *et al.* (2012) in Guangdong, China. It was observed by them that rooting response was in general, best by application of 100 ppm of IBA.

Success of adventitious root regeneration in wax apple is usually greater in air-layering than in stem-cutting. Tangkichote *et al.* (2012) in Thailand made a study on air-layering of wax apple by inoculation with bolete (*Boletus colossum*) spawn. Inoculation with 10 – 14 gram of spawn was observed by them to be most effective in formation of roots.

Budding is done in wax apple on rootstock of the same species. However, rootstocks belonging to other species of *Syzygium* are also employed and in Malaysia, rootstock of the species, *Syzygium densiflorum* which is called wild rose apple is also used.

Micro-propagation of wax apple has been studied by some workers. Litz (1984) had cultured immature adventitious embryos from excised ovules of immature fruits in MS medium. Proliferation of axillary buds had been observed by him in the media with Benzyl Adenine of 2-10 mg per litre concentration, depending on developmental stage of the embryos. Media that contained only 3-10 ppm of 2,4-D could be seen to have induced root information. Embryonic callus had been observed from adventitious embryos of 1-2 cm fruitlets on 2, 4 - D of 1-2 ppm concentration. Wattanawikkit *et al.* (1999) had cultured axillary buds on Woody Plant Medium (WPM) supplemented with various concentrations of BA. Highest average shoot number had been produced when the growth regulator was used at 9 ppm concentration while the medium without any growth regulator had produced highest average shoot length after 120 days of culture. Maximum rate of root induction to 75 percent had been obtained at 4 ppm of IBA for 60 days. Dipping plantlets in 5 ppm of IBA solution before transplanting had brought about 90 percent survival of them.

Culture

Normally, wax apple trees grow and produce satisfactorily without care but manuring, weeding and irrigation are of great help for rapid growth and robustness. Where organized plantation is desired, the trees should be set at 8 - 10 metres apart.

The status of nitrogen, phosphorus and potash content of the leaves of wax apple trees grown in the South 24-Paraganas district in West Bengal had been studied by De *et al.* (1996) at monthly interval from December to May. The results revealed that steady increase of phosphorus took place upto the month of February while increase of nitrogen and potash had continued upto March. Subsequently, the nutrients had sharply declined with advancement of summer season. Evaluation of potassium, calcium, manganese, iron, copper, zinc and bromine of wax apple fruits had been made by Oliveira *et al.* (2006) in Brazil using the Energy Dispersive X-Ray Fluorescence technique and it came to notice that some of the elements had been at appreciably high level.

With studies aimed at making improvement in the quality of wax apple fruits, Mandal and Mazumdar (2000) sprayed 1.0 percent solution of potassium chloride or zinc sulphate or with 50 ppm solution of IBA or GA, after 12 – 15 days of fruit-set grown in West Bengal. Increase in weight, specific gravity and size of the fruits as well as their dry matter, sugars, acidity and the vitamin C contents had been observed by them which varied according to treatments and in general, zinc sulphate solution proved to be highly effective. Shu Zen Hong *et al.* (2001) had examined the effects of light, temperature and sucrose on colour, weight, diameter and soluble solids content of the skin of the pink-colored wax apple fruits by culturing skin discs. The results of their study indicated that light with 20°C and 6 percent sucrose solution caused highest soluble solids concentration and anthocyanin content.

Experiments on pollination of wax apple had been done by Chantaranothai and Parnell (1994) and it has been concluded by them that the species was self-compatible and self-pollination had been more common. The effect of hand-pollination of flowers of several varieties of wax apple had been evaluated by Wanichkul and Natheung (2011) at the Kasetsart University, Thailand. They observed that during 10 days after pollination, fruit-set showed significant difference. Hand-pollination with Phetchsairung cultivar as pollinizer brought about highest fruit-set.

Pests, Diseases and Physiological Disorder

Attack by insect pest is not serious in wax apple trees and their fruits but disease infection by a number of fungal species is often met with. Wang *et al.* (2010) has stated severe decline of wax apple trees caused by the fungal species, *Fusarium solani* in northern Taiwan. A new fungal species, *Pestalotiopsis samarangensis* has been reported by Maharachchikumbura *et al.* (2013) in Thailand. The symptoms had been stated to be small, circular, black, sunken spots that appeared on fruits. The spots later enlarged rapidly, became more sunken and resulted in a soft decay of the fruit pulp. According to them, wounding of the fruits triggered the symptoms and hence, careful handling of the fruits during transport and storage had been advised by them.

Lu and ChinHo (2011) had stated cracking disorder of wax apple fruits in Taiwan. Biochemical changes of the fruits due to this physiological disorder had been analyzed by them.

Harvesting

Wax apple trees raised from seeds come to flowering after 3 - 7 years and those propagated by vegetative means flower at earlier maturity. Trees may flower twice and sometimes three times in a year in more tropical situation. In India, the trees start flowering from the end of February and continue upto April. These fruits are ripened to become ready for harvest from April to July depending on climatic condition. According to Morton (1987), in Sri Lanka, the fruits become ripe from March to May, in Java, from June to August and in India, during May - June.

Harvesting of fruits is done with a suitable pruner along with some twigs as described under the chapter on *Dimocarpus longan*. Some growers shake branches one by one to cause dropping of fruits and collect the dropped fruits. Damage to the fruits should be avoided in harvesting as damaged fruits are liable to fungal infection.

Yield

In mature condition, a wax apple tree produces more than 700 fruits. In rich soil, yield goes very high.

Varieties

There are many variations in fruit characters of wax apple which may not have recognition as varieties. Shape of the fruits in these types is although bell-shaped, they may be somewhat roundish, elongated or swelled at the centre. Similarly, colour of the fruits may be almost white, pale green or reddish in part or full or purple or deep purple. Red-fruited and purple coloured fruits are grown in Thailand and trees bearing such fruits are now grown in West Bengal and other parts in India. These fruits are sweeter than the pale-green coloured fruits. In Taiwan, fruits known as Black Pearls are very famous, the surface colour of the fruit of which is purplish-red. Wax apple fruits from south-east Asia widely vary in size, shape and colour. Some of these varieties are now grown in southern part of West Bengal.

Mode of Consumption

Wax apple fruits are highly relished as fresh and no peeling of them becomes necessary. The fruits appease much just after harvest and become less tasty when stored. All parts of the fruit are consumed except the core near the bottom. In Indonesia, the fruits are sometimes used to make salads (= rujak) and are also preserved by pickling. In Malaysia, the green fruits are taken with salt and spices and are cooked as sauce. They are also stewed with apples.

Quality and Nutritive Value

As per analytical report issued by the Food Standards of Australia and New Zealand, nutritional value of 100 gram of wax apple fruits is as follows. Moisture 0.7 g. Fat : 0.2 g. Fibre :1.9 g.; Sugars : 4.5 g.; Calcium : 13 mg.; Iron: 0.8 mg.; Magnesium: 5 mg.; Potassium: 38 mg.; Thiamine : 0.2 mg.; Riboflavin : 0.4 mg.; Niacin : 0.5 mg and Vitamin –C : 8 mg.

Morton (1987) has quoted the food value of 100 g of wax apple as follows Moisture : 91:40 – 92.96 – 92.96 g.; Protein : 05 g.; Calcium : 0.01 g.; Phosphorus : 0.03 g.; Sulphuric acid : 0.17 percent and Citric acid : 0.15 percent.

(Index : g = gram, mg = milligram)

The present author (Mazumdar, B. C., unpublished) analyzed fruits of wax apple belonging to the red-fruited type and the Thailand variety grown in the South 24-Paraganas in West Bengal. Results revealed that weight per fruit was 45.0 and 51.4 g and soluble solids were 5.9 and 6.3 percent respectively for the red-fruited and the Thailand variety. Number of seed in the former was noted to be 1 and weighed 1.5 g while most of the fruits of the letter variety seedless and some contained 1 seed in fruit the weighed 1.0 g.

Mandal and Mazumdar (1997) determined the constituents of wax apple fruits of a common variety grown in West Bengal. The fruits having whitish to light green colour of fruits were found to have the following composition. Weight of fruit: 29.83 g.; Specific gravity: 0.95.; Stalk-stylarlength: 3.67 cm and Dry matter: 6.26 percent. Total titratable acidity as equivalence of citric acid 0.26 percent and Vitamin C: 20 mg.

The proximate composition of 100 gram of wax apple fruits has Govt. of India has been presented by the National Institute of Nutrition (I.C.M.R.) in Hyderabad as follows. Moisture 93.9 g.; Protein 0.1 g.; Fat: 0.4 g.; Carbohydrates: 3.7 g.; Crude fibre: 2.2 g.; Minerals 0.1:, Calcium 17 mg.; Phosphorus: 3 mg.; Iron: 0.1 mg.; Thiamine 0.01 mg.; Riboflavin 0.02 mg.; Niacin 0.2 mg and Vitamin C: 30 mg. The report states that 100 gram of fruits on consumption would provide 19 K calorie (= 79.49 K Joule) of energy in the body.

(Index : g = gram, mg = milligram)

It has been stated by Shu Zen Hong (1999) from Taiwan that position of fruits on the tree had a marked effect on the quality of them. They observed in the pink-fruited type that the fruits on the lower trunk were heaviest and largest among the 9 positions which were compared. The fruits borne at the upper inner side were smallest but these fruits developed more intense red colour. It has been opined by them that for high soluble sugars content of the fruits, harvesting should be done from inner positions.

Medicinal Value

Flowers and other parts of the wax apple tree are known to have many medicinal properties. As regard fruits, those have been stated to have antibiotic and also antifungal properties against *Staphylococcus, Candida* and *Mycobacter* species. Fruits and other parts are used in traditional medicine against diabetes and other ailments in many tropical and sub-tropical countries since ancient time. Antioxidant properties of the fruits have also been stated in Hawaii in the U. S. A. (Reynertson *et al.*, 2005).

The fruits are used in diarrhoea for the presence of calcium antagonistic activity (Ghayuret *et al.,* 2006).

Phytochemicals

A number of phytochemical compounds have been derived from different parts of wax apple trees. Amror *et al.* (2005) stated spasmolytic flavonoids from the fruits. The chemical composition of essential oils had been analyzed by Gao *et al* (2012) and they reported that five compounds could be considered as the main components of the essential oils.

Constituents in Developing Fruits

The quantitative value as regard weight, specific gravity, stalk-stylar length and diameter of the fruits at maximum bulge as well as percentage of dry matter, total, reducing and non-reducing sugars, titratable acidity as citric acid equivalence and the vitamin C content in them had been determined by Mandal and Mazumdar (1998) at four stages of their growth at interval of 11 days till the fruits attained maturity for harvest, using samples of the common white-fruited type grown in the South 24-Paraganas district in West Bengal. They observed that the constituents had tended to increase quantitatively with development of fruits but the trend had not been clear in regard to vitamin C content in them.

The fluctuation that occurred in the content of carbohydrate, soluble protein, amino acids and the glutamine synthetase activity of wax apple fruits borne by flooded trees had been assessed by Yumeiet *et al.* (1999).

Post-harvest Storage

Wax apple fruit could be stored well for 3-4 days or more at ordinary atmosphere. But the fruits are shrivelled when stored long and this could be averted by storing them in moist environment. Cold storage of the fruits could be done at 2°–3 °C for 6-8 weeks, maintaining a relative humidity of 85-90 percent in the storage atmosphere.

Utilization

A number of preserved products are made utilizing wax apple fruits. In south-east Asian countries, the greenish fruits are cooked as a sauce and are also stewed with apple (Morton, 1987).

Canning of the fruits in syrup is done in many countries. Sometimes the seed(s) of the fruit is taken out on coring from below, *i.e.,* stylar-end and then canned so that the fruits resemble bell-like appearance. Pickling of wax apple fruits is a common practice in Indonesia. In Puerto Rico, fruits are much utilized to make wine. Processing strategies of wax apple fruits in China has been discussed by Hong Yand *et al.* (2013) in China. In West Bengal, some sweetmeat makers make a special preparation by mixing fine pieces of wax apple in sweetened curd or sweetened and thickened milk. Raisins, nuts, cherries, vanilla or chocolate are blended with the product and it is chilled. The chilled sweetmeat is served with biscuits.

Family: Myrtaceae

Species: *Syzygium uniflora* Linn.

Synonym: *Eugenia uniflora* Linn.

Taxonomic Serial: 27224

Common Name

Most widely used common names of the species, *Syzygium uniflora* in English are, Suriname cherry, Brazil cherry and Florida cherry. In other languages and countries, the fruit is called by different names and some are as follows.

Brazilian: Pitanga, Pitanga da praia *etc.* Spanish Pitanga, Pendanga, Guinda, Arrayán, Cerezaquadrada, Cerezo de Cayena, Cerezo de Suriname *etc.* Portuguese: Pitanga da praia, Pitanga do norte, Pitangamulata, Pitangarosea, Pitanga vermulha, Pitangaroxa, Pitangueiro do mato, Pitangueia *etc.* French: Cerise à côtes, Cerise côtelee, Cerisiercarré, Cerise créole, Cerisier de Cayenne *etc.* Italian: Ciligodi Cayenne. German : Suriname kirsche, Suriname kirschmyrtle, Cayenne kirsche *etc.* There is no particular name of Suriname cherry in any Indian language. Where it is grown, it goes by the name as Suriname cherry. The names, Brazil cherry and Pitangaare also used in some places.

Importance

Suriname cherry has great ornamental importance apart from its value in production of edible fruits. The new shoots of the shrub are wine-coloured and attractive. The sub-globose, 8-ribbed, bright coloured small fruits which hang in clusters are charming and the flowers are fragrant. For these reasons, Suriname cherry is preferred for planting as a hedge or as a screen or to form topiary in ornamental gardens. Adaptability of the shrub to a wide range of soil conditions, high resistance against pests and diseases and fast growth are among other attributes of it.

Description of Plants Parts

Suriname cherry is an evergreen shrub or a small tree and when fully grown up, it assumes a compact and somewhat conical appearance. It reaches a height of 7-8 metres and may be greater in deep and fertile soil. Leaves are simple, borne opposite, ovate to lanceolatein shape and with entire margin. A leaf measures 2-5 cm in length and 1-3 cm in width. New leaves are bronze or copper-red coloured. They become deep green and glossy when mature and turn red in winter. A resinous flavour is given off by the leaves. Flowers are white, sweet-scented and have 4 petals. They are borne on long, slender stalks with a conspicuous central cluster of white stamens and yellow anthers. Flowering takes place several times in a year. Fruits hang in cluster on slender stems. They are 2–4 cm in diameter and 8-ribbed. In young stage, fruits are green but later, their colour changes to yellow, orange, scarlet, maroon and in some types, almost black coloured. They are borne in large number. A fruit is embedded with a large seed but sometimes two seeds are also found in a single fruit.

Origin

Origin of the species is countries in the eastern coast of south America, ranging from Suriname to southern Brazil and extending to northern Argentina, Bolivia, Paraguay and Uruguay.

Geographical Distribution

The plant is much grown in Argentina, Bolivia, Brazil, Paraguay, Uruguay and other south American countries and especially at the peripheral parts of the forests in those countries. It is also grown in China, Guam, south-east Asian countries, Australia, New Zealand, West Indies, southern U.S.A., Mexico and other countries.

In India, Suriname cherry is not much grown but, it is met with in and around Bangalore, Mumbai, Baroda, Coonoor and is also found in some parts of Uttar Pradesh (Hayes, 1970; Singh and Saxena, 2008). It is presumed that the Portuguese settlers first introduced the plant in Goa.

Climatic Requirement

Tropical climate is most favourable for Suriname cherry but it is also grown well in mild subtropics where summer and winter are mild. The plant is able to withstand low temperature for a short period. High rainfall increases yield and full sunshine during growth of fruits is necessary for proper development of colour in them.

Soil Requirement

Soil is not an important factor but the drainage should be good. In deep, well-textured and fertile soil, the plants becomes taller, spready and highly productive.

Propagation

Seed propagation is most commonly done in Suriname cherry. Seeds should be sown without storage as their viability decreases when stored. Germination takes place in 3-4 weeks. Stem-cutting and air-layering are also successful but they are not much employed as the trees raised from seeds are capable to retain most of the characters of the mother plant without much variation.

Grafting and budding are done in some countries to take the advantage of rootstocks. Bezerra *et al.* (1999) in Pernambuco in Brazil studied budding and grafting methods in Suriname cherry. The seedling rootstocks which were 6, 9 or 12 months old were grafted by cleft or whip method, or budded by shield or patch method on to 12, 15 or 18-month-old rootstock seedlings. They observed that cleft or whip grafting on 9 or 12-month old rootstocks resulted in a high percentage of graft-union and shield-budding gave higher success than patch budding. In Florida in the U.S.A., cleft grafting is commonly done.

Culture

Suriname cherry grows well without receiving culture but irrigation in summer season is of much help. The shrub responds very well to training by *open- centre* or *vase* method and also pruning. As fruits are usually borne in very large number, thinning-out of fruits at early stage of their growth increase size and quality of them.

Pests and Diseases

Pests and diseases are not serious problems in Suriname cherry. Sometimes, scale insects and some minor pests are although found but they do not cause much damage.

Harvesting

Flowers appear in the tree two or three times in a year and in large number in winter. Fruits ripen in 3-4 weeks from flowering. Best ripening takes place in the trees and the fruits do not ripen any more after they are harvested. Hence, fruits of best quality are obtained which are harvested after they are fully ripened. This may be judged on observing their surface colour, *i.e.*, when the fruits have assumed scarlet red or blackish colour, depending on varieties.

Fully ripe fruits however, become very soft for which they may be difficult to handle. Hence, harvest of the fruits before full ripening appears to be a practical approach.

Yield

Yield of Suriname cherry greatly varies and the total yield of a tree from pickings done at different times of the year may be 3–5 kg. Yield is markedly increased by efficient training and pruning practices and judicious application of manures, fertilizers and irrigation.

Varieties

Named varieties of this fruit are not there but some types are known belonging to the red-skinned group. However, much variations are noticeable in plant and fruit characters of Suriname cherry in different parts where it is grown.

Mode of Consumption

Taste of the fruit is sour to sweet and sweetness goes high with ripening. Ripe fruits are taken as fresh but addition of salt, ground black pepper and spices are added to increase palatability and sugar is added by many for calcification.

Quality and Nutritive Value

Carbohydrate content of Suriname cherry is fairly high in some types. According to Wikipedia, 100 gram of pulp of the fruit contains carbohydrate: 7.49 g.; fat: 0.4 g.; protein: 0.8 g.; vitamin A equivalent: 75 mcg.; thiamin : 0.03 mg.; riboflavin 0.4 mg.; niacine : 0.03 mg.; vitamin C: 26.3 mg.; calcium : 9 mg.; iron 0.2 mg.; magnesium: 12 mg.; phosphorus: 11 mg.; potassium: 103 mg.; sodium: 3 mg and provides 33 Kilocalories (= 138 Kilojoules) of energy on consumption.

(Index : g = gram, mg milligram, mcg = microgram).

Medicinal Value

Volatile oils of the fruit are known to have anti- diabetic, anti-tumoric, anti-hypertensive and many other properties. The fruit has been on use as a folk medicine in Nigeria (Oguntimen and Elakovich, 1991). Leaves are used in Paraguay for lowering cholesterol, weight, uric acid, blood pressure *etc.* (Ferro *et at.,* 1988). In Brazil, leaf infusion is used to reduce fever and digestive disturbance.

Phytochemicals

Fruits as well as leaves of Suriname cherry are rich sources of many essential oils which are highly valued in pharmaceutical and other industries. Bark of the stem contains 25 percent or more tannin which is extracted and used in industries. The seed also contains appreciable amount of tannin.

Undesirable Effect

Pulp of the fruit has no harmful effect but the seeds are toxic in containing high amount of resin. It is found that dogs develop severe diarrhoea on consumption of the fruits along with seeds.

Utilization

Although fruits of Suriname cherry are taken as fresh, they are also processed to make many preserved products. Jelly and jam are commonly done as the fruits are rich sources of pectin of high jelly-grade. To make jam, either the fruits are used singly or in combination with other fruits. Juices and drinks are also made from the pulp juice and the juice is used to make ice-cream as well. In Brazil, an alcoholic beverage is made on fermenting the juice, extracting from ripe or the semi-ripe fruits.

Family: Oxalidacear

Species: *Averrhoa bilimbi* Linn.

[Synonym: *Averrhoa obtusangula* Stocks]

Taxonomic Serial: 506370

Common Name

According to Morton (1987), the strictly held English names of *Averrhoa bilimbi* are "Cucumber tree" and "Tree sorrell", which had been bestowed by the Britishers in colonial times. However, it is also called in English as Bilimbi and this is the commonly used name in India also.

The names of bilimbi used in other countries are as follows.

Sri Lanka : Bilincha, Bimbiri, Biling. Dominican Republic : Vinagrillo. Malaysia: Belimbingasam, Billing-biling. Indonesia : Belimbing wuluh, Belimbing sayur. Thailand: Taling pling. Vietnam : Khetau. China : San Lian, Mu Hu Gua, Huang Gua Shu. France : Carambolier bilimbi. Jamaica: Bimbling plum. Cuba: Grosella china. Brazil: Azedinha, Biri-biri. Argentina Pepinode Indias.

In India, the common name is Bilimbi in different parts. In south India however, the names used are, Ilimbi, Ilumbanpuli, Chilimbi, Irumban puli, Chemmen puli, Keerichakka *etc*.

Botanical Family

Oxalidaceae, to which *Averrhoa bilimbi* is placed is also known as wood-sorrel family although some botanists prefer to put it into a separate family, naming as Averrhoaceae. The family belonging to the Order, Oxalidale is small and consists of 8 genera. The species in the family are herbaceaous plants, shrubs and rarely trees which are small. A great majority of 900 species come under the genus, *Oxalis*. Plants of the family have divided leaves and the leaflets show sleep movements, *i.e.*, they spread open in light and close in darkness. Inflorescence is cyme and the flowers are bisexual and actinomorphic. Perianth consists of a calyx of 5 sepals and a corolla of 5 petals, both of which are distinct. Stamens are 5 and in two series with the outer series opposite to petals. They are sometimes reduced to staminodes. Gynoecium consists of single compound pistil of 5 carpels, 5 styles which are distinct and a superior ovary with 5 locules. Fruit is a loculicidal and sometimes dehiscent capsule or berry.

Genus: Averrhoa

The genus, *Averrhoa* is named after Averroes, who was an astronomer and philosopher from Al-Andalus. The genus comprises many species among which bilimbi, *i.e., Averrhoa bilimbi* and carambola, *i.e., A. carambola* are cultivated for fruit. Some other species are *A. dolichocarpa, A. leucopetala, A. microphylla, A. minima, A. sinica etc.* The genus is native to south-east Asian country and the members have pinnate leaves.

Importance

For high sourness, fruits of bilimbi have less importance than those of the close relative, which is carambola. However, the tree is much grown in Indonesia, Myanmar, Sri Lanka, Thailand, Malaysia, Singapore and some other countries in the south-east Asia to secure fruits. The tree is also much grown in south India where the sour fruits are used for culinary purpose. The fruit has many medicinal properties and has much use in traditional medicine in some countries. Attractive appearance of the small tree is also a desirable feature.

Description of Plant Parts

Height of the bilimbi tree reaches upto 5-12 metres. This may be longer in tropical and shorter in subtropical climates. Leaves are 3-6 cm long, entire, compound, alternate, lanceolate, imparipinnate and in cluster at branch extremities. Leaflets are 14-38, alternate or sub-opposite, oblong and hairy. Flowers are 1.5-2 cm long, fragrant, yellowish green or purplish. They are borne in small, hairy panicles which emerge directly from the trunk. Heterostyly is observed in flowers, *i.e.*, some trees produce flowers which have long styles and in some trees, flowers produced are of shorter styles.

Fruit is 4-10 cm long, ellipsoid, obovoid or nearly cylindrical and faintly 5-sided. It is capped by a thin, star-shaped calyx at the stalk-end and tipped with 5 hair-like remnants at the apex. Colour of the fruit turns from bright green to yellowish-green and is nearly white when ripe with glossy outer skin which is thin. Pulp is very sour and crunchy. Seeds are 6-7, flattened, disc-like, 6 mm wide, smooth and brown. The fruit resembles gherkin.

Origin

Exact origin of the species is not known and Singh and Saxena (2008) have indicated to be of Indo-Malayan origin. However, a great many authors have stated that bilimbi has originated somewhere in the south-east Asia. Moluccas and Indonesia as the possible origin has also been suggested by many.

Geographical Distribution

Bilimbi is cultivated or naturally grown in all south-east Asian countries and particularly in Indonesia, Malaysia, Thailand and Singapore. In south Asia, it is grown in Myanmar, Bangladesh, Sri Lanka and India. The tree is much grown in south India where its fruit has popularity for high sourness and is used in culinary preparations. Outside Asia, bilimbi has been introduced in Queensland in Australia, Zanzibar, Cuba, Trinidad, Puerto Rico, Venezuela, Columbia, Ecuador, Brazil, southern Florida in the U.S.A. and many other countries.

Climatic Requirement

Bilimbi is highly tropical in its climatic requirement. If the tree is grown in subtropical climate, protection from cold and wind becomes necessary. The tree is less resistant to cold than carambola, which is the close relative of it. Young trees of bilimbi are more sensitive to cold than those which are mature. High rainfall is necessary for better performance of bilimbi and ideally, rainfall should be evenly distributed throughout the year. The tree requires full sunshine and growth is slow in shady or semi-shade situation.

Soil Requirement

Soil is not an important factor for bilimbi and the tree grows even in limestone and sand. However, fruiting potential is very high in rich and well-drained soil. The tree prefers slightly acidic soil and a soil pH of 5.5–6.5 is considered very satisfactory.

Propagation

Seed propagation in bilimbi is largely practised. Seeds germinate easily within 4-6 weeks at 20° - 25° C. Grafting and budding are less successful but air- layering is practised in Indonesia since long time. Trees raised from vegetative propagation are found to be less vigorousthan when grown from seeds. Rema *et al.* (1997) reviewed the vegetative propagation methods in bilimbi.

Culture

Bilimbi trees are grown without any culture or application of manure. The tree is usually grown as filler trees in orchards of other fruit crops. However, manure is found to make high vigorous growth and it increases yield. In Taiwan, an NPK fertilizer mixture of 10: 10: 10 is known to be used. Sharma *et al.* (2012) stated dose of 100, 50 and 100 gram of nitrogen, phosphorus and potassium per plant.

Pests and Diseases

No serious pests or diseases have been reported in bilimbi. Sometimes some fungal diseases are known to occur in inflorescence and fruit due to the attack of *Fusarium, Penicillium, Cercospora* and *Colletotrichum* species of fungus.

Harvesting

Seed borne trees start flowering from 4 to 6 years. In India, the tree starts flowering in February and gives fruit over a long period to continue upto December. Fruits naturally fall off when ripe but are damaged as the skin is very thin. Hence, picking fruits or clusters by hand is advised.

Varieties

Any known variety of bilimbi is not known but much variations in tree and fruit characters are recognized. Fruits of trees having different degrees of sweetness are also met with.

Fruits of two types are usually found in India, which are sour and sweet (Singh, 1963 ; Nand, 1970). In south India, the sour type of fruits are more popular. A somewhat sweet variety is found to grow in West Bengal but its sugar content does not exceed 5 percent. Chinese strains are appreciably sweet (Singh, 1963). In Taiwan, some cultivars, *viz.*, Tean, Ma, Min Tao and Dah Pon bear large and sweet fruits. In Florida in the U.S.A., the variety, Newcomb Seedling has been stated (Saulset *et al.*,1986).

Mode of Consumption

Fruits of bilimbi are much sour due to high acid content and hence, they are usually made into sauces and curries rather than taking as fresh. The green and uncooked fruits are prepared as a relish and served with rice and bean in Costa Rica. They are also cooked with fish and meat.

Quality and Nutritive Value

Morton (1987) has cited in her book that 100 gram of bilimbi fruits containing 94.2 - 94.7 g of moiture has 0.61 g of protein, 0.6 g of fibre, 0.31-0.4 g of ash, 3.4 mg of calcium, 11.1 mg of phosphorus, 1.01 mg of iron, 0.035 mg of carotene, 0.10 mg of thiamine, 0.026 mg of niacin and 15.5 mgof vitamin C.

From the National Institute of Nutrition (I.C.M.R., Govt. of India) in Hyderabad, is has been reported that (Gopalan *et al.*, 2018) 100 gram of bilimbi fruit having 94.4 percent of moisture contains protein - 0.5 g., fat - 0.3g., carbohydrates - 3.5 g., 3.5 g., crude fibre - 1.0g., minerals - 0.3 g., calcium - 15 mg., phosphorus - 10 mg., iron -1.2 mg., carotene - 18 mg., thiamine - 0.09 mg., riboflavin - 0.04 mg., niacin - 0.6 mg., vitamin C - 32 mg, potassium - 130 mg and copper 0.05 mg. On consumption of 100 gram of fruit, the body gets 19 Kcalorie (= 79.5 K Joule) of energy.

(Index : g = gram, mg = milligram, mcg = microgram)

Medicinal Value

Leaves, stem and fruit of bilimbi are known to have many medicinal properties. Paste made with leaves is efficacious in itches, swelling, bites of poisonous animals and veneral diseases. Fruit has astringent property. It is stomachic and effective in obesity. Anti-ulcerogenic and anti-cancerous effect of the fruit has also been indicated. A syrup made from the fruits is used in some countries against inflammation, rectal bleeding and hepatitis.

Undesirable Effect

Although bilimbi fruits have medicinal properties for many diseases, high content of oxalic acid in them not only makes the fruits very sour but is also harmful.

In unripe fruits, oxalic acid ranges from 10.5 to 14.7 mg per gram although it reduces to 8.45 to 10.8 mg per gram when the fruits are ripe. This dicarboxylic acid which is usually present as calcium salt is considered to be a non-nutritive component which interferes with calcium absorption in the body by forming insoluble salts with calcium. Intake of high oxalate in the body may result in formation of urinary stone as well.

Post-harvest Storage

Bilimbi fruits have a shelf-life of 4-5 days under ordinary atmosphere but pre-chilling the fruits just after harvest can increase the shelf-life for 2–3 days more.

It is presumed that fruits may be kept well for 2-3 weeks in cold storage condition, maintaining a temperature of 3°–5° C and a relative humidity of 85-90 percent in the storage atmosphere.

Utilization

Bilimbi fruits are very commonly made into chutneys in India with or without addition of sugar but a number of other preserved products are also made on reducing the sourness by dipping the fruits in lime water for few minutes. Pickling the fruits is very common in south India and West Bengal. In Indonesia, fruits are sun-dried which are called asam sunti. On drying, the fruits are shrivelled and turn blackish. In Malaysia, sweet jam is also produced. In Costa Rica, immature fruits are processed as a relish. In many countries, syrup is made with the juice of the fruit pulp. Flowers are also preserved in syrup.

Family: Oxalidaccae

Species: *Averrhoa carambola* Linn.

[Synonym: *Averrhoa acutangula* Stokes]

Taxonomic Serial: 506371

Common Name

In English, common names of *Averrhoa carambola* are Star fruit and Carambola. However, other names such as Five corner, Starpickle, Foreign peach *etc.* are also used in English. It may be stated that the name, Carambola has been derived from the Sanskrit word, Karmaranga which means food appetizer.

In some other countries, the species goes by the common names which are as follows.

Indonesia : Belimbing manis. Vietnam : Khe, Kheta. Laos : Nak Fuang. Thailand : Ma fueang. Malay : Belimbing batu. Guam : Bilimbine. China : Yang tao. France: Carambolier. Germany : Sternfrucht. Portugal : Carambola. Guyana : Five finger. Trinidad : Coolie tamarind. Brazil: Camerunga.

The common names of carambola in different parts of India are Kamrakh, Kamranga, Karmal, Karamara, Kordoitenga, Tamarta Irimpanpuli, Thambaratham *etc.*

Importance

Unusual shape of the fruits is a special feature of carambola and it is said that fruits of carambola may not be very pleasant to taste but pleasant to the eyes. Prolific golden yellow coloured fruits, dark green leaves, pink flowers borne in great mass which smell sweet and symmetrical shape of the tree are attractive features for which carambola is also preferred for planting in decorative gardens. Carambola fruits, especially those types which are sweet to taste make a great appeal to consumers.

However, sweet-sour taste of the common types of carambola fruits is a likeable feature to the people in south India where it is taken as fresh or in many culinary preparations. The juice extracted from the fruits is made into drink in western countries.

The fruit and other parts of the tree are used in traditional folklore medicine in south-east Asian countries. Manda *et al.* (2012) and Avinash *et al.* (2012) made exhaustive reviews on medicinal properties and various other aspects of *Averrhoa carambola.*

Description of Plant Parts

The tree or shrub of carambola is slow growing and is 5-12 metre tall. Branches are drooping and the wood is white which turns reddish. Leaves are produced in axils at the end of twigs and deciduous, spirally arranged, alternate imparipinnate, 15-20 cm long with 5-11 nearly opposite leaflets, and are ovate or ovate-oblong. Leaflets tend to fold at night and also if shock is given abruptly. Flowers are pink or lavender coloured, bell-shaped, 6 mm wide and arranged in small cluster. Petals are 5 with whitish edge. Fruit is 5-15 cm long, ovate to ellipsoid with 5 prominent longitudinal ridges which rarely may be 4, 6, 7 or 8. In cross-section, it resembles a star shape, Fruit (Plate-21) is oblong, 7–9 cm wide, 5-15 cm long, has a thin waxy surface and gives off oxalic acid smell. Each fruit has 10-12 flat, brown seeds.

Origin

Exact origin of carambola is controversial. A great many botanists put forward that the tree has originated somewhere in south-east Asia and most probably, in Indonesia or Malaysia. Hayes (1970) quoted that carambola is a native of the Moluccas or of the Malayan region. However, origin of the species in Sri Lanka, India or Bangladesh has also been suggested by some botanists.

Geographical Distribution

Carambola is extensively grown in all south-east Asian countries and Malaysia is stated to be the "global leader in carambola production." At present, the tree is however, much grown in other countries, notably in southern China, Myanmar, Bangladesh, India, Sri Lanka, Caribbean islands, parts of Africa, central America, Brazil, Colombia and in other countries. In India, carambola is much grown all over south India and in eastern Indian States. According to Hayes (1970), carambola is grown in India since ancient time.

Climatic Requirement

Carambola is a tropical tree and grows best in hot and humid climate. However, the tree grows in subtropical climate with suitable protection in winter but the growth and productiveness are better in tropical environment. Ideal temperature for satisfactory growth of the tree is 20° to 35°C and at temperature below 18°C, growth is stunted. Younger plants are susceptible to low temperature and are killed in frost although mature trees can withstand (-) 3°C for a short period.

High rainfall distributed evenly throughout the year is most favourable for better growth of the trees, their higher yield performance as well as production of fruits of better quality. It is observed in Australia that quality and flavour of the fruits are best where annual temperature is 1800 mm or more. Full sunshine has also a great necessity for the tree for production of fruits of better colour, flavour and taste. The tree grows well upto an altitude of 1200 metres in the tropics but the productivity and quality of fruits are much hampered when the tree grows at very high altitude even in tropical climate.

Soil Requirement

Soil is not considered to be a very important factor for carambola as the tree grows satisfactorily on sand, heavy clay or limestone but leaves become pale yellow when grown on limestone. However, the trees grow at a faster rate and bear profusely in deep and well-drained clayee loam. Mildly acidic soil with a *p*H of 5.5 – 6.5 is preferred by the trees. Drainage is a very important factor but soil should be adequately moist throughout the year and never gets dried.

Propagation

A number of propagation methods are employed for multiplication of carambola trees. Seed propagation is by far, most common but as the seeds are recalcitrant and remain viable only for few days, freshly extracted seeds should be used. Seeds should be plump and obtained from fully ripe fruits. Seedlings should better be raised in polythene bags and transplanted after about a year. Trees raised from seeds take longer time to come to fruiting which may be 4 to 6 years.

Root formation in stem-cuttings is difficult in carambola but use of auxin in suitable concentrations to induce root formation has been observed by Bora and Das (1998). Protrusion of roots in air-layering takes place more easily but air-layered plants make slower growth. Shield-budding, Forkert-budding, veneer and side-grafting on seedling rootstocks of carambola or bilimbi are practised in Java, Indonesia and other south-east Asian countries. In India, nurserymen prefer inarching on rootstocks raised from carambola seedlings. Grafted trees attain bearing maturity within 2-3 years.

Plantlet prodution adopting micro-propagation techniques has been stated by a number of workers. Kantharajah *et al.* (1992) described a method using MS medium supplemented with BA and NAA. Regenerating plants from callus culture has been stated by Amin and Razzaque (1993). Li Ji Hong *et al.* (1993) in China observed that MS medium supplemented with BA and NAA was best for shoot tip culture of carambola *in vitro.*

Culture

If plantation of carambola is required to be set up, trees should be planted at a spacing of 7 - 8 metres in both directions but as the trees make vigorous growth in rich soil, planting distance should be increased in such soil.

Carambola being a minor fruit, no culture is practically adopted in raising the trees but liberal application of manure, fertilizers, irrigation and adoption of suitable training and pruning practices are of great benefit. Chadha (2012) stated that it would be a good practice to manure the trees once a year with 40-50 kg farmyard manure per tree and also nitrogenous and phosphatic fertilizers. The doses should depend on age and size of the trees beside soil fertility. Morton (1987) cited that at the Research Center in Homestead, Florida in the U.S.A., carambola trees which were 2.4 - 3 metres tall responded well to application of 0.5 kg of N, P, K and Mg in the ratio of 6:6:6:3 given to the trees 3 or 4 times in the year.

She suggested that in case the leaves showed chlorotic symptoms, iron, zinc and manganese should be applied as corrective measure.

Under West Bengal condition, De *et al.* (1996) undertook studies in regard to accumulation of nitrogen, phosphorus and potash in carambola leaves sampled at monthly interval from December to May. It was observed by them that all the nutrients had decreased with advancement of summer season following increase that took place in the winter months. In an observational study, Mazumdar and co-workers (unpublished) noted that quality of fruits, their colour development and shelf-life were improved by foliar application of potassium sulphate, zinc sulphate and magnesium sulphate solutions to the trees.

Carambola trees require adequate sunshine for higher yield and production of fruits of better quality. Accordingly, training the trees to *open-centre, i.e., vase* system at early stage of their growth would be of great help to admit more light in the inner parts of the trees.

Thinning of fruits may become necessary for production of better quality fruits. Liberal irrigation in summer season ensures vigorous growth of the trees and higher yield performance.

Pests and Diseases

Carambola is almost free from attack by serious pests and diseases. Among insect and animal pests, fruit flies, fruit moths, ants and birds cause some damage. In Malaysia, damage is usually high due to the attack of fruit flies and especially by *Dacus dorsalis*. Trapping with methyl eugenol as an attractant has been found to be effective to control the insect pest in that country. Among fungal diseases, anthracnose caused by *Colletotrichum gloeosporioides* and leaf spots due to the attack of *Phomopsis* and *Phyllosticta* species or *Cercospora averrhoae* are noted in the trees and copper fungicides are effective to control them.

Harvesting

Carambola trees raised from seeds give fruits from 4 to 6 years after sowing while the grafted trees start production from 2 to 3 years after planting and may even be earlier. In highly tropical zone, the trees flower and fruit at every month (Hayes, 1970). Under Indian condition, stray fruits are borne by the trees at several times but main season of harvest is August to October and then again in December - January, depending on locality, duration of rainy-season and winter temperature.

Carambola fruits are non-climacteric and are best ripened in the trees. The fruits do not ripen any more after harvest but they are harvested slightly earlier than when fully ripe. After harvest, the fruits are softened and become slightly yellowish when maintained under ordinary atmosphere but sugar content in them does not increase. When the fruits are ripe, they naturally fall off from the trees but harvesting is done before attainment of full-ripening stage and this is done by hand picking of the individual fruits. Carambola fruits which are yellowish have a greater demand in the market then those which are of green colour. Bhattacharyya *et al.* (1988) had attempted de-greening of the fruits harvested from the trees growing in the South 24-Paraganas district of West Bengal. Just after harvest, the fruits were steeped with brine water ranging from 1 to 24 percent concentration. Although dipping in some of the treatment solutions caused some yellowing, many fruits were shrivelled and their ridges had burning appearance particularly when dipped in solutions of higher concentrations of salt. Hence, dipping in saline water does not appear to be a good practice to de-green the fruits.

Yield

Yield of carambola fruits greatly varies and under Indian condition, it may be 50 to 100 kg per tree per year (Singh and Saxena, 2008). In Java, Indonesia and other highly tropical countries, yield is much higher and goes to 150 to 180 kg per trees in a year. Pollination is a problem in carambola flowers due to heterostyly and this affects yield of the trees considerably. Hand-pollination of flowers increases yield to a high extent (Hayes, 1997).

Varieties

Two distinct classes of carambola are grown. These are, trees which bear relatively small fruits and those which bear larger fruits. The smaller fruits are sour to taste and they also emit a smell of oxalic acid. Large fruits are sweet or moderately sweet or have little sourness and they are mildly flavoured. Hayes (1970) has commented in his book that in India, trees which bear sour fruits are mainly grown and in China, trees that bear sweet fruits are largely grown. Ping Sheng (1999) has mentioned a variety in China which is of high quality and is a sport of a sweet variety. Pareek and Sharma (1999) have stated that a number of Chinese strains of carambola claimed to be very sweet in taste. Of late, many sweet or moderately sweet varieties are grown in the southern and eastern parts of India, the trees of which have been introduced from south-east Asian countries.

Some of the famous varieties in other countries are Dah Ponin Taiwan, Fwang Tung in Thailand, Maha in Malaysia, Demakin Indonesia, Arkin in Florida in the U.S.A. *etc.* In Florida, the variety, Golden Star, which is evolved as a seedling selection is claimed to be of high quality.

Mode of Consumption

Fully fruits of carambola are highly relished as fresh. They have a taste which has a resemblance of apple, pineapple and kiwi fruits blended together. Sour fruits are taken on mixing with salt, sugar and spices. Fruit is cut into sections, sugar or salt is sprinkled on them, maintained for sometime and is taken sprinkling spices on them. In south India, carambola is much used in preparation of curries and pulses. In West Bengal, Assam and Orissa, soury preparations are made with pieces of carambola fruits with addition of sugar or molasses. Fruit slices are also added in cooking small fishes.

Quality and Nutritive Value

Sourness is a distinguishable feature in carambola fruits and Morton (1987) stated that experiments done in Florida in the U.S.A. revealed that in fruits of the Golden Star variety, total acid content was 12.51 mg per gram of pulp in the freshly plucked mature green fruits and in mature yellow fruits it was 13 mg. According to her, among other acids, oxalic and tartaric acids were noted to be predominant acids in carambola fruits. She has presented food value of 100 gram of carambola fruits as follows.

Calorific value : 35.7.; Moisture: 89-91 g.; Protein: 0.38 g.; Fat 0.08 g.; Carbohydrate: 9.38 g.; Fiber: 0.8 -0.9.g. Ash: 0.26 : 0.26 – 0.40 g.; Calcium : 4.4 6.0 mg.; Phosphorus: 15.5 - 21.0 mg.; Iron: 0.32-1.65 mg.; Carotene 0.003-0.552 mg.; Thaimine: 0.03-0.038 mg.; Riboflavin : 0,019 -0.03 mg.; Niacin: 0.294 - 0.38 mg.; and Ascorbic acid (analyses made in Cuba and Honduras): 26.0-53.1 mg.

(Index : g = gram, mg milligram).

Physico-chemical constituents of carambola fruits had been determined by a number of other workers as well. Heredia *et al.* (1998) analyzed that main carbohydrate in the fruit pulp was sucrose. Mandal and Mazumdar (1995) under West Bengal condition observed that average weight, specific gravity and stalk-stylar length of the fruits were respectively 71,0.097 and 8.1 cm while the total, reducing and non- reducing sugars and the total titratable acidity (equivalence of citric acid) of the pulp were 2.84, 1.98, 0.82 and 1.28 per cent respectively. The volatile components in carambola were determined by Pino (1997 b). Sinha Roy (1996), Mandal (1997) and Mandal and Mazumdar (2000) determined weight, dry matter, total, reducing and non-reducing sugars, pectic compounds, titratable acidity (equivalence of citric acid) and vitamin C content of carambola fruits by spraying the trees borne by them with solutions of KC1, $ZnSO_4$, or with IBA or GA_3.

Medicinal Value

Carambola has many medicinal properties in its root, leaves and fruit. In traditional medicine in India, fruit is recommended to cure bleeding piles, billiousness, diarrhoea, eye inflammation, vomiting and for many other purposes. Fruits are also known to have ani-ulcerogenic, anti-bacterial and cytotoxic properties. Fruit is rich in anti-oxidant and potassium. Seeds are beneficial in asthma. Avinash *et al.* (2012) have reviewed the medicinal properties of carambola.

Constituents in Developing Fruits

Changing pattern of physical and chemical constituents in carambola fruits during their development and maturation has been studied by a number of workers. Siti Halijah Ali and Jaafar (1992) ovserved that total soluble solids, pH, sugars and ascorbic acid in fruits increased with maturity while acidity decreased, Mandal (1997) and Mandal and Mazumdar (1998) studied that total duration of fruit development of carambola fruits had been around 56 days. Analyses done by them at four, equally spaced stages of development revealed that the constituents studied increased with development of fruits and markedly from 42 days after fruit set. The acidity level increased highly upto 42 days and decreased subsequently. Sinha Roy (1996) determined pectic fractions in the fruits during their development, maturation and ripening.

Narain *et al.* (2001) had studied physico-chemical composition of the Golden Star cultivar of carambola fruits at green, half-ripe and ripe stages of their maturity. It has been revealed that weight of the green mature fruits was significantly different from those of half-ripe and ripe fruits, *p*H of the fruits increased with maturity, ripe fruits were significantly less acidic than green mature and half-ripe fruits and calcium content was higher in ripe fruits.

Undesirale Effect

Both *Averrhoa bilimbi* and *A. carambola* belonging to the family, Oxalidaceae have their demerit in that they contain appreciable amount of oxalic acid. Harmful effect of this dicarboxylic acid in kidney and urogenital system has been stated in the Chapter on A. *bilimbi*. High intake of carambola fruits may in fact, be fatal to persons who have urinary disturbance and are prone to formation of stone in kidney. As mentioned in A. *bilimbi*, oxalic acid also interfers with calcium absorption in the body.

Post-harvest Storage

Carambola fruits are found to keep well for 6 - 8 days under room temperature condition. In cold atmosphere, fruits may be stored for 4 to 6 weeks at a temperature of 5 - 10°C maintaining a relative humidity of 85 - 90 percent. Morton (1987) stated that in storage trials in Florida in the U.S.A., carambolas picked at the first signs of yellowing kept in good condition for 4 weeks at 10°C, 3 weeks at 15.56°C and2 weeks at 21.1°C. She added that waxing the fruits extended their storage life and preserved the vitamin value. Post-harvest storage of carambola fruits had been studied by other workers as well (Sankat and Balkissan, 1992 O'Hare, 1993; Miller and McDonald, 1997).

Utilization

A number of products like jelly, jam, preserve, juice *etc.* are made with carambola fruits. In Malaysia, the fruits are stewed with sugar and cloves with or without blending with apple. In China, slightly under-ripe fruits are salted, pickled or made into jam. In Taiwan, fruits are canned in syrup. In Hawaii in the U.S.A., the juice of sour carambola fruits is mixed with gelatin, sugar, lemon juice and boiling water to make drink. Wine is also made with the juice of fruits. In Java, flowers of carambola are utilized and are added to salads.

Family: Phyllanthaceae

Species: *Baccaurea ramiflora* Lour.

[Synonyms: *Baccaurea sapida* (Roxb.) Mull. Arg.; *B. cauliflora* Lour.; *B. flaccida* Mull. Arg,; *B. oxycarpa* Gagn.; *B. pierardi* Wall.; *B. propingua* Mull. Arg.; *Pierardia sapida* Roxb.; *P. flaccida* Wall.]

Taxonomic Serial: 506428

Common Name

The commonly used names of the fruit are Burmese grape, Lantern tree fruit, Mafai, Baccaurea *etc.* in English. In Chinese, the common name is Mu nai guo, in Mayanmar, Kanzaw, in Vietnam, Giau gia dat, in Thailand, Mafai or Mak fai pa, in Cambodia, Phunai and in Bangladesh Kusumtenga.

In India, the names are Latka, Lotkon, Bubi *etc.* in Bengali, Lateku in Assamese and in Tripura, the name is Khuchumu bathai. There is no specific name in Hindi and it is usually referred to as Khatta phal.

Botanical Family

The species comes under the Phyllanthaceae family belonging to the order, Malpighiales. The order is divided into 40 families or more and the target family is Euphorbiaceae. The family, Phyllanthaceae includes 60 genera.

The family is divided into two sub-families, which are Antidesmatoideae and Phyllanthoideae. The former contains six tribes and the latter, four. The genus, *Baccaurea* comes under the tribe, Scepeae of the sub-family, Antidesmatoideae. The genus, Phyllanthus belonging to the tribe, Phyllantheae comes under the sub-family Phyllanthoideae.

The species of the phyllanthaceae family are herbs shrubs or trees and some are climbers and succulents. Leaves of the plants are usually alternate and rarely opposite. Inflorescence is usually auxiliary and flowers are unisexual except for a few species. Fruit is a schizocarp, drupe or berry.

Genus: Baccaurea

The genus contains more than hundred species which grow from the south-east Asian centuries to the western Pacific islands. The trees of the genus bear flowers in clustered spikes or racemes and some cauliflorous. The fruits are firm, pulpy and contain upto six seeds. The generic name, *Baccaurea* in Latin refers to the golden yellow colour of the fruits.

Importance

Burmese grape fruit has a fondness to the people in north-eastern State but is not so popular in other parts of India. In some part of north-eastern India, it is grown as a rainy-season fruit crop either as a monocrop or as inter-crop in orchards. The tree also grows naturally in the rain-forests of north-east India, Bangladesh and in south-east Asia.

Apart from the rose-colored delicious pulp, the fruits and other parts of the trees have many other uses. Stewing of the fruits is very common. Juice of the fruits is used in making wine. Fruits are rich source of pectin and hence, are used to make jelly and jam (Rahman *et al.,* 2014).

An important attribute of the seeds of the fruits is existence of annatto dye in them. Annatto is an orange-red coloured condiment present in the testa of the seeds. It is used to add colour and also flavor in butter, other foods, silk, cotton *etc.* The plant is cultivated in Bangladesh not only for the fruit but also to secure annatto dye. (Abdulla *et al.,* 2005).

Roots of the trees are used to add flovour in culinary preparations and to secure strong fibers to make rope. The oil extracted from the seeds is used in many industries (Dutta *et al.*, 2013).

Description of Plant Parts

Burmese grape tree is evergreen, perennial, slow growing and reaches a height of 20-25 metres. Crown of the trees is round and spreading. Stem is greyish brown with thin bark. Leaves are obovate-oblong, entire, alternate, membranous, green and papery. They are 10-15 long and 4-9 cm wide. Flowers are dioecious and apipetalous. Male inflorescence is densely papillose and 12-15 cm long. Female inflorescence is 35 cm long.

Fruit is oval, 2-4 cm in diameter and of yellowish, pinkish or purplish colour. It is glabrous, juicy, sweet-sour in taste and has pleasant flavour. Fruits are borne directly on trunk. Seeds are 2-4, flat-elliptic or rotund and resemble loquat. They are reddish and purplish and have white arils.

Origin

The species has been stated to have originated in the eastern part of India (Pareek and Sharma, 1993). Goyal *et al.* (2013) from the North Bengal University in Siliguri have stated south-east Asian country is the origin of the species.

Geographical Distribution

Burmese grape is cultivated or it is naturally grown in the rain-forests in the south-east Asian countries, Myanmar and Bangladesh. In India, it is much grown and is also naturally grown in northern West Bengal, Assam and other States in the north-eastern part. It is found to grow to some extent in the sub-Himalayan tract, Nepal, Sikkim and Bhutan.

Climatic Requirement

The tree is tropical and grows best at a temperature of 24° to 28°C and annual rainfall of 1500 to 3500 mm. Tolerance limit to low temperature in high of the tree, which may be a characteristic feature of some eco-types. It may grow and produce upto an altitude of 1300 metres in the tropics. The tree requires bright sunshine for better performance of fruits.

Soil Requirement

The trees grow and produces in wide types of soil provided the drainage is assured. Moderately acidic soil with a pH of 5-6 is favorable for the tree.

Propagation

Burmese grape is conventionally propagated by seeds. Seeds germinate easily and the seedlings are usually raised in perforated poly-bags, filled with a mixture of loamy soil rotten cowdung and sand in many parts of northern West Bengal. The emerged out seedlings become ready for transplating within 45 days.

In south-east Asian countries, air-layering, budding and grafting are also done. Abdulla *et al.* (2005) experimented with use of auxin in propagation by stem-cuttings in Bangladesh. Under Chittagong condition in Bangladesh, they observed that maximum rooting percentage of stem-cuttings had taken place by the use of 0.4 percent of IBA. Survival of the cuttings was also found to be highest at this concentration of auxin.

Culture

The fruit crop responds well without receiving agro-inputs. However, many Burmese grape growers in the northern West Bengal and Assam apply organic manures and in hot weather, irrigations are also applied and they secure high yield. As per instructions of the present author, some growers in the northern West Bengal sprayed weak solution of molybdenum sulphate twice on four trees and fruit size was found to be markedly increased due to that.

Some growers in the northern part of West Bengal make annual pruning of the trees in early winter and they secure high yield. Lopping of branches is also done at low height to encourage new shoots.

Pests and diseases are not serious in Baccaurea and any insecticide and fungicide may be sprayed to the trees when insects or disease symptoms would be noticed.

Harvesting

The trees produce flowers from February to March and may continue upto the month of April. Fruits are ready for harvest after 80–90 days from flowering and are harvested in rainy-season. Harvesting of fruits may be continued upto October. In the high rainfall areas in Assam and Meghalaya, Burmese grape trees produce fruits in several times in the year but the peak season of harvest is rainy-season.

Fruits are best ripened in the trees and do not ripe any more if unripe fruits are allowed to ripen under postharvest condition.

Yield

Exact yield of fruits from a tree is not known. However, according to many growers in the north–eastern States, a tree may produce 40 to 50 kg of fruits and among these, 80–90 per cent attain frull ripening.

Varieties

Well-recognized varieties of Burmese grape are not known but many types are met with which greatly differ on plant and fruit characters.

Evaluation of twelve germplasms of Burmese grape in respect of physico-chemical characteristics has been made by Deb and Bhowmick (2013) under Cooch Bihar condition in northern West Bengal. It was observed by them that harvest–maturity was least in Accession No.9. Fruit size, weight and juice content of the fruits were greater and acidity was lesser in Accession No.11.

The study undertaken by Rahman *et al.* (2014) in Bangladesh as regard fruit characters, yield contributing factors and yield behaviour of five promising genotypes of Burmese grape deserves interest. The results of their study done under Sylhet condition revealed wide variation among the genotypes. Proportion of edible matter and the soluble solids content were found to be highest in the line BS–Jai–001, weight and size of fruits were recorded to be highest in the line, BS–Jai–003 while the fruiting intensity and yield had been observed to be maximum in the line, BS–Jai–005.

Mode of Consumption

Traditionally, Burmese grape berries are eaten fresh. The fruits are broken by peeling off skin and the pulpy yellow arils are eaten. Seeds are usually consumed with pulp. Some people prefer to boil the fruits, cool and take the arils with salt and black pepper.

Medicinal Value

Many medicinal properties of the Burmese grape fruit and tree are known. Fruit is used in skin disease. Sap of bark, root and wood have benefit in treating arthritis, abscess and other ailments. Goyal (2013) studied anti–oxidant properties of the fruits.

Constituents in Developing Fruits

With growth of fruits, it is observed that constituents like length, diameter, weight and precentage of pulp go on increasing till maturity while acidity decreases. The non–reducing sugars content in the fruit juice does not change much during growth of the fruits. Colour of the fruits conspicuously changes from green to yellowish, reddish or purplish colour with maturity and ripening.

Changes in some physical and chemical components of Burmese grape fruits during their growth have been assessed by Bhowmick *et al.* (2013) under Cooch Bihar condition in the northern West Bengal. Studies conducted by them from the 30th to 86th days after fruit-set revealed that fruit size and juice content of fruits increased with growth of fruits while peel thickness and seed weight decreased. Increase had also occurred in soluble solids and sugars content and concomitantly, acidity level had decreased. The authors suggested that best time of harvest of the fruits would be 80–85 days after fruit-set to achieve best dessert quality.

Undesirable Effect

Consumption of Burmese grape fruits does not cause any harm in the body but too much consumption may cause belching, stomach pain, vomiting and uneasiness.

Post-harvest Storage

Bermese grape fruits can be kept well under room temperature condition for 5–7 days. Storage period may be much increased by boiling the fruits, cooling them and mixing with common salt.

Utilization

Stewing is a common practice to utilize the berries. For this purpose, they are slowly heated mixing with common salt. After heating, the water is drained and the fruits are mixed with spices.

Fermentation of the juice of Burmese grape fruits to make a delicious wine is done in some south-east Asian countries. The wine is of golden colour and has high popularity. Fermentation is done completely or partially. Either the juice of the berries singly or mixed with juices of other fruits are used for fermentation.

Goyal *et al.* (2013) in Siliguri in the northern West Bengal made an interesting study to evaluate some compounds like total phenols, flavonoids, flavonols, pro-anthocyanidin as well as DPPH-scavenging activity of Burmese grape juice and wine in regard to ageing in comparison with the juice. The range of the compounds in the wine and the juice has been tabulated by them.

The oil extracted from the seeds of Burmese grape has demand in some food industries as well as in pharmaceutical, cosmetic and soap industries. The oil is used as fuel and hence, it is nicknamed as bio-diesel.

Dutta *et al.* (2013) in Assam made a detailed study in regard to composition of the oil. They analysed iodine value, moisture content, density, saturated and unsaturated fatty acids and other constituents of the oil and provided results that would be of help in oil industry. Of late, oil extraction from the seeds of Burmese grape has much increased among the rural people in north-eastern India.

Family: Phyllanthaceae

Species: *Phyllanthus acidus* (L.) Skeels

[Synonyms: *Phyllanthus distichus* Muell. Arg.; *Cicca acida* Merr.; *Cicca disticha* L.].

Taxonomic Serial: 28366

Common Name

The commonly used names of the species *Phyllanthus acidus* in different countries are as follows:

In the U.K., the fruit is known as Otaheite gooseberry, Star gooseberry, Malayan gooseberry, Tahitian gooseberry, West Indian gooseberry, Damsel, Wild plum *etc.*

The common names in other countries are as follows. Malaysia: Chermala, Kemangor, Cherme *etc.* Java: Tijerme, Cherme *etc.* Thiland: Mayem. Vietnam: Tam duot, Chum rout *etc.* Venezuela: Cerezo agrio. Spain: Grosela. Mexico: Manzana estrella. Bangladesh: Nor.

In different parts of India, the fruit is known as follows. Maharastra: Harpharori. Tamil Nadu: Aranelli, Catarum *etc.* Karnataka: Rayaranelli. West Bengal: Nor, Hariphal. U.P., Punjab and Haryana: Chhota aonla, Harpharari.

Botanical Family

The species comes under the family, Phyllanthaceae, which has been described under the species, *Baccaurea ramiflora*.

Genus: Phyllanthus

As stated under the species, *Baccaurea ramiflora*, the family, Phyllanthace is grouped into two sub-families. The genus, *Phyllanthus* comes under the sub-family, Phyllanthoideae which contains four tribes.

The genus, *Phyllanthus* is the largest in the Phyllanthaceae family and consists of 1200 species. The species belonging to the genus are both annual and perennial and comprise as herbs, shrubs, climbers, succulents and floating aquatic plants. Their floral characters are of wild types. Many species have flat types of stems resembling leaves which are botanically known as cladodes.

The species of *Phyllanthus* abound in all tropical and sub-tropical parts but largely in tropical countries. Many species of the genus are of great economic importance, medicinal properties and a notable fruit plant, emblica (= aonla) comes under this genus apart from Otaheite gooseberry.

Importance

Fruits of Otaheite gooseberry have not much importance for human consumption for their very high sourness. But in the south Indian States, it is highly liked in culinary preparations, particularly for high sourness. In West Bengal, eastern Indian States and in Bangladesh, fruits are used to make pickles, preserves, drying and in making chutney. Ayurveds recommend the fruit for medicinal properties. The tree has importance for planting in ornamental gardens and avenue planting.

Description of Plant Parts

Otaheite gooseberry trees are somewhat between trees and shrubs in appearance. Height goes upto 8.5 metres or more. Crown is bushy, dense and are highly spreading. Branchlets are produced at the top of branches. They are greenish or pinkish in colour and 20–30 cm long. Leaves are ovate or ovate–lanceolate and are arranged in two rows along small branches (Hayes, 1970). Leaves are 3–7 cm long, pointed, green and thin. At the leaf base, there are two pointed stipules.

Inflorescence is panicle and 6–12 cm long. Flowers are small, mainly male and other flowers are female, besides some which are bisexual. Fruit (Plate-22) is oblate in shape, 1.5–3 cm wide, pale yellow and have 6–8 ribs. Surface of the fruit is waxy and pulp is fleshy. Stone is very hard, lightly embedded in the centre and contains 4–6 seeds.

Origin

Exact origin of the species is not known. A section of taxonomists is of the opinion that the coastal area of north–eastern. Brazil might be the origin. Other viewpoints are, the species might have originated in Malagasy or in the caribbean countries.

Geographical Distribution

Otaheite gooseberry is grown in most of the tropical and subtropical countries in the world. Some of the important countries where it is much grown are Malaysia, Indonesia, Laos, Guam, South Vietnam *etc.* The species has spread to the Caribbean islands, southern Mexico, Venezuela, Suriname, Columbia and other south America countries where its performance is still better. In India, the tree is grown all over the plain land and abundantly in the southern States. However, it is also much grown in West Bengal and the eastern States.

Climatic Requirement

Tropical climate with hot, humid condition and high rainfall is most favourable for Otaheite gooseberry. However, it grows fairly well in mild subtropical climate. The tree is hardy and thrives upto an altitude of 900 metres in the tropics.

Soil Requirement

Soil is not important for the tree except that drainage should be good. Soil which remains moist all over the year is best suited for vigorous growth and higher productivity of the tree.

Propagation

Seed propagation is most common for Otaheite gooseberry. Scarification of the hard seed–covering by rubbing between two sand–papers is effective to loosen the covering and thereby, the method facilitates germination. Soaking the extracted seeds in lukewarm water for few hours also ease germination process. Stem–cuttings by the use of softwood cuttings is practised in many countries. Air–layering gives abundant adventitious roots when practised in rainy-season. Propagation by budding on rootstock of same species is done in some south-east Asian countries.

Culture

Organized plantation and culture in Indian is not common for the tree. The tree is grown here and there from seeds but in some south-east Asian countries, small plantations are raised.

For higher productivity and better quality fruits, pruning is much helpful. This should be done by heading–back and thinning–out systems. Application of organic manure in winter season is found to increase productiveness. If regular plantation is desired, a spacing of 7–8 metres should be maintained. Spacing should be less if afforestation is done.

Pests and Diseases

Any serious pest or disease of the tree is not observed. However, a caterpillar attack has been reported from Florida in the U.S.A. The caterpillar feeds on the bark heavily and also feeds on leaves which are young. As a result, high leaf fall takes place. From Indonesia, attack by the caterpillars of the insect species, *Parallelia* is reported. Otaheite gooseberry fruits are also much damaged by fruit fly which pierces the fruits.

Harvesting

Fruits of Otaheite gooseberry ripen best in the trees and they do not ripen in post-harvest condition if harvested earlier. Maturity for harvest of the fruits is determined when colour of the fruits becomes yellowish and some fruits naturally drop off.

In high rainfall area under south Indian condition, the trees produce two crops in a year, which are in April–May and in August–September but few fruits are produced almost all throughout the year. In West Bengal and eastern India, fruits are produced from March to May. Conventional method of harvest is to shake the branches so that fruits drop and to collect the dropped fruits. Fruits are hard and hence, are not damaged when dropped.

Yield

Yield of fruits of Otaheite gooseberry is not precisely known. However, yield is greatly increased in high rainfall area, rich soil and if proper training practice be adopted. Training by *open-centre, i.e., vase* system in which the branches much spread out increases yield considerably.

Varieties

Any named variety of the fruit crop is not known. However, much difference is noted in regard to growth pattern of the trees, foliage characters and size and shape of the fruits.

Mode of Consumption

In south India, either intact fruits or slices made from them are used for cooking, removing the hard stones of the fruits. In eastern India, the fruits are consumed as fresh by mixing with salt and spices or cooked with sugar to make chutney. In some south-east Asian countries, juice of the fruits is extracted, blended with juices from other fruits, mixed with sugar and used as drink. In Indonesia, leaves of the tree are known to be cooked and taken.

Quality and Nutritive Value

Mandal and Mazumdar (1997) analyzed ripe fruits of oteheite gooseberry, randomly obtaining samples from trees grown in the South 24–Paraganas district of West Bengal. Average results of 100 gram of fruit samples of their study revealed the following.

Weight: 5.02 g.; Specific gravity: 1.71 g.; Stalk–stylar length: 1.5 cm.; Titratable acidity (citric acid equivalence): 2.83 per cent; Vitamin C: 6 mg per cent. Reducing, non–reducing and total sugars: 0.6–2.5 percent respectively.

An analytical report from Thailand presented the following constituents (Saraju Devi and Paul (2011)).

Carbohydrate: 6.4 g.; Fibre: 8 g.; Ash: 0.51 g.; Protein: 0.155 g.; Fat: 0.52 g.; Calcium: 5.4 mg.; Phosphorus: 17.9 mg.; Iron: 3.25 mg.; Carotene: 0.019 mg.; Thiamine: 0.025 mg.; Riboflavin: 0.13 mg.; Niacin: 0.029 mg.; Vitamin C: 4.6 mg.; Moisture: 91.9 g.

Nutritive constituents determined from 100 gram of otaheite gooseberry fruits containing 91.2 per cent moisture have been presented by the National Institute of Nutrition (I.C.M.R., Govt. of India) in Hyderabad as follows (Gopalan *et al.,* 2018).

Protein: 0.7 g.; Fat: 0.6 g.; Minerals: 0.5 g.; Crude fibre: 1.1 g.; Carbohydrates: 5.9 g.; Calcium: 6 mg.; Phosphorus: 6 mg. and provides 32 Kcal (= 133 KJ) of energy in the body.

Majumdar and Mazumdar (2001) determined pectin content of the fruits.

(Index : g = gram, mg = milligram, cm = centimetre)

Medicinal Value

Oteheite gooseberry has plentiful medicinal properties. Fruits are known to tone up liver, purify blood and stimulate appetite. Rupal and Narasimhacharya (2011) stated anti–oxidant potential of the fruits is case of fluoride toxicity.

Roots of the tree are useful to alleviate asthmatic distress. They are used to reduce cough on inhalation of steam boiled with water. They have also laxative action. Paste made from roots is used to treat psoriasis. Leaves have antimicrobial property and drinking of their decoction reduce blood pressure and have sudorific property. Poultice made with leaves relieves pain in sciatica and joint pain. Latex of the plant has puragative action. Seeds are useful in cough and cold.

Phytochemicals

Lignans, *e.g.*, phyllanthin is present in leaves, roots, stems, bark and fruits. Compounds like saponins, tannins, amine compounds *etc.* are present in different parts of the tree.

Undesirable Effect

Fresh fruits being very sour may cause acidity, flatulence, heartburn *etc.* if taken too much. Saponins, tannins and other toxic substances present in the fruits may produce harmful effect if too much fruits are taken at a time.

Post-harvest Storage

Fresh fruits keep well for 4–6 days under ordinary room temperature condition. Storing the fruits with moistened cloth and in cool condition increase shelf life. At 10°–20° C and 85–90 per cent relative humidity, fruits may be stored well for 25 days or more.

Utilization

Otaheite gooseberry fruits are much utilized to make pickles in eastern India. In some countries, fruits are candied and preserved in syrup or in brine and jelly is also made, though not very common.

Family: Phyllanthaceae

Species: *Phyllanthus emblica* Linn.

[Synonyms: *Emblica officinalis* Gaertn.; *Phyllanthus mairei* H. Lév.; *Cicca embica* (L.) Kurz; *Emblica arborea* Raf.; *Diasperus emblica* (L.) Kuntze; *Dichelactina nodicaulis* Hance; *Phyllanthus glometatus* Roxb. ex Wall nom. inval.; *Phyllanthus mimosifolius* Salis.; *Phyllanthus taxifolius* D. Don].

Taxonomic Serial: 504352

Common Name

In English, *Phyllanthus emblica* goes by the names as Indian gooseberry, Emblica, Myrobalan plum, Malaccan tree *etc.* In other countries the names are as follows.

Thailand: Mo kham pom. Vietnam: Chu me, Kam lam *etc.* Malaysia: Melakkha, Popok melakka *etc.* Laos: Makkham bom. Chila: Anmola, Yougan, Ganlanshu *etc.* Nepal: Amala. Bangladesh: Amlaki.

The species is known in India since ancient time. In Sanskrit literature, the fruit of the species has been referred to by various names and some of which are Dhatri phal, Amalika, Abaleha, Amrut, Sriphalam, Vayastha, Divya *etc.* In different parts of India, the fruit is now known as follows.

Northern India: Aonla or Amla. Gujarat: Maharashtra and Goa: Aonla. Kerala: Nellikka. Tamil Nadu: Nelli. Andhra Pradesh and Telangana: Rasi Usiri. Odisha and West Bengal: Amlaki, Assam and Manipur.: Heikru, Amlaki, Amla *etc.*

Genus: Phyllanthaceae

At present, the species has been placed under the genus, *Phyllanthus*. Characteristic features of the genus have been described under the species, *Phyllanthus acidus*.

Importance

In India, fruits of aonla, *i.e.*, emblica are regarded as the queen of all fruits for many attributes of them, notably vitamin C, plentiful medicinal properties and other benefits. In the Hindu religion, emblica has been given a significant status. It has been mentioned in the Hindu mythology that the emblica plant had its birth from the drop of amrita (= ambrosia) which fell on Earth during samudra–manthan, *i.e.*, churning of the sea by gods and demons (= asooras).

In the ancient texts of the Hindu religion, it has been mentioned that Lord Vishnu dwelt under the emblica tree. The fruits, twigs and leaves of the tree are considered sacred amongst the Hindus and are used in worshiping gods. In Himachal Pradesh, emblica tree is worshiped in the month of *Kartik*. The fruit is honoured in Buddhism also.

Emblica fruit serves as an excellent material in preparing a delicious fruit product, known as *morabba*.

Above all, a remarkable medicinal preparation is done with fruits in India since ancient time, which is known as *Chyavanprasha*–a product which is claimed to be a panacea.

The tree of emblica has a great merit in that, it is adequately hardy and adapts in soil which has less agricultural suitability. Wood of the tree is used to make soft type of cane furniture, firewood *etc.* Seeds are used to extract oil which has value in industries. Leaves of the tree are fed to pet animals as fodder of high food value.

Description of Plant Parts

Emblica tree has a graceful appearance. Pathak and Patkal (1993) have presented description of the tree parts.

Height of the tree is 8–18 metres but it becomes more tall in rich and well-drained soil where it may reach upto 30 metres. Trunk of the tree is crooked. Branches are highly spreading, glabrous or finely pubescent. They are 10–20 cm long and are usually deciduous. Bark is smooth and pale green. Trunk and branches of the tree shed strips of bark which is known as de–barking as takes place in guava and other species of the family.

Small leaves are simple, sub-sessile, light green in colour and are densely set along branches. They are oblong, about 2 cm long and 3 cm wide. Leaves resemble leaflets of pinnate leaves. The tree is deciduous particularly in sub-tropical climate where it sheds leaves and branchlets but the tree usually does not become completely bare without leaves and branches.

Emblica tree is monoecious and male and female flowers are borne on the same inflorescence. Proportion of male flowers is high. Karale *et al.* (1991) in Rahuli, Maharashtra studied flowering and sex–expression of emblica. They stated that the branchlets produced 59.1 per cent purely male flowers and 35.4 per cent both male and female flowers while 5.45 per cent had not borne any flower. High sex ratio of male to female flowers as 26 was observed by them.

Male flowers in emblica are light in colour and female flowers are green but later they become brown. Cross–pollination is very necessary to set fruits as inadequate pollination may result 70 per cent of the female flowers to drop shortly after flowers are set.

Fruit of emblica (Plate-23) is round to oblate according to varieties. It is light green in colour in young stage but in mature stage, becomes whitish or yellowish.

Fruit is indented at the base and each fruit has 6–8 vertical strips or furrows on its surface. Surface of the fruit is smooth and hard even on ripening. Skin is thin, transluscent and adherent to pulp. Stone is very hard and tightly embedded in the centre of the pulp. It is hexagonal in shape and contains 6 small seeds.

Origin

Origin of emblica is controversial. A large section of botanists is of the view that central part of India has been the origin while other botanists consider nativity of the species in some part in south-east Asia. Area ranging from Nepal, India and Sri Lanka has also been indicated as the origin by some taxonomists. Sham Singh *et al.* (1967) and Singh and Saxena (2008) stated that tropical Asia had been the centre of origin of emblica. Chadha (2012) commented upon that emblica could be indigenous to the Indian subcontinent.

Geographical Distribution

Emblica is grown in many subtropical and tropical countries. Khan (2009) from the V.I.T. University, Vellore in his review has stated that among many subtropical and tropical countries, emblica is grown in India, Pakistan, Sri Lanka, Malaysia and many other countries.

India has undoubtedly anchored a vast area in emblica culture and this tree of versatile use is grown almost everywhere in this country, upto an elevation of 1200 metres in the Himalayas and upto 1500 metres in the Nilgiri hills in south India.

Uttar Pradesh and other parts of northern India are by far, the most important areas where the trees are abundantly grown, either in man–made plantations or as wild. In Uttar Pradesh, the districts of Pratapgarh, Varanasi, Rai Bareli, Jaunpur, Azamgarh, Sultanpur *etc.* have the pride where emblica is grown in abundance and with superior varieties. From Uttar Pradesh, the tree has spread to Punjab, Haryana, Madhya Pradesh and Rajasthan. Thereafter, the tree has migrated to southern India, and later to Bihar, West Bengal and other eastern States.

Pathak mentioned that in 1999–2000, emblica has covered an area of 49,620 hectares in India and yielded a production of 1,50,500 tonnes.

Climatic Requirement

Emblica is primarily a subtropical tree and its performance is best in places where winter is cool but not freezing, summer is dry and rainfall is moderate. In tropical climate, the tree grows in good vigour but its quality and productivity are low. Under subtropical situation, the tree is deciduous but when grown in the tropics, it remains evergreen. Highly suitable climate for emblica seems to be north–eastern part of Uttar Pradesh as stated earlier. Although frost damages feeder roots of the tree which are largely near the surface soil and it occurs in some parts in the said agro-climatic zone but severity of the frost is not high there and that it does not occur all throughout the winter season. Trees raised from seeds are found to have greater tolerance to frost than those grown from budded and grafted trees.

Temperature tolerance of the tree is from 10° to 45° C but it is able to survive upto 0° C for some days. Warm summer temperature is said to be favourable for floral-bud differentiation, *i.e.*, anatomical flowering (Chadha, 2012). Annual rainfall of 630–800 mm is desirable for the trees. High relative humidity during growth of fruits is conducive and in absence of that, fruit drop occurs heavily (Chadha, 2012).

Soil Requirement

Emblica grows in a wide range of soil but heavy clay and poorly drained soils are not desirable. High organic matter in the soil is favourable for higher productivity of the tree. Growth and performance of the tree are not hampered much in slightly acidic (pH–6.5) to moderately alkaline (*p*H–9.5) soil but in higher soil *p*H, deficiency of some nutrients are found. The tree adapts well in low soil moisture content for which it is recommended for planting in low rainfall areas.

Propagation

A number of methods are adopted to propagate emblica in different parts of the world which are as follows.

(a) Seed Propagation

Propagation of emblica by the use of seeds is easy and the resulting trees are hardy, vigorous and long-lived but it has two drawbacks. These are firstly, the trees that are raised from seeds do not retain all characters as with the mother trees, *i.e.*, the vegetative characters are segregated as well as the fruit quality is impaired. Secondly, the trees take long time to attain maturity to produce flowers that is, their juvenile period becomes long. Nevertheless, seed propagation is very common in emblica in many parts of the world.

To propagate by seed, they should at first be extracted from the fruits. In the centre of each fruit, a hard, hexagonal stone remains lightly embedded in the pulp and within that stone, six small seeds remain present.

To take out the seeds from the stones, fully ripe fruits are to be taken and these should be dried in sun for 2–3 days. By that, the fruit pulp will be softened and then each fruit should be cut lengthwise and crosswise with a knife. On doing so, the hard stone of each fruit could be taken out. The pulp adhering the stones should be wiped out by rubbing thoroughly with a cloth. The hard stones should be broken then with a wooden mallet or a nut-cracker and the six seeds present within each fruit should be removed out.

It is advisable to sort out the bold seeds for propagation. For this purpose, water should be taken in a container and to this, little amount of common salt should be dissolved. Seeds should be kept in this solution for two minutes. Only the bold seeds which will sink at the bottom should be taken for propagation, discarding those which float. One kilogram of seeds could be obtained from 1–1.25 quintal of fruits. The seeds should be sown just after extraction.

A study undertaken by Nayak (2006) in Dharwad, Maharashtra suggested that least number of seeds were required for 50 per cent germination when the seeds were scarified by dipping 30 seconds in concentrated sulphuric acid. The seeds also germinate easily by soaking them in water for 3–4 days with replacement of water at 12–16 hour interval.

Germination has been observed to be accomplished in a still better method and in easier way by soaking the seeds for few hours in 200–500 ppm solution of gibberellic acid or in a weak solution of thiourea (Dhankar and Kumar, 1996; Gholap *et al.,* 2000 a). Under the condition of Agra in Uttar Pradesh, Bhadauria *et al.* (2000) studied that seeds of emblica which were treated with biofertilizer, *Azopirillium* slurry for 24 hours increased germination to 79 per cent as against 50 per cent that had occurred in untreated control.

Just after soaking the seeds in water or with any germination promoting chemical, they are to be sown without delay. Seeds may be sown in beds (Chadha, 2012) and to make the beds, soil should be thoroughly pulverized and mixed well with adequate amount of well-rotten cow–dung, little sand and wood–ash. The beds may be of any size but they should be raised from ground level to 6–10 cm. Paths should be maintained between beds for drainage and walking. Seeds should be sown at 1–2 cm depth and after sowing, beds should be sprinkled every day.

After 10–15 days of sowing, the seedlings will become 2–4 cm long. At this stage of growth, the beds should be thoroughly watered and each seedling should be carefully uprooted with a stick. The seedlings thus, uprooted should be transplanted in a similarly made, second nursery–bed at 2 cm depth and maintaining a spacing of 8–10 cm. After transplanting, beds should be regularly irrigated lightly.

After 1–1 ½ months, the seedlings should again be uprooted and transplanted in small pots. The potted plants should be maintained in semi–shade situation with regular watering. They should be planted at the plantation site during rainy-season.

Instead of sowing in beds, the seeds may also be sown in perforated polythene bag (30 cm × 15 cm) filled with mixture of pulverized garden soil and well-rotten cow–dung and maintained over a layer of sand.

A better method may be to sow seeds in both ends opened polythene sleeves, filled with a uniform mixture of garden soil, well-rotten cow–dung, leaf–mould, wood–ash, sand and neem–cake. Polybags containing seedlings should be maintained vertically over a layer of sand and regularly watered (Mazumdar and Ali, 1998).

(b) Budding and Grafting

A number of budding and grafting methods are adopted in emblica in India and other countries. Some of these methods are stated below:

(i) Raising of Rootstock

Preparation of rootstock plants is pre-requisite for budding and grafting operations. Rootstocks are prepared from seedlings of same species, *i.e.*, emblica. To raise rootstock seedlings, seeds should be obtained from fruits borne by robust and hardy trees which grow as wild.

Seedlings for rootstock should be made in the same way as described above. The seedlings should be maintained in pots and in partial shade. Budding or grafting should be done on the rootstock seedlings after 5–6 months of their growth when their stem thickness would be 7–10 mm. But in most cases, it is advisable to do the operation during the end of rainy-season. Singh and Saxena (2008) stated that emblica could be successfully propagated by budding when carried out in the month of June.

The time of budding or grafting differs according to locality. Nayak (2006) from Dharwad, Maharashtra observed in his studies that maximum success of grafting to 87.5 per cent had occurred when the operation was done in the month of November and success was least and had been 16.52 per cent when carried out in June. In parts of northern India, budding and grafting give high success when these are done in September–October or in February–March.

(ii) Selection of Scion Materials

Scion materials that are to be budded or grafted depend on the variety of emblica. Nevertheless, in choosing the materials from the selected variety (ies), condition of the trees should be critically considered. The trees(s) from which scion material should be obtained should be highly vigorous and free from the attack of pests and diseases.

It is sometimes observed that some trees of emblica have biennial or irregular bearing habit, *i.e.*, they do not produce flowers every year or do not produce flowers in equal abundance in every year. Trees having such demerits should not be used.

(iii) Budding

There have in fact, been a number of budding methods that give success in emblica and many of these methods are adopted widely in India and other countries.

Under north–Indian condition, patch–budding, ring-budding, Forkert-budding and modified Forkert–budding, are followed by the nurserymen. It is generally observed that patch–budding and Forkert–budding give more than 85 percent success.

Budding may be done from the time when the rootstocks have become 6–months old. But 1–year old plants are found to serve as better materials for bud–take and growth of the sprouts. Which time of the year should be considered best for budding depends on rainfall in the locality, advent of winter season and other conditions. In north India, the period from middle of May to September has been found to give success in patch–budding and modified Forkert–budding.

At the Allahabad Agricultural Institute, Shanker obtained high success by adopting chip-budding methods. In this method, the bud-piece contains a chip (= small piece) of wood (hence, the name) which is kept attached with the inner side of the bark. This wood chip saves the bud-piece from breakage during the operation. Slightly older rootstocks that are 1–1 ½ year old appear to be more suitable materials.

Possible comparison of different methods and time of budding have been made by a number of workers (Pathak *et al.,* 1991; Keskar *et al.,* 1991; Singh and Parmar, 1998).

In an observational trial, the present author compared efficiency of the T-budding method with a modified method designed by him. The methods described by the author earlier (Mazumdar, 2007 a) is a little modification of T-budding method and had been observed by the author to give success in other plants. The methods termed as 'cross-budding' consisted of giving a cross-like cut on the rootstock instead of "T"-shaped cut to make the matrix. The bud-piece is taken out from the scion-shoot by giving cuts from both ends so that it comes out giving oval appearance at both ends of it. On the rootstock, the matrix is made by giving cross-cuts on its bark, instead of a T-shaped cut. To facilitate easy movement and proper placement of the bud-piece on the matrix, small oblique cuts are also given on both sides of the upper end of the flat cuts on the cross-cut of the rootstock and this may be done with a small pliers or a nail-clipper. T - budding and the modified T-budding methods were compared by the author (Mazumdar, B.C.) at the Calcutta University Farm at Baruipur in West Bengal, using several emblica plants which were 8–10 months old. Results obtained indicated that the designed (modified) T-budding method had not given any extra benefit nor the rate of success had been greater than T-budding, although time taken to carry out the operation had been somewhat lesser.

It should be noted that in any of the budding operation which is performed, the rootstock stem above the operated part should be cut off when the sprout has grown to at least 4–5 cm and this should be cut in instalment at interval of 3–4 days, giving V-shaped cuts.

(iv) Grafting

Grafting is usually a less common method in emblica as compared to budding. Percentage of success is found to be low in approach-grafting (= inarching). Choudhary *et al.* (2016) in the Agricultural University at Parbani in Gujarat studied softwood grafting in different emblica varieties. It was observed by them that minimum days for sprouting of buds and maximum graft success of 91.1 per cent had occurred in the grafts of the NA-7 cultivar.

(c) Stem-cutting

Stem-cuttings are ordinarily not successful in emblica. Success had been reported by Sen *et al.* (1990) by pre-conditioning stock plants and use of growth regulators in semi–hardwood cuttings.

(d) Micro-Propagation

Some workers have achieved success in micro-propagation of emblica and in most cases, MS medium had been observed to be the best culture medium (Sehgal and Khurana, 1985; Kant and Verma, 1997; Gupta *et al.,* 1997; Verma and Kant, 1999; Goyal and Bhadauria, 2007).

Culture

Emblica is a hardy fruit tree and does not require much care. But cropping the tree with cultural practices enhances its productiveness. Some of the aspects of growing the tree are stated below.

(i) Planting

Plants are well-established when set in rainy-season although spring planting may also be done, if 2–3 light irrigations could be provided after planting.

Plants should be set at a spacing of 6–7 metres but it should be widened to 10–11 metres in poor soil. Planting should be done in pits of 1–1.25 metre cubes, well-filled with organic manure, wood ash *etc.* along with 0.5 kg of neem cake. If the soil is alkaline, 5–7 kg of gypsum and 20–25 kg of sand should also be thoroughly mixed. Pits should be regularly watered every day and any depression in a pit, if noticed should be filled with soil–manure mixture. Planting should be done in the filled-in pits after 10–15 days.

Unfruitfulness appears to be a problem in some varities of emblica. Hence, to ensure pollination, planting several varieties together is advisable to secure pollen grains.

(ii) Manuring

Emblica trees respond very well to the application of manures and fertilizers. A general recommendation is to apply 10–15 kg of farmyard manure, 100 gram of nitrogen, 50 gram of phosphorus and 80 gram of potash to a 1–year old plant. The dose should be increased every year in same proportion and upto the 9th year. The dose received in the 9th year should be maintained in subsequent years till the plantation exists. The entire amount of farmyard manure and phosphorus along with half of nitrogen and potash should be applied before flowering in January–February and the remaining part in August–September. Sarkar *et al.* (1985) studied on nutritional aspect of the Banarasi varity of emblica and stated the benefits of nutrients.

(iii) Irrigation

Emblica trees are able to withstand drought for a considerable period and hence, regular irrigation is avoided by growers in mature trees. However, fortnightly irrigation in summer and during development of fruits is of great help for the grown–up trees. Irrigation is necessary at early stage of growth of the plants. If planting is done in January–February, light irrigation should be given at weekly interval for a month and fortnightly irrigation for the subsequent two months. If possible, drip irrigation at early stage of growth of the plants should be done as this is of much help.

(iv) Training

Branches of emblica tress are long and drooping and fruits are borne heavily on these branches. Hence, encouraging long and drooping branches is desirable for higher productiveness. This could be done by adopting the system of training which is known as open-centre, *i.e.*, *vase* system.

However, breaking of branches is a problem in emblica trees which is popularly known as *limb breakage*. Breakage becomes greater when fruits are borne heavily. Hence, supporting the branches with bamboo poles which is known as propping should be done in such case. Many horticulturists are therefore, of the opinion that instead of central–leader, modified central–leader system of training would be better in emblica, in which case the branches do not become very long and the height of the tree does not assume very tall appearance. This should be done by regulated trimming of the branches and top of the trees during the first few years of growth.

Artificially bending of the drooping branches is also a good practice in emblica as is done in guava (Hayes, 1970). To do this, long branches should be tied with ropes, pulled downward and the other ends of the ropes should be firmly fixed to the ground with strong iron hooks. This should be done in such a way that the bent down branches look like arches.

(v) Pruning

Regular annual pruning is not necessary in emblica. But dead, dried, broken, infected and weak branches as well as those branches that closely overlap one another should be regularly removed when visible.

Sometimes over–bearing may become a problem in emblica. In such situation, there is more chance of breaking of branches and the size of the fruits become smaller. To avoid very high production of fruits, thinning of some fruits is desirable. If it is desired, it should be done at very early stage of growth of the fruits. Thinning, in other words, forced dropping of small fruits may even be done by spraying solutions of plant growth regulators and notably with weak solution of 2–chloroethyl phosphonic acid.

Some emblica trees are found to produce insufficient shoots. Pollarding seems to be a good practice in such trees but it should not be done every year.

Sometimes some abnormal swellings, *i.e.*, bulged symptoms are noticed in some parts of the trees. Such swelled parts should be cut off whenever visible and just after cutting, fungicidal paste should be applied on cut surfaces.

Regular removal of water–suckers that arise at the base of the trees should not be ignored as these unwanted growths draw up much water and nutrients from the soil but do not produce fruits. As a result, the fruit producing shoots are deprived.

(vi) Soil Management

Extensive inter–cropping may be done in emblica plantation with application of inputs during first few years of growth. Weeding and mulching would raise productivity of the fruit crop.

(vii) Top-working

Where emblica trees grow abundantly as in parts of Uttar Pradesh, many trees are found to grow here and there as wild and in jungles. These trees produce fruits which are not of desirable quality. Fruits of inferior quality are also grown from many seed–propagated trees. It is desirable that these trees should be upgraded, in other words, changed to varieties that would bear fruits of desirable standard. This can be done by a horticultural practice which is known as top–working.

The method to be employed to top–work a tree has been described in the Chapter on *Ziziphus mauritiana*. The same procedure should be adopted in emblica also.

Pests, Diseases and Physiological Disorder

The common insect pests, diseases and the disorder in emblica are as follows.

(a) Insect Pests

Most serious insect pest in emblica is no doubt, bark–eating caterpillar (*Indarbela tetranois*). This insect pest is serious in many fruit crops also, *e.g.*, in *Psidium guajava,* many citruses, *Syzygium cuminii etc.*

The caterpillars make tunnels inside the bark and feeds on sap. As a result, the shoots become hollow and hence, they break easily. The insect attacks largely on new shoots that come out in April after dormancy. Effective control measure is to plug the holes when visible with plaster of paris or cotton soaked with kerosene. To control the adult insects, the trees may be sprayed with 0.05 per cent Endosulfan or 0.8 per cent Dimethoate.

Shoot gall–maker is sometimes a serious pest in emblica. Caterpillar of the insect (*Betousa stylophura*) bores into the aerial part of shoots after rainy-season and feed on the inner tissue. As a result, growth of apical part is arrested and the side–shoots that come later develop galls. To control, the affected parts should be cut off and burnt. Spraying 0.05 per cent Monocrotophos before and during rainy-season may be done to control the insect. But it may be noted that gall insect is in fact, very difficult to control.

(b) Diseases

Leaf rust is the serious disease in emblica which is caused by the fungal species, *Ravenillia emblice*. The fungus infects after rainy-season and the symptoms are brown pustules that develop on leaves and fruits which later turn black. Leaves and fruits are shed. To control the disease, Indofil M–45, or Dithane M–45 (0.3 per cent) may be sprayed to the trees at fortnightly interval after rainy-season.

Fruit rot (blue mould) occurs in mature fruits or after harvest. The disease is due to the attack by species of *Penicillium* fungus. Fruits show water–soaked spots on them which become swelled. Initially the spots are yellowish and later they become brown and black. The disease infection may be prevented by pouring in solution of common salt or borax after harvest but complete prevention would hardly became possible.

(c) Physiological Disorder

Emblica fruits are sometimes threatened by a non–pathogenic, physiological disease which is known as necrosis. The symptoms are at first not visible because they start to develop from the centre of the fruit. Gradually, the rotting extends outside and the symptoms appear on fruit surface. Hence, the disorder is also called internal necrosis of emblica fruits. Symptoms look yellowish at first and later they become brown to black. The dark spots become gummy and corky.

A variety of emblica known as Francis (Syn. Hathijhool) is known to be most susceptible to this disorder but fruits of the famous variety, Varanasi are also susceptible. Spraying the trees with 0.6 per cent borax solution three times at 10–day interval during September–October is recommended which may prevent the appearance of symptoms to some extent.

Harvesting

Emblica trees raised from seeds attain maturity to produce flowers usually not before 10 years. Budded trees produce flowers after 4 or 5 years of growth.

The trees start flowering from the beginning of spring season and extends upto the early part of summer. In most parts of south India, emblica trees produce flowers twice in a year, which are in February–March and then again in June–July. Early flowering of February–March gives a small harvest of fruits. In high rainfall tropical zones in south-east Asian countries, some flowers and fruits are produced almost all throughout the year.

Emblica fruits are harvested in winter but the actual time of harvest depends on varieties and prevailing weather condition. Under the condition of Uttar Pradesh, fruits of the Desi variety which are small in size and very early maturing become ready for harvest during the middle of October to early part of November. Early varieties, *e.g.*, Varanasi, NA–9, NA–10 *etc.* attain harvest–maturity from the end of November to early December. Fruits of the mid–season varieties such as Amrit (= NA–6), Neelam (= NA-7), Francis (= Hathijhool) *etc.* are harvested in December–January. Fruits of the late–maturing variety, *viz.*, Chakaiya are harvested from the end of January to the middle of February.

Determining proper maturity of fruits for harvest is a very important operation in emblica. This is because, prematurely harvested fruits are of poor quality, very sour, very hard in texture and do not ripen after harvest. Delayed harvesting is also not desirable as the fruits become more soft, lose colour and have low shelf-life.

Proper maturity for harvest of fruits is judged by observing their sizes and colour. The growers are of the opinion that fruits should be harvested during early or late hours of the day and not during noon.

Harvesting is normally done by collecting the fruits which have naturally fallen from the trees every day. Shaking the branches vigorously to force dropping of fruits is done by many growers. This is not desirable as vigorous shaking of branches may cause the branches to break which are highly breakable. Harvesting may also be done with a sickle–shaped, movable knife which is attached with a cloth bag and tied to a bamboo pole as described in the Chapter on *Dimocarpus longan*.

Yield

Yield of fruits depend mainly on varieties but soil fertility, rainfall *etc.* are also important factors. Yield may be 100–200 fruits per tree and 15–20 tonnes in a hectare area of plantation. A fruit weighs 50–70 gram.

Varieties

Although plentiful types of emblica are grown in different parts of India, named varieties of the fruit crop are not many. Most of the commercially important varieties are selected or evolved in Uttar Pradesh and at the Narendra Deva University of Agriculture and Technology at Faizabad in that State.

(1) Banarsi (syn. Varanasi)

Fruits of this emblica variety are highly popular. The tree of the variety appears to have originated in the Varanasi district of Uttar Pradesh. Fruits of the variety mature very early like those of the Desi variety. Colour of the fruit surface is light green and this becomes yellowish on ripening.

Fruit weighs 25–30 gram. Shape of the fruit is oval which is 3–4 cm long and 4.5–5 cm wide. Fruit skin is thin. Six segments of the fruits are paired. Pulp is fibrous, moderately juicy and is transluscent. Stone is thin. Yield performance of the variety is relatively low for poor fruit–set.

Shape of the tree is semi–spreading. Trees have biennial or irregular bearing habit but this can be overcome to a considerable extent by interplanting with other varieties in the same plantation.

Juice of the fruit is moderately astringent. Total soluble solids of the juice is 10 per cent and vitamin C content (on pulp basis) is around 600 mg per 100 gram of pulp. Fruits of this variety are highly preferred to make *morabba*, which is a preserved fruit product.

(2) Francis (syn. Hathijhool)

Fruits of this variety are of mid-season maturity. Size of the fruit is very large, shape is roundish to oval and bulged at the apex. Surface colour of the fruit is yellowish. Tree is vigorous and of regular bearing habit. Fruit is preferred to make *morabba*.

Fruits of this variety have a drawback in that, they are susceptible to be affected by the physiological disorder, known as necrosis which has been described earlier.

(3) Chakaiya

This is a late-maturing variety of emblica. Fruit is smaller and weighs about 28 gram including stone weight of 3 gram. The trees are heavy yielder and of regular bearing habit. Fruit juice has 10.6 per cent total soluble solids. Vitamin C content of the fruit is high and is stated to be 793 mg per 100 gram of pulp.

Trees of this variety are very much suitable to act as pollinizers and hence, are preferred for interplanting with trees of other varieties to ensure pollination in them.

The above three varieties are highly popular in Uttar Pradesh and also at the country level. Other important varieties of Uttar Pradesh are as follows:

(4) Krishna (syn. Narendra Aonla-5)

The variety appears to have originated as a seedling selection of the Banarasi variety and is an early maturing variety. Fruit colour is yellowish with red blush. Size of the fruits is similar to those of the Banarasi variety but weight is heavier and stone is bold. Vitamin C content of the fruit is high and is stated to be 824 mg per 100 gram of pulp and the total soluble solids of the juice is 11 per cent. Productiveness of the tree is low.

(5) Narendra Aonla-10 (syn. Agra bold)

This is stated to be a seedling selection of the Banarasi variety. Tree is not very tall. Fruits are large-sized and flattened in shape. They mature early in the season which is from the middle of October to early November. Fruits are susceptible to the disorder, necrosis like those of the Hathijhool variety but the disorder is of mild type in fruits of this variety.

(6) Kanchan (syn. Narendra aonla-4)

This is also considered to be a seedling selection of the Chakaiya variety. The tree is tall and fruits mature late in the season. Trees are of regular bearing habit and are of high productive capacity. Fruit weighs 30 gram and has high fibre content. Vitamin C content of the fruits has been stated to be about 785 mg per 100 gram of pulp. Fruits are preferred for pulping. The tree is suitable for growing in semi-arid zone.

(7) Narendra Aonla-6

Like the Kanchan variety, this variety is also known to be a seedling selection of the Chakaiya variety. Tree is tall, semi-spreading, late maturing and a branchlet of the tree produces more than 10 female flowers. Fruits are medium to large in size, attractive in appearance, have low fibre content and are suitable to make candy and sweet products.

(8) Narendra Aonla-7

The variety is considered to be a selection of the Hathijhool variety. Trees are tall, precocious and thus, attain maturity to produce flowers after three years of planting. Yield potential of the tree is very high. Fruit weighs 40 gram or more and has very high vitamin C content. Vitamin C of the fruit has been stated to be high and 977 mg per 100 gram of pulp and total soluble solids, 12 per cent. Fruits have suitability for processing.

(9) Narendra Aonla-9

Fruits of this variety mature early which is from middle of October to early November. Vitamin C content of the fruits is fairly high.

In addition to the above varieties, there are many other famous varieties also. Some may be named as Anand-1, 2 and 3 of Gujarat, Bhavasagar (BSR-I), Lakhshmi-52 *etc.*

It may be stated in this connection that although not recognized as varieties, a number of types of emblica are grown in many parts of India and some of them produce good quality fruits.

A wild type is known which grows as wild in the low altitude area in the Himalayan mountain. The tree is cold hardy and a heavy yielder. Some types which go by the names as green-tinged fruits, red-tinged fruit, white-streaked fruits *etc.* (Sham Singh *et al.,* 1967) are used for drying and making products.

Mode of Consumption

Morabba (preserve) and other preserved products are although made from the fruits of emblica, fresh fruits are also relished with salt and black pepper. Taking slices of the fruits after dipping in salt water or syrup is also preferred by many.

Quality and Nutritive Value

Emblica fruit is considered as the queen of all fruits in respect of vitamin C content. However, fruits are known to have many other qualities.

Physico-chemical studies on quality and nutritive value of emblica fruits have been done by many workers. Singh *et al.* (1984) compared fruits of Banarsi and Desi varieties in respect of physical characteristics, soluble solids, acidity and vitamin C. Yield behaviour and fruit quality of three emblica varieties have been studied by Sharma *et al.* (1989) under Haryana condition. Mazumdar (1986) reported results of analytical studies done with four types of emblica grown in West Bengal. Evaluation of five varieties in regard to potentiality of fruits in making preserved products has been made by Singh and Pathak (1987).

Morton (1987) has presented in her book the report of analytical studies of emblica fruits done in Thailand. She commented upon that analytical studies done in various countries had shown wide differences on vitamin C content of emblica fruits. A study undertaken by Guha and Mazumdar (1996) made it apparent that vitamin C content of emblica fruits differed according to layer of fruits as well. Shankar *et al.* (1999) evaluated seven types of emblica fruits in Tamil Nadu.

From the V.I.T. University in Vellore in Tamil Nadu, Khan (2009) reviewed the retrospective studies on emblica and put forward the results on composition of the fruits. Nutritional value and chemical constituents of fruits have also been reported by Ekta Singh *et al.* (2011) from the Banasthali University in Rajasthan.

From the National Institute of Nutrition (I.C.M.R. of the Govt. of India) in Hyderabad, nutritive constituents of emblica fruits as observed in 100 gram of pulp containing 81.8 per cent moisture has been stated as follows (Gopalan *et al.,* 2018).

Protein: 0.5 g.; Fat: 0.1 g.; Carbohydrates: 13.7 g.; Crude fibre: 3.4 g.; Minerals: 0.5 g.; Calcium: 50 mg.; Phosphorus: 20 mg.; Iron: 1.2 mg.; Sodium: 5 g.; Potassium: 225 mg.; Carotene: 9 mcg.; Ribiflavin: 0.01 mg.; Niacin: 0.2 mg.; Vitamin C: 600 mg.; Choline: 256 mg. It has been stated that 58 Kcalorie (= 242.67 KJoule) of energy is obtained in the body on consumption of 100 gram of fruit. Besides, nutritive value, 100 gram of fruits has been stated to provide 5.8 gram of Insoluble and 1.5 g of Soluble dietary fibres in the body.

(Index : g = gram, mg = milligram, mcg = microgram).

Physico-chemical constituents of a type of emblica fruits grown in West Bengal have been studied by Mandal and Mazumdar (1995).

Presence of tannin in a fruit is known to protect its vitamin C against loss and it may be mentioned that as early as 1954, Jain and Lal reported 5.02 per cent tannin in emblica fruits. Thus, much of vitamin C in emblica fruits is protected from loss in post-harvest condition due to the presence of tannin.

Medicinal Value

Plentiful medicinal properties of emblica fruits had come to the knowledge of the human beings in ancient time. India could be said to be pioneer in which country, the various health giving properties have been extensively studied and mentioned in *Charak–Samhita* and other medico–scientific treatise that came to light.

Emblica fruit is highly merited for its efficacy to keep at bay and cure the dreadful disease, scurvy for its high vitamin C content. Apart from this effect, fruits have multitudes of other medicinal properties also. Some of these may be stated as therapeutic effect on disorders of liver, kidney, heart, skin, hair, teeth *etc.* Not only fruits but leaves, milky sap of the tree, root and seed oil are known to have therapeutic properties against dysentery, diarrhoea, diabetes, skin diseases, asthma *etc.* Krishnaveni and Mirunalini (2010) mentioned that in ayurvedic and unani medicines, almost all parts of the emblica tree are used.

Extensive studies have been done to examine the effect of emblica in treating health abnormalities of diverse types, such as renal importment (Tariq, 2007), diabetes (Qureshi *et al.,* 2009), pancreas infection (Sidhu *et al.,* 2011), high cholesterol (Jacob *et al.,* 1988), some types of cancers (Ngamkitidechakul *et al.,* 2010) *etc.* Contraceptive property of vaginal application of fruit paste of emblica has been stated (Mazumdar, 1989).

Plenty of other research results on therapeutic role of emblica have been on record and the exhaustive review made by Khan (2009) in this line is worth mentioning.

Phytochemicals

A large number of chemical compounds have been identified and isolated from fruits and other parts of trees of emblica and few are mentioned here.

Morton (1987) has mentioned in her book compounds like tannin, phyllemblin, gallic acid, fatty substances, ellagic acid, emblicol *etc.* Ellagitannis, pedunculagin, kaemferol and other compounds from the fruits have been mentioned by Bhattacharya (1999). Zhang *et al.* (2001), Nisha *et al.* (2004) and Bajpai *et al.* (2005) reported quercetin, phenols, ellagic acid, kaempferol, geraniin, furosine and other compounds. Zhang *et al.* (2001) reported phyllanemblinins A–F and new ellgitannins. The important compound, gallic acid has been determined by Sawant (2010) from dried fruits of emblica.

Constituents in Developing Fruits

Acidity, tannin, fibre and some other compounds decrease with development of fruits. Balasubramanyan and Bangarusami (1988) reported that vitamin C content of the fruits tended to increase steadily till full maturity of the fruits.

Undesirable Effect

Emblica fruits have apparently no adverse effect but it has been reported that the fruits contain 296 mg of oxalic acid per 100 gram of pulp (Gopalan *et al.,* 2004). This amount of the dicarboxylic acid is fairly high and is liable to interfere with calcium absorption by forming insoluble salts of calcium.

Post-harvest Storage

Under room temperature condition, emblica fruits can be kept well for 7–10 days. In cold storage at 6°–9° C at a relative humidity of 85–90 per cent, fruits can be stored for about two months.

Storing fruits in brine water is a household method. Mazumdar (1986) experimented with steeping freshly harvested emblica fruits in common salt solutions of 0.1, 1.0, 1.5, 2.0, 2.5 and 3.0 per cent concentrations for 7 days with change of the solutions and replacement with freshly made solutions at alternate days. It was observed that vitamin C content of the fruits was lost by the treatments and the extent of loss was greater with higher the concentration of the brine water. Storage of emblica fruits by spraying with solutions of some chemicals has also been stated by Yadav and Singh (1999).

Utilization

By far, emblica fruits are excellent materials in making *morabbas* (preserve) of delicious taste. High quality *morabba* is now prepared particularly in the north–eastern part of Uttar Pradesh, which has high demand in the country and is also exported. Pickle and candy are also made from the fruits. In West Bengal, a product known as '*mukh-shuddhi*' is made by slicing the fruits, drying in sun and mixing with salt and spices. Morton (1987) stated that in many Indian homes, fruits are cooked with sugar and saffron for giving to children in the morning.

Family: Rhamnaceae

Species: *Ziziphus mauritiana* Lamk.

[Synonym: *Rhamnus jujuba* L.]

Taxonomic Serial: 28595

It should be noted that the Chinese jujube, which is *Ziziphus jujube* Mill. is not the same species as Indian ber, *i.e., Z. mauritiana*. The tree of the former species is small, upright growing, deciduous and can withstand very low winter temperature.

Indian bers have on the other hand, several species. Among these, commercially most important species is *Ziziphus mauritiana* Lamk. Other Indian species which are of lesser importance are several in number. Among those, the common ones are as follows.

(i) *Ziziphus rugosa* Lamk.
(ii) *Z. vulgaris* Lamk.
(iii) *Z. nummularia* (Burm. f.) Wight & Am.
(iv) *Z. oenoplia* Lamk.
(v) *Z. rugosa* Lamk.
(vi) *Z. sativa* Gaertn.
(vii) *Z. xylopyrus* Willd.
(viii) *Z. fumiculosa* Buch.–Ham.
(ix) *Z. spina* Christi Willd.
(x) *Z. oxyphylla* Edge.
(xi) *Z. glabrata* Heyne.

Common Name

In English, ber is referred to as Indian date, Indian plum, Indian jujube, Chinese date, Sour jujube, Dunks *etc.* In French, the name is Cottony jujube, in German as Filzblattrige jujube and in Spanish, it is called Ponsigue.

In Sanskrit, ber is known as Badri, in Hindi and Punjabi as Ber, in Marathi as Bor, in Tamil as Elandhai, in Telugu as Regu, in Bengali as Kul, in Assamese as Bogari and in Manipuri as Boroi.

Botanical Family

Ber comes under the family, Rhamnaceae belonging to the Order, Rosales. The family is also called Buck-thorn family or Ber family. The species of the family are trees and shrubs and some are vines. There are about 55 genera and 950 species in the family. They usually abound in sub-tropical and tropical countries.

Leaves of the species are simple and alternate, spiral or opposite. In many genera, leaves are modified into spines. Flowers are symmetrical and have 5 sepals and 5 petals. Petals are yellowish, whitish, greenish, bluish or pink. Stamens are 4 or 5 and the ovary is superior. Fruits are usually berries but may be drupes or nuts.

Genus: Ziziphus

The genus *Ziziphus* contains 40 species or more. They are shrubs or trees and have spines. The plants are grown best in subtropical climate. Leaves are entire, alternate with distinct basal veins and measure 2–5 cm in length. Flowers are small and greenish yellow in colour. Fruit is drupe, globose, oblong or ovate. Fruit peel is brownish, pinkish, red or deep red in colour. Some species are deciduous.

Importance

In India, ber is called "Poorman's fruit" (Sharma *et al.,* 2012). The tree has many importance. Fruits of some varieties are choicable. Timber is strong, hard, durable and is used in making agricultural implements, boat ribs *etc.* Leaves are used to rear lac insects and to feed camels, goats and poultry birds. Many good varieties of ber have been developed in India by breeding procedure and area under this fruit crop is increasing by growing these varieties. There is a report that economic status has much improved of the small and marginal farmers by growing ber in eastern Rajasthan.

Description of Plant Parts

Ber is a spiny, bushy shrub or tree. Tree is evergreen but where winter is prolonged and severe, it becomes deciduous. Height of the tree or shrub is 2–6 metres but may be upto 12 metres. Tap-root is deep. Trunk is 40–50 cm in diameter and the bark is rough. Branches are dense and drooping. Old stem assumes blackish colour.

Leaves are simple, alternate, 2–9 cm long, 1.6–4 cm wide, ovate and are finely toothed at their margins. They are dark green at upper surface and whitish and pubescent at the under surface. Flowers are 5–8 mm across and are borne in leaf axils in clusters. They are greenish yellow, greenish or whitish. Petals are 5. Fruits (Plate-24) are 2-6 cm long. It is a drupe and contains a hard, oval or oblate and rough stone. Stone contains two sub-globular or ellipsoid seeds which have hard skin.

Origin

There are several viewpoints as regards origin of ber. Some botanists consider India as the origin. Another section of botanists suggest Indian to Malaysia is the origin. The region from southern Asia to eastern Africa and including some islands in the Indian ocean has also been suggested by some as the origin of ber. It has further been proposed that the region from southern China to Afghanistan, Malaysia and Queensland might be the origin of ber.

Geographical Distribution

Ber grows in south-east Asian countries, northern Australia, India, Bangladesh, Sri Lanka, South Africa, Mexico, South American countries and many other countries. It has naturalized in many countries. In India, ber grows all over the country and upto an elevation of 1000 metres.

Climatic Requirement

Ber is highly productive in subtropical and tropical climates but quality of fruits is better in subtropical situation. It grows in annual rainfall of 125–2000 mm but thrives and produces more suitably in areas of 300–500 mm annual rainfall. The tree is "extremely drought hardy owing to its deep root systems and other xerophytic characters" (Chadha, 2012). Temperature range of ber is 37° C–48° C as maximum and 7°–13° C as minimum. The tree is sensitive to frost.

Soil Requirement

Soil is not an important factor for growing ber and the tree grows satisfactorily in almost any type of soil. However, growth and productiveness are better when it grows in sandy loam. In some countries, alkaline soil is considered favourable. After the tree has been established, it is able to withstand soil salinity upto 21 mm hos/cm (Chadha, 2012). Drainage is by far, very important for the tree.

Propagation

Unlike Chinese jujube, Indian ber is much easy to propagate. The tree propagates easily from seeds but the resulting trees do not retain all characters of the mother trees. Seed-borne trees grow slowly and their productiveness is low. Seeds remains viable for 2–3 years. Removal of seeds from the stony endocarp is difficult and this may be done by hammering with a mallet. Chemical scarification of the endocarp by dipping in concentrate sulphuric acid for few minutes is also helpful.

It India, ber is commercially propagated by budding which is very successful. However, grafting is also done in some parts of the country. In budding and grafting, rootstocks are raised from seed-borne plants. Wild species of ber, *e.g., Ziziphus rotundifolia, Z. oenoplia, Z. nummularia, Z. rugosa etc.* are hardy rootstocks. *Z. rotundifolia* rootstock brings about dwarfness of the scion tree.

A number of budding methods are successful in ber but T-budding is most commonly adopted. Budding gives success in all seasons except winter and when it is done after 3–4 months of growth of the rootstock plant. Season of budding differs according to zones. In Punjab and Haryana, bud-take is high when done in August and also in June. June budding ensures new growths in short time. In Rajasthan, success of budding is high when it is done in July, in Maharashtra, favourable time of budding for high success is April–May and in Gujarat, budding before onset of rain is recommended.

Grafting is not much practised in India and if it is done, whip or whip and tongue grafting may be adopted. Stem–cuttings are ordinarily not successful but success is reported to be obtained by treatment with 200–400 ppm of NAA (Pareek, 1983). Stool (= Mound) and air–layering have been found to be successful (Pareek, 1983; Morton, 1987).

Micro-propagation of ber has drown attention of scientists (Rathore *et al.,* 1992; Mathur *et al.,* 1995). Ibrahim *et al.* (2012) stated that *in vitro* plant regeneration in Indian ber was achieved through *de novo* formation of meristems, cultured on MS medium with IBA, NAA or BA. According to them, whole plants were developed when the adventitious shoots were transferred to half–strength MS medium with NAA.

Culture

When mass planting at any site is desired, ber trees should be set at a spacing of 6–8 metres and it should be 10–12 metres in poor soil. The tree is not grown with much care except in organized plantations. However, application of any amount of organic manure is beneficial for higher yield. Application of 1–2 irrigations in summer season is also of great help.

If no culture is adopted, it is at least training and annual pruning that are recommended in a ber tree for higher yield performance as well as to secure fruits of larger size. Training is a very important operation to impart proper shape of the tree. It should be done during first few years of growth of a tree and *open-centre, i.e., vase* systems is satisfactory in most cases. Wide crotch angles should be maintained between branches. Lopping of branches or young stems encourages production of new shoots. In dwarf varieties which produce branches near the ground level, coppicing should be done at low height.

Pruning should be done every year after the harvest of fruits in winter. The dead, diseased or dried shoots or branches should at first be completely cut off at their bases. Where there is over–crowding of shoots in the tree, thinning–out of some shoots should be done. On the other hand, heading–back of shoots should be done where growth is sparse. Ber produces flowers and fruits on new shoots and pruning has its necessity to produce such new shoots that would produce flowers in next year.

Ber is an axillary bearing tree like guava. This means that it produces flowers and fruits in the leaf–axils. It is observed that the shoots which grow vertically upwards produce flowers only at their upper parts while the lower parts of the vertically growing shoots remain barren or are less productive. On the other hand, shoots which grow more or less in a horizontal manner or like arches produce flowers on almost all parts of the shoots. Hence, while pruning operation is done, care should be taken that vertically growing shoots should be pruned and horizontal or arch like shoots are encouraged to grow.

Water–suckers whenever found should be cut off at their bases as they are unproductive but draw moisture and nutrients from the tree. In case of over–production of fruits, some fruits should be thinned-out when they are young. Top–working is highly recommended in ber nowadays. This is done with the objective to transform a tree of inferior variety or type into a desirable variety. The method of top–working as narrated in other chapters should be adopted to top–work ber trees.

Pests, Diseases, Physiological Disorder and Phanerogamic Parasites

(a) Insect Pests

Among several insects that attack ber, the following are more destructive.

(i) Fruit-fly

The insect attacks fruits at their young stages and fruits are damaged. Several species of the insect attack fruits, among which *Carpomyia vesuviana* Costa is most serious. Eggs are laid by the insects on the surface of fruits when they are young. On hatching, the larva goes inside the fruit. After 10–20 days, the larva comes outside the fruit, leaving a hole in the fruit. Larva bores into the soil and pupate there. After two weeks, the adult fly comes out. Fruits which are attacked drop off.

Fruits which are thin–skinned are more commonly attacked. Fruits of some varieties, *e.g.*, Sanaur, Ilaichi, Safeda, Umraon, Dudhia, Meharun *etc.* are somewhat resistant to the pest. To control, tilling the orchard soil is to be done regularly. Application of 0.1 per cent Aldrin to the soil kill the pupae. During fruiting condition, the trees should be sprayed with Dimethoate, or Sevin or Malathion.

(ii) Bark-eating Caterpillar

The insect (*Indarbela quadrinotata*) lays eggs during rainy-season. A moth lays upto 500 eggs. On hatching, the caterpillars go on feeding the bark. The caterpillars pupate in summer for 3–4 weeks and the adult moths come out. To control the pest, Endosulfan 35 EC or Monocrotophos or Methyl parathion should be applied. Holes made by the caterpillars should be plugged with cotton soaked with kerosene.

(iii) Leaf Chafer

The beetles (*Adoretus pallens*) which are 10–15 mm long come at night (nocturnal). In rainy-season, they come in huge number and voraciously feed on leaves. To control, Malathion or Endosulfan or Parathion may be sprayed. Light–trap may be used to collect the beetles at night.

(iv) Mites

They form scale-like galls on leaves and twigs and are most active in rainy-season. To control, trunk and main branches should be wrapped with cloth soaked in crude oil emulsion so that the mites cannot cross the cloth. Dicofol (0.02 per cent) and Dimethoate (0.045 per cent) may be sprayed alternately.

(v) Mealy bugs and Scale insects

They suck the sap from tender leaves and young shoots and are more common in south India. To control, Pongamia oil (1 per cent) or neem oil emulsion may be sprayed.

(vi) Thrips

They feed on leaves which become curled. Phosphamidon or Carbaryl may be sprayed to control.

(b) Animal Pests

Birds, bats and squirrels cause heavy damages to fruits. Nets over trees are effective to control. Scare–crows should be installed at several places.

(c) Diseases

Some of the common diseases on ber are as follows:

(i) Powdery Mildew

White powdery mass of condia of the fungus (*Oidium zizyphii* var. *indicae*) are observed on young leaves and fruit surfaces after rainy-season. Leaves and fruits drop off. To control, sulphur dust is effective. Karathane (0.05–0.1 per cent) may also be sprayed from August to December to control.

(ii) Rust

Brown and rusty patches appear on lower surface of leaves due to infection by the fungus (*Phakospora zizyphi*). Leaves fall off. Wettable sulphur (2 gram per litre) should be sprayed to control when symptoms are visible.

(iii) Leaf–spot

The fungus (*Alternaria* species) produces dark brown patches on leaves which fall off. To control, Mancozole (2 gram per litre) may be sprayed in winter.

(iv) Fruit Rot

A number of fungal species, notably belonging to *Alternaria* and *Colletotrichum* are responsible for rotting of fruits. Grey spots appear on fruits. Dithane Z–78 (0.2 per cent) may be sprayed to control.

(v) Black Spot

Black sooty spots are found on lower surface of leaves. Leaves fall off. The fungus which is the causal organisum (*Isariopsis zizyphi*) may be controlled by spraying Mancozit (2 gram per litre) or Baycor/Topain–M (1 gram per litre).

(d) Physiological Disorder

Cracking of fruits is found in many ber trees. Although infection by powdery mildew disease may also be a cause for fruit cracking, symptoms of attack by this fungal species are not found on leaves and fruits. Fruit cracking as a physiological disorder may take place due to water imbalance of fruits, lack of soil moisture, improper soil pH or faulty nutrition of the trees.

(e) Phanerogamic Parasite

Ber trees are very often attacked by phanerogamic, *i.e.*, flower producing parasitic plants. Among other species, those species belonging to the genus, *Cuscuta* are very common. The plants are leafless, slender with yellow stem which tightly entwine the branches of the ber tree. The plant is an obligate (total) parasite. By means of haustoria, the parasitic plants suck sap from the host plants.

Control of the parasitic plant is actually difficult. Whenever the parasitic plants are found on fruit tree, they should be pulled out by hand. Dicotox, which is a 2, 4-D preparation, Methoxone *etc.* may be applied but with care and in low dose because the chemicals may cause damages to the host trees also.

Harvesting

In case of budded trees, flowers appear after third or fourth years of growth and trees raised from seeds produce flowers from fourth to sixth year. However, varieties, locality and duration of winter season play influence in regard to maturity of the trees to produce flowers. Ripening of the fruits also depends on agro-climatic condition and varieties.

Under north Indian condition, if the winter is prolonged in some years, harvest season of ber is of relatively short duration. Harvest of ber is done between the middle of February and upto the middle of May in north India, while it is between January and March in western Rajasthan and from the middle of December to March in Gujarat (Pareek, 1983).

Under south Indian condition where the winter is mild, harvest season of ber lasts from the end of November and upto April. In Maharashtra, ripening time of fruits lies between the end of September to early November. In West Bengal and other eastern States, ripening of fruits occurs from the middle of January and continues upto the middle of April, depending on varieties and duration of winter season.

Ber fruits are best ripened in the trees but they are harvested before full ripening for easy handling and storage. All fruits in a tree do not ripen at the same time. Hence, harvesting is done in 4 or 5 installments. For harvesting, the branches are shaken or beaten with bamboo sticks. The method is unripe not desirable because many fruits also fall off by doing so. Picking the fruits by hand, climbing on a self-supported latter appears to be an ideal method.

Yield

Yield of ber tree greatly differs according to varieties, soil condition and other factors. In general, a tree bears 50–100 kg of ripe fruits and yield per hectare may be 6000–7000 kg. Yield is higher in late ripening varieties.

Varieties

There are many varieties of Indian ber and these have been mostly evolved as seedling selections. Singh and Saxena (2008) have mentioned some famous varieties. These are Narma and Karaka of Varanasi, Umrao or Umri of Delhi region, fruits of which are large with high shelf-life and the small fruited early variety, Gola of Delhi, fruits of which are though excellent in quality but have very low shelf-life. The said authors have mentioned other good varieties of Delhi, *viz.*, Dandan, Kheera and Chanchal, the famous Kotha or Katha variety of Ahwar and the Mehroon variety of Jalgaon in Maharashtra. They have also brought to light a variety of Pune which bears seedless fruit but of very small size.

Pareek (1983) made a list of commercially grown early, mid–season and late varieties of ber grown in Punjab, Haryana, Maharashtra, Uttar Pradesh, Rajasthan and Gujarat. Subsequently, Chadha (2012) added to the same list some varieties of Tamil Nadu and Andhra Pradesh.

It may be stated that under Indian condition, the famous early varieties of ber are Gola, Shamber, Kaithali, Seo, Narma *etc.*, the mid–season varieties are Dandan, Meharun, Banarsi, Sanaur *etc.* and the late–ripening varieties are Umran, Ilaichi, Jogia, ZG–2–3, Aliganj *etc.*

Many other varieties or types are found to grow in different parts of India. A highly popular variety of ber of West Bengal known as Narkeli deserves mentioning. Fruits of this variety have prominent nipples at their stylar end, attractive appearance, negligible acidity and have coconut flavour. Hayes (1970) mentioned ber varieties grown in Assam and among those, the variety known as Jherberi cited by Dutta has been quoted by him which has high popularity. It has been stated that seedlings of the Jherberi variety make good rootstocks of ber (Anon., 2005).

Mode of Consumption

Fruits which have sweet pulp are relished as fresh. Those having considerable sourness are taken with salt and spices. Sometimes they are crushed with green chilli, mixed with salt and taken.

Quality and Nutritive Value

Nutritive value of Indian ber fruits has been stated by Pareek (1983) and Morton (1987). Proximate constituents of ber fruits grown in West Bengal have been stated by Mandal and Mazumdar (2007). Subsequently, fruits of the Narkeli variety of West Bengal have been studied by the said authors.

Medicinal Value

All parts of the Indian ber tree are used in folk medicine. Fruits are useful in lung infection, intestinal troubles and have anti–bacterial properties against urinary and renal bacteria. For weight gain, muscular strength and in cancer, fruits are efficacious. Fruits have antioxidant activity. Seeds are helpful in nervous disorder, insomnia, palpitation, nausean, abdominal pain and dyspepsia. Paste of seed is topically used to heal wounds. Decoction made from bark is administered in diarrhoea and dysentery. Leaves are used as poultice on wound. There is a belief that sty on eye is cured by drawing ber leaves very lightly on the sty for one or more days.

Phytochemicals

Plentiful phytochemicals are present in fruits and other parts of ber tree. Some of these may be named as flavonoids, tannins, terpenoids, flavanol, cyanogenic glycosides, phenolics, lignins, sterols, alkaloids, polygalitol, anhydro-glucitrol *etc.*

Constituents in Developing Fruits

According to Pareek (1983), growth of ber fruits could be divided into three distinct phases. This suggests that growth is most active during the first 6 weeks and the last 8 weeks while the rate is slow during the middle 8 weeks. The trend of growth remains almost similar as regard length and diameter of fruits but the total growth rate in diameter is greater during the last phase than in the first phase. Total growth rate occurs to the maximum level during the first phase and growth of fruits goes on continuing till harvest of the fruits. The author cited research references on the changing pattern of constituents in the fruits with progress of their development.

Undesirable Effect

Intake of much quantity of ber at a time may produce symptoms like hyperacidity, belching, vomiting, nausea *etc.* More pulpy fruits are found to be more harmful than the less pulpy fruits. In India, pregnant woman are advised not to take ber fruits. (There is a religious custom among the Bengali Hindus that ber fruits should not be taken before the Saraswati Puja festival occurring during the month of February. Whether any chemical which may cause harmful effect goes out of the fruits with advancement of their maturity however, draws attention in this regard).

Post-harvest Storage

Fruits of ber can be stored for 5–7 days in ordinary atmospheric condition. Occasional sprinkling of water is known to increase storage life. At 7°–10° C and relative humidity of 85–90 per cent, fruits can be stored for 3–4 weeks.

Utilization

A number of products are made with the fruits of ber and among these, the following are of more importance.

(i) Pickling

Making pickles from ber fruits is very common in India. To make this product, small and bright red coloured fruits are more commonly used. Pickles are made either in oil base or in dry form.

(ii) Dehydration

Thick-skinned fruits are preferred to make this product. On removing the stones, the fruits are dried in sun or in oven for several days. Sometimes the peels are partly cut to facilitate thorough drying.

(iii) Drying of Peels

To make this product which is called *kul-kutha*, fruits are peeled and the peels are dried in oven or in sun for a week. When the peels are thoroughly dried, they are ground mixing with spices. Small and dark red fruits are preferred to make the product.

(iv) Pulping

On removing the stones from the fruits, the pulp is made into pieces. The pieces are mixed with water and heated for sometimes and then passed through a sieve. Adequate sugar is added to the pulp, thoroughly mixed and preserved. Sometimes dates on removing stones are mixed with ber pieces and pulped.

(v) Jam

Ber is a rich source of pectin of high jelly-grade and jam is made from the pulp, usually by mixing pulp of other fruits.

(vi) Canning

Canning of ber is made in some countries although it is not done in India. Large-sized and pulpy fruits are preferred for canning.

(vii) Nectar and Leather

These are done to some extent in India.

Apart from the above products, other preserved products are also made from the ber fruits (Pareek, 1983).

Family: Rutaceae

Species: *Aegle marmelos* (Linn.) Correa ex. Roxb.

[Synonyms: *Feroni pellucida* Roth.; *Crataeva marmelos* Linn.]

Taxonomic Serial: 506380

Common Name

In English, the species, *Aegle marmelos* is most commonly known as Wood apple. Other names that are used in English are Bengal quince, Indian quince, Stone apple, Golden apple, Holy fruit *etc.*

The name, bael is also used in English nowadays which is borrowed from the Indian name. It may be mentioned in this connection that in English, another fruit in the same botanical family, *viz.*, *Limonia acidissima* is called Wood apple. It would perhaps be appropriate to call *Limonia acidissima* as Elephant apple and *Aegle marmelos* as Wood apple. Such names were also prevalent in the past and several decades ago.

However, as the name, Bael is more commonly used nowadays for the species, the name has been used in the present text to refer to the species, *Aegle marmelos.*

In French, the fruit is known as Malabar, in German, Belbaum and in Portuguese, Matun. In Indonesia, bael is called Maja, in Malaysia, Pokok maja batu and in Thailand, the names are Matun, Mapin *etc.*

In Sanskrit literature, bael has been referred to by many names such as Sreephal, Vilva, Asholam, Alimangalia, Gandhagarva, Pitaphal, Duraruha, Mahaphalam, Tripatra *etc.* In Hindi, it is known as Bael, in Punjabi as Beel, in Gujarati as Bil, in Marathi as Bael, Bili, Kaveeth *etc.* in Konkani as Gorahamli, in Malyalam as Vilwam, in Tamil as Kuvalum, in Kanada as Kumbala, in Telugu as Maredu, in Oriya as Bilva and in Bengali as Bael.

Botanical Family

Bael comes under the family, Rutaceae, which is commonly called Rue family. The family is also called citrus family as the *Citrus* species which assume high commercial importance come under this family. Rutaceae comes under the Order, Sapindales belonging to the sub-class, Rosidae of the class, Magnoliopsida. There are about 160 genera and 1600 species in the family. Some of the large genera in the family are *Baronia, Agathosma, Zanthoxylum etc.* The family is closely related to the Sapindaceae and Meliaceae families.

Majority of the species in the family are trees and shrubs and species of a few genera, *e.g., Dictamus, Boenninghausenia etc.* constitute herbs. *Citrus* is the most important genus in the family as it consists of many commercially important species.

The plants of the family have alternate or opposite leaves which are simple or compound, smooth, gland-dotted in most cases and without stipules. Inflorescence is usually cymose. Flowers are mostly bisexual and regular. Sepals and petals are 4–5, free or united, glandular and imbricate. Stamens are 8–10 and carpels are usually 4–5 and sometimes united above. Ovary is superior, usually 4–5 celled and the fruit may be hespiridium, berry, drupe, samara, capsule or follicle.

Genus: Aegle

The genus *Aegle* is monotypic, *i.e.,* it has a single species for presenting some well-marked peculiarities that may constitute it as a genus. Flowers of the genus have many ovules in each loculus. Leaves of the plants are trifoliate, stamens of flowers number 30–60, fruits are large, globose and the testa of seed is mucilagenous (Mukherji, 1974).

Importance

Few of the fruit trees have so great importance to various facets of human beings as is bael. All parts of this tree are of use to the human beings in a number of ways, from religion to industries.

In Hindu religion, leaves, fruits and twigs of the tree have indispensable role. The trifoliate leaves of the tree are known to symbolize the trident, *i.e., trishul* of Lord Shiva and the leaves and fruits are essential rituals to worship the Lord. The twigs of the tree are used in oblation. Ripe fruits of superior varieties highly appeal consumers. All parts of the tree have medicinal properties.

Jam, jelly and other delicious products are made from the fruit among which, *morabba* made from the unripe fruit is of excellence. The fruit has its use in adhesive, cosmetic, detergent and other industries. Wood of the tree is used in carving, making knife–handles, turnery, detergent, combs, dyeing *etc.* The tree is hardy and serves as a protective barrier. It is highly adaptable to inferior soils, uneven physiography, tableland and in hills and hence, is suitable for productive utilization of land of low agricultural potential.

Various aspects of bael have been extensively studied in many countries of the world. The reviews made by different authorities have presented the pertinent results of researches done on the fruit and other parts of the tree (Roy and Singh, 1979; Parmar and Kaushal, 1982; Singh and Roy, 1984; Morton, 1987; Mazumdar, 2004 a; Sharma *et al.,* 2011; Sekar *et al.,* 2011). Nowadays, many plantations of bael have come up in many parts of India and the area and production of the fruit crop have greatly increased. Hence, many horticulturists consider bael as a major rather than a minor fruit crop.

Description of Plant Parts

Tree of bael is slow growing. Height of the tree may be 6–14 metres depending on varieties, soil depth and fertility. The tree is deciduous and aromatic. Trunk is short, thick, strong and of light brown colour. Branches are dimorphic and have strong, straight and axillary throns. Lower branches are drooping. Gum exudates from ruptured branches.

Leaves of the tree are alternate, compound and trifoliate. Terminal leaflet is 5–7 cm long, 2–4 cm wide and has long petiole. Leaflets are oval or lanceolate. New leaves are pinkish.

Flowers are bisexual, greenish white, sweetly scented, actinomorphic, ebracteate, hypogynous, stalked and occur in clusters of 4–7 along young branchlets. Calyx is pale yellow, gamosepalous, 5-lobed, and pubescent. Androecium is polyandrous, numerous and basifixed. Style is terminal and stigma is capitate.

Fruit (Plate-25) is spherical, phyriform, oval, ovate, obovate, oblong or somewhat elongated. It has hard and smooth rind, which is greenish, greyish or yellowish in colour and has small dots on outer surface. Size of the fruit greatly varies and may be 5–18 cm in diameter. Weight of fruit may be 50 gram to 2 kg. Pulp is yellowish. Seeds are many. Each seed is about 1 cm long, whitish, hard, flattened to oblong and has hairs on outer surface. Seeds remain embedded in pulp. Srivastava and Singh (2000) have reported floral biology of bael from Faizabad in Uttar Pradesh.

Origin

India has been stated to be the origin of bael (Hayes, 1970; Pathak and Pathak, 1993) and origin in north India has been suggested (Pareek and Sharma, 1993; Singh and Saxena, 2008).

Geographical Distribution

Bael is grown as wild or as semi–wild in almost all parts of India, including the low hilly areas of Himachal Pradesh, upto an altitude of 1000 metres. It is much grown in the northern and central parts of India and in relatively low level in the eastern and southern parts. Systematic plantations of bael have been made in large scale in many parts in the districts of Pratapgarh, Faizabad, Kalyanpur, Mirzapur and Varanasi in eastern Uttar Pradesh. Plantations have of late, been made extensively in the Sunderbans mangrove forest surroundings in West Bengal.

The tree is grown in many other countries notably, Egypt, Pakistan, Nepal, Sri Lanka, Bangladesh, Thailand, Laos other south-east Asian countries and parts of Japan and China. At present, it has been introduced in southern parts of the U.S.A. also.

Climatic Requirement

Bael is primarily a subtropical fruit tree but it grows well in tropical countries also in accordance with varieties and edapho–climatic condition in the tropics.

However, productiveness of the tree and quality of fruits are much higher when it grows in subtropical climate where the summer and winter are distinct.

Temperature tolerance of bael tree is wide and ranges from (-) 10° C in winter and to about 48° C in summer. It grows upto an altitude of 1000 metres in Punjab and Himachal Pradesh. In the tropical mountain in south India, it grows at still higher altitude. In high rainfall area in south India, it grows upto 2000 mm annual rainfall. Long dry season throughout the year is not desirable but dry atmosphere during development of fruits is favourable for fruits of better quality.

Soil Requirement

Soil is not an important factor for bael and it is found to grow well in clayee, sandy and shallow soils of low fertility and at soil *p*H from 5 to 8. However, poor drainage hampers growth and productivity of the tree and causes fruit drop.

Propagation

Seed propagation is most common in bael but all the characters of the mother tree are not retained in the trees that are raised from seeds. However, polyembryony is observed in the seeds. Mazumdar (1990) noted polyembryony in 33 per cent of the seeds procured from fruits borne by tree that were grown in West Bengal and upto a maximum of 4 seedlings were found to be protruded from a seed. Thus, seedlings coming out of adventive embryos, if carefully detached would give rise to seedlings that would maintain all the characters of the mother tree.

Seeds of bael being recalcitrant do not retain their viability for long period. Hore and Sen (1985) reported enhancement of seed viability period by suitable treatments. Importance of sowing media for better germination of seeds has been studied by Chattopadhyay and Mahanta (1989). Nayak and Sen (1999) reported improvement of seed germination by scarifying them with acid, growth regulators or by mechanical means.

Propagation of bael by air–layering has been studied by Mukherjee *et al.* (1986) and they observed that invigoration, etiolation and application of 1 per cent of IBA gave maximum rooting response of upto 90 per cent with high survival rate of 77.7 per cent of the layers.

Budding and grafting are successful in bael when rootstocks of same or related species are used. Maiti *et al.* (1999) reported success by grafting with whip method.

Success of multiplying bael plants by micro-propagation has been reported by some workers. Varghese *et al.* (1993) observed numerous buds by application of NAA and kinetin in the culture media. Plantlet regeneration from leaf explants by the use of benzyl adenine and IAA in the MS medium has been observed by Islam *et al.* (1993).

Culture

If plantation of bael is desired to be set up, a spacing of 8–10 metres should be maintained between the trees and it may be less if vegetatively produced plants are used. Manuring the soil in winter is of great help to enhance yield performance.

If other cultural operations are omitted, training and pruning of the trees should not be ignored. Training should be done at early stage of growth of the trees and the open-centre, *i.e.*, *vase* system is desirable so that the interior parts of the trees get optimum sunshine in their mature condition. Pruning of the trees should be done annually in early part of winter by heading–back and thinning–out systems. Over–crowded shoots should be clipped off at their bases.

Leaves of bael tree often show symptoms of deficiency of nutrients. Shortening of the midrib of a leaf to give it a heart–shaped appearance is noticed as a result of calcium deficiency. Occasional liming the soil becomes necessary to overcome the deficiency of the nutrient. Zinc deficiency symptoms showing stunted growth of internodes, little leaf and leaf fall are sometimes met with particularly when the tree grows in phosphorus rich soil. Spraying the trees with 0.3–0.5 per cent aqueous solution of zinc sulphate two or three imes at 3-4 month interval should be done to get rid of zinc deficiency.

Nitrogen phosphorus and potash content in the laminar tissues of bael have been determined by De *et al.* (1996) under West Bengal condition Analysis of the leaf samples done at monthly interval from December to May showed that there has been a steady increase of the nutrients all along except potash which tended to decline from April to May. Development of the fruits had been found to mark with lesser increase of the nutrients that had occurred during winter season.

Mulching the soil of bael trees during summer and winter seasons conserve soil moisture and boost up yield and quality of the fruits. Planting bael trees of several varieties closely together is believed by the growers to enhance yield and fruit quality.

It should be noted that in order to obtain the trifoliate leaves from the tree to worship the Lord Shiva, resembling the *trident* of the Lord, it is desirable that they should be clipped carefully and tearing off with force or a stick should be avoided. This is because, tearing off leaves or shoots may cause damage to the small branches. Instead of that, a sickle–shaped clipper attached with a cloth bag and bamboo pole as described under the Chapter of *Dimocarpus longan* is a better alternative. With the help of this instrument, clipping of the desired leaves, shoots or tender branches could be done standing on the ground and pressure to the branches could be avoided. Attention should be given to this aspect as large number of leaves and twigs are clipped to worship the Hindu Lord Shiva in all parts of India, besides that, thick branches are cut off to perform *yajna* (= homa).

Pests, Diseases and Physiological Disorders

Pests and diseases are note serious problems in bael. However, some insects, *e.g., Phyllocnistis citrella, Aonidiella aurant*, lemon betterfly (*Papilio demoleus*), mormon (*Papilio polytes*) *etc.* cause damage to the leaves and branches. These can be easily controlled by spraying Malathion (0.05 per cent), Monocrotophos (0.036 per cent), Carbaryl (0.1 per cent), Dimethoate–30 EC (0.2 per cent) or Endosulfan-35 EC (0.2 per cent).

Among diseases, fruit canker caused by the bacterial species, *Xanthomonas bilvae* is common which produces water–soaked spots on leaves and shoots. To control the disease, the insect vectors should be controlled by spraying insecticides. Fungal diseases are not observed in bael. It may be mentioned in this connection that extract made from leaves of bael has in fact, been reported to have antifungal activity (Balakumar *et al.,* 2011). Further, the fruit of bael has also been reported to control rice blast disease (Sharma *et al.,* 2012).

Physiological disorder, *i.e.,* cracking of fruits is sometimes observed in bael. It is presumed that lack of adequate soil moisture may be a cause and mulching to conserve the moisture would be of help. However, zinc deficiency to the trees for this abnormality cannot be ruled out. Chilling injury is another disorder which results in producing brown spots on fruit surface during later part of fruit development and also after harvest if the fruits are exposed to low temperature.

Harvesting

Flowers appear in the bael trees after 5–6 years of growth but optimum harvest of fruits is done after 7 years. Vegetatively propagated trees produce fruits a year or two earlier. Trees produce flowers from the end of April to early June. Fruits take long time, *i.e.,* about a year to become ready for harvest. Peak season of fruiting is end of April to June but the trees produce fruits over a long period and may continue upto September particularly in tropical climate. Before harvest, the trees become leafless and thus, fruits are fully exposed to sun.

Quality of the fruits is best when they ripe in the trees. But they are harvested before full ripening and then fully ripened under postharvest condition. Each individual fruit should be harvested separately. The conventional practice is to tie a short stub at the end of a bamboo pole and fruit–stalk is pulled with the stub standing on the ground. But this may cause the small branches to break. Instead of that, clipping the stalk of each fruit with the plucking implement as stated under the heading on Culture should be used. With this device, fruits could be harvested standing on the ground and without causing damage of fruits and branches.

Just after harvest, the cut ends of fruit–stalks should be thickly coated with lime to avoid infection by micro-organisms.

Yield

Yield of fruits widely varies in accordance with variety, locality, soil type and fertility, amount of rainfall and its distribution in season, age of the trees and other factors. Yield is relatively low in vegetatively propagated trees. It is higher when the trees is trained by open-centre system. Properly done annual pruning of the trees enhances their yield performance and also quality of fruits.

At sixth to seventh year of growth, a tree usually produces 80–150 fruits. It increases in subsequent years and a tree may produce 300–400 fruits in tenth to twelfth years. Under certain conditions, some trees may produce 800–1000 fruits though such a production is rarely met with. Optimum yield is obtained from a tree upto 30–40 years of its growth and in general, 16–18 quintals of fruits could be harvested from an hectare area of organized plantation in a year.

Varieties

Varietal nomenclature of bael has much been done in India but whether all the named varieties should be recognized as true varieties needs to be ascertained. In India, bael is largely propagated from seeds. Hence, extensive variations in fruit and tree characters are met with in the offspring. By careful selection of seedlings and their hybridization, many varieties or types have been produced in different parts of the country. Nevertheless, extensive survey of the trees grown in all agro-climatic zones in the country needs exploration for their use in hybridization to produce highly desirable ba\ trees.

Among the earlier reports, mention may be made to the survey done by Jauhari *et al.* (1969) who described characteristic features of seven genotypes collected at the Fruit Research Station at Basti in Uttar Pradesh. Mazumdar (1975) described characters of five types of bael fruits grown in West Bengal. On examining 24 genotypes collected from different parts of India, *i.e.*, Agra, Kolkata, Delhi and Varanasi, Roy and Singh (1978) described the notable features and they opined that large-sized fruits had better processing quality.

Analysis of few types of bael in regard to their pectin content had been made by Roy and Mazumdar (1989) and Ali and Mazumdar (1991). On collecting several types of bael fruits grown in West Bengal, Kumri *et al.* (1997) determined physical and chemical characters and put forward their distinguishing features. Physico-chemical characters of bael fruits of some varieties or types grown in West Bengal have been studied by Maity *et al.* (1999).

At the Agricultural University at Pantnagar in Uttar Pradesh, Jaiswal *et al.* (1999) studied clonal variations in growth and yield behaviour of 8-year old budded trees of bael. At the same University, Misra *et al.* (2000) assessed early performance of 8 selected genotypes of bael. Mukhopadhyay *et al.* (2002) determined physico-chemical constituents of three types of bael fruits borne by trees grown in the Sundarbans area of West Bengal. They put forward that the fruits which had been grouped as type–A and characterized by oblong shape proved to be superior for many traits.

Under Uttar Pradesh condition in north India, bael fruits of the varieties, *viz.*, Kaghzi, Mirzapuri, Kaghzi Gonda, Kaghzi Etawah and Kaghzi Banarasi are known to have high popularity.

At the Narendra Deva University of Agriculture and Technology at Faizabad in Uttar Pradesh, Srivastava and Singh (2004) studied bael fruits of 4 cultivars, *viz.*, Narendra Bael–4, 5, 7 and 9. They observed that maximum fruit weight of 3.618 kg skull thickness of 1.98 mm, fruit length of 19.5 cm, diameter of 21.25 cm and fibre content of 335.0 gram per kg of pulp was recorded in the fruits of cultivar, Narendra Bael–7 and the values were maximum in fruits of the cultivar, Narendra Bael–4. There is necessity of making critical sensory evaluation of the fruits of bael that grow in different parts of India and making groups under grades.

Varieties of bael having chemical bitter principles in the fruits are also sometimes met with. A Rao *et al.* (2002) in Andhra Pradesh made a comparative study of bael fruits which had bitter or sweet taste.

Mode of Consumption

Unless any processing is done to make preserved product, fruits of bael in their fully ripe condition are relished as fresh. Fruits of varieties which do not contain much astringency, gum or phenolic compounds are highly favoured. In some parts of India particularly in West Bengal, Odisha and other eastern States and also in parts of Bangladesh, drink made from ripe pulp of bael is highly liked. This is made by blending the pulp with water, sugar, lemon juice, salt, black pepper, ginger, *jaljira etc.* in suitable proportion and is called *bela pana*. This is used as a cooling drink in summer. In Indonesia, there is a trend of taking the pulp in breakfast, mixing with sugar and other spices.

Quality and Nutritive Value

Analytical reports presented by the National Institute of Nutrition (I.C.M.R., Govt. of India) in Hyderabad revealed the following constituents in 100 gram of pulp of bael, containing 61.5 per cent moisture (Gopalan *et al.,* 2018).

Protein: 1.8 g.; Fat: 0.3 g.; Carbohydrate: 31.8 g.; Crude fibre: 2.9 g.; Minerals: 1.7 g.; Calcium: 85 mg.; Phosphorus: 50 mg.; Potassium: 600 mg.; Copper: 0.21 mg. and among vitamins, Carotene: 55 mcg, Riboflavin: 0.03 mg, Niacin: 1.1 mg and Thiamine: 0.13 mg. It has been mentioned that 100 gram of the pulp provides 137 Kcalorie (= 573.2 K Joule) of energy in the body.

(Index : g = gram, mg = milligram, mcg = microgram)

Medicinal Value

Bael has much greater importance for its enormous medicinal value than its use as a fruit of horticultural importance. Plentiful medicinal properties of the fruits and other parts of the tree have been amply mentioned in the ayurvedic and unani literature.

Fruits of beal is particularly efficacious in gastro–intestinal disturbances. It is of great help to cure chronic diarrhoea and dysentery particularly when diarrhoea alternates with constipation in a regular manner. The fruit has role to cure dyspepsia, hepatitis, peptic ulcers, removal of intestinal worms and in other abdominal troubles (Jauhari and Singh, 1971; Roy and Singh, 1978; Alam *et al.,* 1990).

Unripe fruits of bael are known to combat rotaviruses (Bajpai *et al.,* 2009) which cause sporadic acute enteritis in infants and also giardia. Its decoction mixed with ginger is of much help in treating piles.

Marmelosin (= Imperatorin), the bioactive coumarin of the bael pulp is dinuretic, laxative and has value as a potent antioxidant.

Leaves of bael tree are used to treat ophthalmia and peptic ulcer. Root is a remedy of belly ache, healing of wounds and its decotion reduces fever. Tannins derived from fruit, rind and leaves are of help in peptic ulcer. Seeds of the fruits have anti–allergic properties. Flavonoids and alkaloids of leaves have antioxidant properties (Bhalla *et al.,* 2012).

Phytochemicals

Many chemical compounds are derived from different parts of the tree. They have therapeutically active properties and have role in prevention and treatment of cancer as well (Baliga *et al.,* 2013).

Leaves of the tree are rich in essential oils, *viz.,* limonene, phellandrene, cineol, citronellal, citral *etc.* and contain alkaloids, sterols, aegelin, tannin *etc.* Fruit is a source of marmelosin and other coumarins, gum *etc.* Oils like palmitic, stearic, oleic, linolenic, linoleic *etc.* are present in seeds. Roots contain the coumarin marmin. They also contain psoralin, scopoletin, xanthotoxin *etc.*

Constituents in Developing Fruits

Ghosh *et al.* (1985) had determined gibberellin content in the fruits during their development. Singh (1986) at Saharanpur studied that the different constituents of fruits had maximum development upto the early part of October and then declined. Pande *et al.* (1986) reported steady increase of sugar and vitamin C in the pulp of fruits during their development but much change in the acidity of pulp could not be observed.

Undesirable Effect

Undesirable effect of taking bael fruit is not heard but excess consumption may cause belly ache, rectal discomfort and uncontrolled evacuation. In some parts of India, there is a belief that taking of bael fruit or bael leaves may bring about miscarriage in woman. However, abortifacient property of bael is not known to be on record.

Post-harvest Storage

Pulp of bael fruits is found to be easily infected by black coloured mould during storage. The fungus invades through the stalk–end and also through the rind if it is cracked. When fully ripe, fruits fall from the trees naturally on ground and are cracked. Hence, harvesting should be done slightly before the fruits have become fully ripened.

Each fruit should be plucked individually along with little portion of stalk. For this purpose, the plucking implement as stated under the heading on harvesting should be used with convenience. Cold storage of fruits has the difficulty in that, it causes a physiological disorder which is referred to as chilling injury.

A practical approach is to pluck a fruit along with a relatively long stalk and just after harvest, cut surface of the fruit–stalk should be covered, *i.e.*, painted with a thick coating of lime. This should be reported after a week. Instead of lime coating, the cut surface of fruits–stalk may also be dipped in molten paraffin just after plucking the fruits.

Utilization

In regard to products made from fruits, *morabba* comes to the mind first. This is made from emblica and other fruits but what is made from bael is perhaps unparallel. Varanasi area is known to be famous in making *morabba* but Pratapgarh, Faizabad, Kalyanpur and other parts of eastern Uttar Pradesh have also great fame in making *morabba* of excellent quality. High quality of *morabba* is also made in some parts of Bihar, Odisha and to a considerable extent in the Birbhum district of West Bengal. It should be noted that *morabba* is prepared from unripe fruits of bael.

Bael fruit contains high amount of pectin (= polygalacturonate)and of high jelly–grade. Hence, good quality jelly and jam are also be made from its pulp but the products become somewhat stiff. Drying and grinding of bael pulp is also a practice which is used for medicinal purpose. Unripe fruits are commonly used for this purpose

Family: Rutaceae

Species: *Casimiroa edulis* Llave and Lex.

Taxonomic Serial: 506385

Common Name

In English, the fruit is known as White sapote, Mexican sapote, Mexican apple or Casimiroa. In Spanish, it goes by the name as Zapote blanco or Matasano and in French, it is called Sapote blanc. The fruit has no specific name in Indian languages.

Genus

The genus, *Casimiroa* has been stated to have been named to honour Cardinal Casimiroa Gomez de Ortega, a Spanish botanist of the 18th century. The genus is small and contains not more than 10 species of trees or shrubs. Among these, 3 shrubs which are *Casimiroa pubescens* Ramirez, *C. pringlei* Engler and *C. watsonii* Engler have been stated to be important (Morton, 1987).

Importance

Fruits of *Casimiroa edulis* are of much importance as they are pleasantly sweet and have negligible acidity. Pulp of the fruit is creamy white in colour and has resemblance of a mixture of custard apple and banana in flavour and taste. Peel of the fruit is very thin and is taken with pulp without its removal.

Wood of the tree which is yellow, compact, fine grained and dense is used to make art objects and small furniture in central American countries. Insecticide is made from the seeds. Fruit and other parts of the tree have medicinal properties.

Description of Plant Parts

The tree reaches a height of 5–6 metres and is evergreen although it may be deciduous when stress is felt by it. Branches of the tree are drooping and the bark is thick. Leaves are alternate, palmately compound with 3–7 lanceolate leaflets which are smooth or hairy. Petioles are 5–8 cm long and pubescent. Newly emerged leaves are reddish in colour.

Flowers are borne in terminal or axillary panicles with 50–150 flowers. Normally, flowers are bisexual but may be unisexual due to aborted stigmas. They are pale green to creamy colour. Pollen grains greatly vary in flowers. Fruit is spherical, oval or ovoid in shape, 5-lobed, 6–12 cm wide and 10–12 cm long. Skin of the fruit is very thin, yellowish or greenish and pulp is creamy white in colour. Fruit contains 1–6 seeds which are hard, oval, 3–5 long and white in colour.

Origin

White sapote tree is believed to have originated in the highlands of Mexico and central America. Nativity of Wooly leaf sapote has been stated to be in the neighbourhood of Costa Rica.

Geographical Distribution

The tree grows as wild or sometimes planted in the central American countries. In Mexico, it is grown as shade trees in coffee plantations. It is grown in Costa Rica, Guatemala, Caribbean countries and introduced in South Africa and New Zealand. Of late, it has been introduced in India and performance is found to be satisfactory in this country.

Climatic Requirements

White sapote trees grow well in tropical and subtropical countries but fruits of better quality are obtained when grown in subtropical climate. It grows upto 1000 metre altitude in tropical mountain. Trees withstand low temperature of upto 10° C for short period and mature trees have greater tolerance. Bright sunshine for long hours is necessary for better quality fruits. Wooly–leaf sapote trees are less hardy to adverse climatic condition.

Soil

Trees grow in any type of soil provided drainage is good. They are drought resistant and grow at a *p*H of 5–7.5. In saline soil, leaves become chlorotic and they fall off.

Propagation

Seed propagation is common to propagate white sapote. Budding and grafting are also done on rootstocks belonging to the same species or on *Casimiroa pringlei*. Success of micro-propagation of white sapote has been reported by Sabour *et al.* (2002) from Egypt by the process of effective sterilization and use of growth regulating chemicals.

Culture

White sapote trees are grown practically without any culture. If organized planting is desired, the trees should be spaced at 6–8 metres. Application of organic manure is of great help to the trees. Irrigation is highly beneficial and deep irrigation is preferable, particularly in summer. Trees may become vary tall with less spread of side–branches. To avoid this, top of the tree and the branches should be cut back. Full sunshine for many hours in the day is required during development of the fruits. Building up ideal candy of the tree by efficient training and pruning practices is a great necessity for the tree.

Pests and Diseases

Among insect pests, fruit-fly causes heavy damages to the fruits. To control, Endosulfan 35 EC 0.035 per cent or, Malathion 50 EC 0.05 per cent should be sprayed. Aphids may attack new shoots and to control, Monocrotophos–36 SL or Dimethoate 30 EC may be sprayed. Stem of the tree is attacked by mealy bugs. To control, Endosulfan 30 EC may be sprayed to control. Banding the base of the trunk with polythene is also effective. Among diseases, anthracnose is often noticed. To control the disease, copper oxychloride–0.25 per cent may be sprayed. Among animals, birds, bats and squirrels cause much damage to the fruits.

Harvesting

Fruits ripen after 6–9 months from flowering. In Mexico, flowers appear in January–February and fruits are obtained from June to October. Fruits are best ripened in the tree. Yonemoto *et al.* (2006) stated index for harvesting time of the fruits in Japan. Harvesting should be done with great care to avoid damage of the ripe fruits.

Yield

Yield of white sapote varies greatly on varieties, soil fertility and other factors. Securing 1000 fruits in a year could be considered to be a good yield.

Varieties

White sapote has many varieties and hybrids. Some may be named as Lemon gold, Maltby, Chestnut, Pike, Fiesta, Wilson, Cuccio, Louise *etc.* A large number of clonal selections have been made in Florida and California in U.S.A. over many years and some may be named as Coleman, Chapman, Dade, Gillespie, Golden, Blumenthal *etc.*

Mode of Consumption

Fruits of white sapote are very sweet and total sugars and sucrose content of the pulp are more than 18 and 10 per cent respectively in ripe fruits. Acidity of the pulp in negligible. Hence, the fruits are enjoyed as dessert. Fruits of some varieties have however, bitterness near the peel and the fruits should be taken cutting off the outer parts of them.

Quality and Nutritive Value

Morton (1987) stated the food value per 100 gram of fresh and ripe fruits as follows.

Moisture: 78.3 g.; Protein: 0.143 g.; Fat: 0.03 g.; Fibre: 0.9 g.; Ash: 0.48 g.; Calcium: 9.9 mg.; Phosphorus: 20.4 mg.; Iron: 0.33 mg.; Carotene: 0.053 mg.; Thiamine: 0.042 mg.; Riboflavin: 0.043 mg.; Niacin: 0.472 mg. and ascorbic acid: 30.3 mg.

(Index : g = gram, mg = milligram)

Medicinal Value

Fruit is known to be soporific and is effective in disorders of central nervous system. Leaves, bark and particularly, the seeds are sedative, soporific and tranquilizer. Crushed seeds are applied to heal sores. In Costa Rica, decoction of leaves in taken in diabetes. The compound zapotin, which is present in the fruit has been stated to be a cancer chemopreventive agent (Maiti *et al.,* 2007).

Phytochemicals

Some of the phytochemicals in white sapote which are of pharmaceutical and industrial importance are zopotin, zapotinin, casimiroin, histamine, isoplimpinellin, flavones, limonoids, coumarins *etc.*

Undesirable Effect

Kernels of the seeds are highly toxic if consumed. They are observed to have cidal effect in some lower animals.

Utilization

In Mexico and central American countries, white sapote fruits are used in making jam, marmalade, squash, pudding, ice–cream *etc.*

Family: Rutaceae

Species: *Citrus limettioides* Tanaka

[Synonyms: *Citrus lumia* Risso *et*]

Taxonomic Serial: 825213

In regard to a specific name, there is some confusion and the following comments have been put forward:

Hayes (1970) quoted, "A similar fruit growing in Europe has been called *Citrus limetta* but Tanaka considers the Indian sweet lime is different and calls it, *Citrus limettioides*." Morton quoted, " It (*C. limettioides*) is often confused with the sweet lemon, *Citrus limetta* Tanaka, which in certain areas is referred to as sweet lime. In some of the literature, it is impossible to tell which fruit is under discussion."

Common Name

In English, the fruit is most commonly called as Sweet lime or Indian sweet lime and is less commonly known as Limetta. In French, sweet lime is called Limettier doux and in Spanish, the name is Lima dulce (Morton, 1987).

In Hindi and in most parts of India, sweet lime is referred to as Mitha nimbu. The name, Kolinchi pazham, Gaja nimble pandu and Gaja nimbe are used for the fruit in Tamil, Telugu and Kannada respectively.

In Bengali, the fruit is called Sarbati lebu. In this connection, it may be mentioned that many fruit sellers sell the sweet lime fruits calling them as Mosambi variety of sweet orange. However, Mosambi is a different fruit and it is a par excellence horticultural variety of sweet orange, *i.e., Citrus sinensis* Osbeck. (This variety is considered perhaps the best variety of sweet orange in India. For nutritional superiority and much less acidity, fruit juice of the Mosambi variety is recommended to the ailing persons). As the fruit of sweet lime greatly resembles with that of Mosambi in respect of shape, colour of peel, segments and juice colour, it is mistakenly called Mosambi.

Among fruits of all species of the genus, *Citrus*, those belonging to the Mosambi variety are only characterized by a ring like circular marking at the digital end of each fruit. This marking at the stylar end of each Mosambi fruit is called *areole*. Mosambi fruit is in fact, recognized by the circular, areole marketing. The marking is however, not deep and it fades out with maturity of the fruit. In a fully mature fruit, the areole marking is not very clearly observed.

Genus: Citrus

The genus, *Citrus* comes under the botanical family, Rutaceae, sub-family, Aurantioideae and the tribe, Citreae. The species in this genus have great importance in commerce. They are known to have originated in Australia, New Guinea and Caledonia although it is not confirmed. Exact number of species under the genus is not known and one of the season is, the species of the genus easily hybridize naturally between other species. Hence, which one is a species and which one is a hybrid or type becomes controversial.

The hybrids of the genus are grouped as major hybrids, *e.g.*, sour lime, or minor hybrids, *e.g.*, citrange, Rangpur lime *etc.*, or hybrids of the related genus, *Fortunella*. Some of the *Fortunella* hybrids are calamondin, orangequat *etc.* Some taxonomists are of the opinion that only 3 species coming under the genus, *viz.*, citron, pummelo and mandarin orange are not hybrids.

The species of the genus which have much commercial importance are brought under some groups, notably, mandarin orange and sweet orange, sour lime (= lime) and lemon and as pummelo (= shaddock) and grapefruit. Some hybrids, particularly Kinnow (= Kinow) is also of great importance.

The species of the genus, *Citrus* are trees or shrubs and all are evergreen. Height of them ranges from 4 to 10 metres, leaves are alternate, glossy with plenty of oil glands and have winged petioles in almost all species. Branches are thorny, flowers are solitary or in corymbs which are small. Flowers are scented, white, 1–5 cm in diameter and have 5 or in few cases, 4 petals. Stamens are many. Fruit of the *Citrus* species is hesperidium. Length of the fruit ranges from 5 to 25 cm and diameter from 5 to 15 cm. Shape of the fruit is globose, round or oval.

Peel of citrus fruits is pericarp and has three layers. Outermost layer is exocarp (= flavedo or zest). Middle layer is mesocarp. From the lower, *i.e.*, innermost layer hairs arise which extend into the locules where they provide nourishment. Segments of the fruit which are called liths enclose juice vesicles. Citric acid is predominant acid and is highly abundant in the juice of all citrus fruits.

Importance

Fruits of sweet lime are thin–skinned and they contain abundant juice. Juice is transparent, moderately sweet and has very low acidity Nevertheless, the juice has the drawback that it becomes bitter when exposed to air for sometimes which is due to the compound, limonin present in it. The trees of sweet lime are hardy and adapts well in adverse condition and drought.

Description of Plant Parts

Sweet lime trees are fairly large and spreading. The tree is evergreen. Leaves are not deep green and petioles have very narrow wings.

Origin

Morton (1987) stated that exact origin of sweet lime could not be known and according to her, it might be a hybrid between a Mexican type of lime and a sweet citron. She had also quoted that nativity of the species in India had been suggested by the Mediterranean botanists.

Geographical Distribution

Sweet lime is much grown in Egypt, Libya, Tunisia, Mexico, south American countries, India, Bangladesh, Sri Lanka and south-east Asian countries. In India, it is grown to some extent everywhere upto an altitude of 800 metres and much in the eastern States.

Climatic Requirement

Sweet lime trees grow in tropical and subtropical climates but productivity and fruit size are greater when grown in the tropics. Trees are found to be more vigorous in tropical climate but the sweetness of juice is greater in fruits when grown in the subtropics. Trees do not grow well if winter temperature goes below 5° C.

Soil Requirement

Soil is not an important factor for sweet lime trees and it is able to grow in soil of low horticultural potential. However, deep soil and that of near neutral reaction are favourable for higher productiveness of the trees.

Propagation

Seed propagation is common for growing sweet lime trees. But stem cutting is successful with medium–hardwood and hardwood cuttings and preferably by the use of auxin at relatively higher concentration. Air–layering is highly successful when it is done in rainy-season and gives rise to plentiful adventitious roots.

Harvesting

Harvesting continues from the end of summer and beginning of autumn and the peak season of harvest is later part of rainy-season. Maturity of the fruits for harvest should be determined when the deep green colour of their surface fades and turns to yellowish green. Timely harvest is an important factor as juice becomes granulated if harvesting is delayed.

Yield

Yield of sweet lime fruits varies on a number of factors. Poor fruit setting occurs in sweet lime under some conditions. Accordingly, yield becomes very low in such situations. Singh and Dhuria (1960) stated that the causes for poor fruit–set are production of high percentage of male flowers, occurrence of heterostyly and also self–incompatibility.

Morton (1987) pointed out that Indian scientists have observed the benefit of self–pollination to cause high fruit–set. She had also brought to notice the observations of the Indian scientists that cross–pollination of sweet lime flowers with pollens of sweet orange or those of grapefruit flowers resulted in higher retention of fruits, besides increasing fruit size.

Varieties

Morton (1987) named and described the following three varieties.

Indian: The strain bears oblong, ovoid or nearly round (= spherical) fruit with rounded base and small nipples at the apex and also slightly ribbed occasionally. Peel of the fruit is thin, greenish to orange–yellow in colour, smooth with prominent oil glands. Fruit usually bears 10 segments. Juice vesicles are highly juicy and the juice has some bitterness. The tree is shrubby in appearance with thorny branches. Leaves have prominent oil glands with faintly winged petioles. Trees bear profuse fruits at the later part of rainy-season.

Columbia: A clonal selection which is mentioned by Reuther *et al.* (Ref.: *Citrus Industry.* Volume I, revised, 1967).

Soh Synteng: Fruit is a strongly acid variation in Assam. New shoots and flower–buds are briefly pinkish.

The said author stated that several strains of sweet lime existed in India which differed in respect of fruit shape and yield of trees.

Mode of Consumption

Juice of sweet lime could be relished when it is fresh, *i.e.*, just after extraction from fruits. Fresh juice is faintly sweet and has negligible acidity. Unfortunately, if the juice is exposed to air for an hour or two or the fruits are cut off and then left, the juice become *bitter*. Bitterness is sometimes very high such that the juice cannot be drunk. The reason for bitterness of the juice is due to appearance of a compound, *viz., limonin* in it. Some precursors of limonin remain already present in the juice and when the juice remains exposed to air for sometime, the compound, limonin is formed by the action of enzymes. Hence, it is called, delayed enzymatic conversion of limonin. However, in fruit juice of some varieties or types of sweet lime, the bitterness is low or very low on exposure of the juice to air.

Morton (1987) stated that in "West Indies and central America, the fruits are commonly enjoyed out–of–hand. The stem–end is cut off, the core is pierced with a knife and the juice is sucked out."

Quality and Nutritive Value

Although qualitative constituents of the allied fruit, *Citrus limetta* has much been studied, those of sweet lime have scantily been studied. Mazumdar in 1973 in the orchard of the Allahabad Agricultural Institute sprayed sweet lime trees with different concentrations of 2, 4–D solutions before one month of harvest. The report published (1985) revealed that different physical and chemical constituents of the fruits had in general, increased on receiving the spray solutions of upto 16 ppm concentration and at the higher concentrations, undesirable results were produced. Gopalan *et al.* (2018) from the National Institute of Nutrition in Hyderabad stated that sweet lime fruit juice on consumption of 100 gram provide 1.3 and 1.4 gram of insoluble and soluble dietary fibres respectively.

Medicinal Value

Sweet lime juice is believed to reduce blood sugar in persons afflicted with type–2 diabetes. There is a belief that the juice has benefit in urinary tract infection.

Family: Rutaceae

Species: *Citrus mitis* Blanco

[Synonyms: *Citrus microcarpa* Bunge; *C. madurensis* Lour.; *C. microcarpa* (Bunge) Wijnands]

Taxonomic Serial: 511076

The species is considered to be an intergeneric hybrid. Morton (1987) stated that it had been given the hybrid name *i.e.*, X *Citrofortunella mitis* J. Ingram and H.E. Moore.

Common Name

In English, *Citrus mitis* fruit is commonly called as Calamondin or Chinese orange. Other names in English are Golden lime, Panama orange, Scarlet lime, Cala mandarin, Mush lime *etc.* In Thailand, it is called Ma–nao–wan. In India, the fruit is called Hazara in north India. Other names in India are Khatta narangi, Chhota narangi, Chhoto kamla, Tok kamla *etc.*

Importance

Calamondin has the following attributes. The tree is an excellent material for planting in ornamental gardens. It assumes attractive appearance when the fruits are borne. Leaves are creamy green in colour and the small fruits which are 3–5 cm in diameter resemble exactly similar to mandarin orange. The tree bears fruits profusely and hence, it is called *Hazara* in north India which denotes that it bears one thousand fruits (although it is not so). During blooming condition, excellent flavour is given off by the flowers. For these merits, the tree is chosen for planting in gardens. In household compound also, the tree is preferred to grow in large pot.

The fruits when ripened contain appreciably high amount of pectin of superior jelly–grade and hence, excellent marmalade is prepared from them. Although juice of the fruit is sour but the peel is somewhat sweet.

By regular trimming, the trees are sometimes set as bushy rows in ornamental gardens.

Description of Plant Parts

The tree is like shrub, evergreen and attains a height of 4–6 metres. Branches are profuse and arise at low height of the tree. They are less thorny. The tree produces deep tap-root. Leaves are alternate, dark-green on upper surface and yellowish at the under surface. They are glossy, broadly oval, aromatic and have narrowly winged petioles.

Flowers are whitish. Petals are 5, scented, 2.5 cm wide and are borne singly or in terminal clusters. Fruits are round to oblate, 2.5–3.5 cm wide but sometimes they are upto about 5 cm in diameter. Rind colour is tangerine when the fruit is ripe. It is very thin and is loosely held to the segments. Segments are 6–10, orange coloured, highly juicy and the juice is very sour. Fruit resembles small-sized mandarin orange in appearance. Seeds are small and 8–10 in a fruit. Seedless fruits are sometimes noticed.

Origin

Nativity of calamondin is claimed to be China and from there, it spread to Indonesia. South-east Asian origin of the tree is also stated.

Geographical Distribution

Calamondin is much grown in China, south-east Asian countries, India, South Africa, Caribbean countries, U.S.A., Mexico, Brazil, Equador and other south American countries.

Climatic Requirement

Calamondin is a subtropical citrus tree but unlike most of the *Citrus* species, it is more cold–hardy. Low winter temperature upto 10° C is tolerated by the tree. For appearance of tangerine–reddish colour of the rind, bright sunshine becomes necessary.

Soil Requirement

Soil is not an important factor but drainage should be good for vigorous growth of the tree. Deep soil is necessary for the tree as it is fairly deep–rooted.

Propagation

Seed propagation is common in calamondin. Seeds are polyembryonic and a seed gives rise to 3–5 seedlings. Careful detachment of the plants arisen from vegetative embryos would give rise to plants, true–to–type with the mother plant. Semi–hardwood cuttings produce roots by application of auxin. Budding is a common practice for the tree in other countries and is done by T–budding. Grafting is usually not done for propagation.

Culture

If the tree is intended to be grown in pot, large–sized pot of greater height should be preferred. Fruiting performance goes very high by application of organic manure. Balanced fertilizer (20–20–20) gives very good result for growth of the tree and for fruit production. The tree should be watered lightly as over–watering is harmful.

Training the trees by *open-centre, i.e., vase* system at early stage of growth make them more spready and highly fruitful. The tree are necessary to be pruned annually in winter by heading–back and thinning–out systems. Shade should be avoided to the tree for higher production and colour development of fruits. Canopy management of the trees is an important aspect for graceful appearance.

Pests and Diseases

Damage of the fruits in ripe condition by the attack of fruit-fly is very common. The insect can be controlled by spraying Malathion 50 EC–0.1 per cent or Endosulfan 35 EC–0.035 per cent. Some other insect pests which are found to attack other citruses are also found to appear in calamondin. Disease is not serious in the tree. Birds pick fruits in ripe condition and scaring device should be adopted to drive them.

Harvesting

Calamondin fruits are available all throughout the year. Colour development of the fruits is best when they are fully ripe. For preparation of marmalade, fully ripe fruits should be used. Harvesting should be done with a secateurs or pliers alongwith some twigs. As the rind of the fruit is thin, plucking should be done with care to avoid damages of the fruits.

Mode of Consumption

Juice of calamondin fruits being very sour is not used as drink but it is taken by many on mixing it with juices of oranges, shaddock, pomegranate *etc.* Fruit is also used by mixing with cranberry.

Quality and Nutritive Value

Calamondin juice has a nice flavour. Morton (1987) stated the nutritive value of the fruits. According to her, on whole fruit basis, moisture, protein, fat, carbohydrates, ash, calcium, phosphorus, iron and citric acid of the fruits had been 87.08–87.12, 0.86, 2.41, 3.27, 0.54–0.64, 0.14, 0.07, 0.003 and 2.81 per cent respectively. The energy value had been stated by her as 380 calories per kilogram of fruits.

Medicinal Value

Juice of calamondin fruits is used for some ailments in the south-east Asian countries. Morton (1987) stated that the juice had the property to banish itching, promote hair growth, bleaching freckles, cure acne vulgaris and pruritis vulvae and relieve cough. She commented upon that the fruits crushed with the saponaceus bark of *Entada phaseoloides* could be used for shampooing the hair, or the juice of the fruit might be applied on scalp after shampooing. According to her, diluted juice on warming is used as a laxative.

Post-harvest Storage

Calamondin fruits could be kept in good condition for two weeks at 8.89°–10° C maintaining 90 per cent relative humidity in the atmosphere and weight loss of the fruits would only be 6.5 per cent by that. Waxing the fruits would be able to retard loss of vitamin C content of fruit juice for a considerable period (Morton, 1987).

Utilization

Fruits could be best utilized in making the fruit product, marmalade. Rind of the fruit is thin and whole fruit is used to make marmalade instead of peeling. Marmalade made from a mixture of calamondin fruits and unripe papaya has popularity in Hawaii in the U.S.A. A sour sauce is made with juice of the fruit. Peel of the fruit is dried and used as a flavouring agent in cooking foods.

Family: Rutaceae

Species: *Limonia acidissima* Linn.

[Synonyms: *Feronia limonia* Swingle; *F. acidissima* Linn.; *F. elephantum* Correa; *F. limonia* (L.) Swingle; *Schimus limonia* L.]

Taxonomic Serial: 506392

Common Name

The fruit is known in English as Wood apple, Elephant apple, Monkey apple or Curd apple. (N.B. It should be noted that another fruit, *viz.*, *Dillenia indica* is also called Elephant apple in English). In French, the fruit is known as Citron des mois, Pomme de nois *etc.* In Java, the fruit is known as Kawista, Kawis *etc.*, in Malaysia Vilam kai, in Laos, Ma–fit and in Sri Lanka, Divul.

In Sanskrit, the names of the fruit are Kapittha, Dudhiphala, Dadhistha, Surabhiada, Bilwa, Graphiphala *etc.* In Hindi, it is known as Kaitha, Kabeet and Kaithbel, in Malayalam as Viram kai, in Marathi as Kavath, in Telugu as Vellaga pandu, in Tamil as Vilam palam and in Bengali, the fruit is called Kath bael or Koeth bel.

Genus: Limonia

The genus, *Limonia* of the Rutaceae family comes under the sub-family, Aurantioideae and the tribe, Citreae. It contains 43 species which are mostly polygamo–monoecious trees. Leaves are alternate, mostly deciduous, imparipinnate, the petioles and rachis are winged, leaflets are 5–7, opposite and sub-sessile. Lamina is ovate–obovate, thorns are axillary, short and the young branchlets and leaves are minutely appressed.

Inflorescence of the tree is axillary or terminal and the flowers are perfect. Calyx is small, flat, caducous and petals are spreading and oblong to ovate. Stamens are double in numbers than petals and filaments are dilated. Anthers are large, linear–oblong and ovary is globose. Fruit is large, globose with hard, woody shell.

Importance

Elephant apple fruits are relished in their fully ripe condition by mixing the pulp with molasses or sugar. A number of products are made with the pulp among which, a stiff jelly is common (Hayes, 1970). To make the jelly softened, soft pulp of other fruits is mixed in preparing the product. Chutney is made with pulp having sweet taste by mixing pastes of ginger, green chilli and this is highly popular. Beverages are made from the pulp in some countries. The fruit and other parts of the tree are used in folk medicine. Cosmetic, medicinal and anti–fungal properties of the fruit have also been stated (Zarga, 1986; Ghosh *et al.*, 1989; Macleod *et al.*, 1989).

Extract made from the stem bark of the tree has been reported to have insecticidal property (Bandara *et al.*, 1989). Elephant apple is used as a rootstock of citrus and it causes early flowering of the scion. The tree has religious importance in Hinduism and is sometime planted in temple yard.

Description of Plant Parts

The tree is slow growing, erect and attains a height of 8–10 metres. It is evergreen or semi–deciduous. Leaflets are 5–7 and each leaflet contains 25–30 long leaved. Leaves are pinnate, alternate, dark green, leathery and minutely toothed. Bark is rough, ridged and have sharp thorns.

Flowers are yellowish green, tinged with red and are fragrant, Calyx is small and 5–6 leaves. Petals are 5–6, free, 5 mm long and spready. Stamens are 10–12 and ovary is glabrous and sessile. Fruit (Plate-26) is a hard berry, which is round or globose. Pericarp is hard and woody, grey coloured and 5–9 cm in diameter. Fruits are of two types, one with sweetish pulp and the other of sour pulp. Pulp is brown, has a sweetish smell, mealy and resinous. Seeds are small, soft and embedded in pulp.

Origin

Origin of the species is not confirmed but is believed to be in the region from India and Sri Lanka to Java and Malaysia in the south-east Asia.

Geographical Distribution

Elephant apple grows in India, Bangladesh, Sri Lanka to all countries in the south-east Asia. It is also grown much in Mexico and northern parts of south American countries. In India, it grows almost everywhere and upto an altitude of 500 metres.

Climatic Requirement

The tree grows satisfactorily in dry tropical and subtropical climate and upto an altitude of 450 metres. Mean annual temperature of 20°–30° C and annual rainfall of 800–1200 mm are most favourable for the tree. In tropical climate, the tree remains evergreen.

Soil Requirement

Elephant apple grows in wide types of soil but light soils are much favoured. Slightly acidic soil having a pH of 5–6 makes better growth of the trees.

Propagation

The tree is easily propagated by seeds, which give normally 90 per cent germination. Germination takes place within a fortnight. Seedlings have the drawback in that, their mortality rate is high at early stage of growth. Trees grown from seeds do not also maintain all characters of the mother trees.

Vegetative propagation by stem–cuttings, budding and grafting are commonly done in south-east Asian countries. Root regeneration in leafless semi–hardwood cuttings and leafy softwood cuttings by application of auxin had been observed by Poi and Mazumdar (1989) under West Bengal condition.

In vitro proliferation of shoots had been stated in elephant apple by Handique and Bhattacharjee (2000) with the use of B–5 medium supplemented with kinetin and benzyl adenine. The hypocotyl and internodal segments from *in vitro* grown seedlings of elephant apple were cultured by Hiregoudar *et al.* (2005) on MS medium supplemented with BA or ADE or kinetin. Optimum response was recorded on the medium containing 2 micro-M of BA.

Culture

The tree is hardy and is found to grow well without culture. But application of organic manure in winter and 1–2 irrigations in summer greatly help in enhancing productiveness.

Elephant apple tree grows very tall and hence, the trunk and the branches are liable to be broken by high wind. To avoid this, trees should be properly trained at their young stage so that they do not become very tall when they become grown up. Training should be done by open-centre or by modified central leader system during the first 4–5 years of growth. Annual pruning in winter by trimming of some shoots is of great help for higher production in next year.

In assessing the nutritional requirement of the elephant apple trees, De *et al.* (1996) analyzed nitrogen, phosphorus and potash content of the peripherally borne leaves borne by trees grown in West Bengal. It was observed by them that the nutrients tended to increase in the trees from December to May and at higher rate during April to May.

Pests, Diseases and Physiological Disorder

Insect pests and diseases are not serious problems in the tree and some of these are same as are known to attack bael. Cracking of fruits is sometimes noticed which is a physiological disorder and water stress in the fruits may be a factor. Premature fall of fruits is sometimes noticed in trees and liberal irrigation of them may reduce the incidence.

Harvesting

Elephant apple fruits are ripened best in the trees. Pre-ripened fruits do not ripen any more after harvest. Ripening of fruits for harvest should be judged by observing their colour, *i.e.*, when the rind assumes more greyish in appearance. Harvesting should be done in instalment. Individual fruit with some stalk attached to it should be harvested and preferably with the implement as narrated under the Heading on longan. The cut–end of the stalk should be thickly painted with slaked lime to prevent fungal attack of the fruits after harvest which usually occurs.

Yield

Yield of the tree varies on a number of factors and a tree may produce 150–200 fruits.

Varieties

Named varieties of elephant apple are not known. Two types are usually grouped by the growers. These are fruits of relatively sweet pulp and the other type having sour pulp of fruits. However, a number of types of elephant apple are found to grow in different parts of India. They differ on height of the trees, morphology of leaves and branches and also fruit size, which may be very small to fairly large. Fruits of several types are also grown in West Bengal (Plate-26).

Mode of Consumption

Fully ripe fruits of the sweet type are relished by mixing with molasses or sugar and usually by mixing with pulp of other fruits.

Quality and Nutritive Value

Quality of fruits differ according to trees from which they have been obtained. Sinha Roy and Mazumdar (unpublished) analyzed fruits in November procuring samples from several parts in the South 24–Paraganas district and the western hilly areas in the Purulia district of West Bengal. It was observed that the fruit weight ranged from 175–250 gram, length from 8–10 cm, diameter at maximum bulge from 7.8–8.8 cm and the shape of fruits differed from almost spherical to ovate or tapering at the distal end. Carbohydrate and protein content of the pulp were recorded as 16.6–18.7 and 6.4–7.6 per cent respectively. Quality of the fruits sampled from the hilly and low rainfall areas in the Purulia district was found to be much superior.

From the National Institute of Nutrition (I.C.M.R., Govt. of India) in Hyderabad, 100 gram of elephant apple pulp having 64.2 per cent moisture in it have been stated to have the following nutritive constituents (Gopalan *et al.,* 2018).

Protein: 7.1 g.; Fat: 3.7 g.; Carbohydrate: 18.1 g.; Crude fibre: 5.0 g.; Minerals: 1.9 g.; Calcium: 130 g.; Phosphorus: 110 g.; Iron: 0.48 mg.; Magnesium: 41 mg.; Copper: 0.21 mg.; Manganese: 0.18 mg.; Zinc: 0.46 mg.: Chromium: 0.007 mg.; Carotene: 61 mcg.; Thiamine: 0.04 mg.; Riboflavin: 0.17 mg.; Niacin: 0.8 mg and vitamin C: 3 mg. Consumption of 100 gram of pulp of the fruit has been stated to provide 134 Kilocalorie (= 560.656 K Joule) of energy in the body.

(Index: g = gram, mg = milligram and mcg = microgram).

Mazumdar (2007c) quoted some proximate constituents of elephant apple fruits borne by trees grown in West Bengal.

Medicinal Value

Fruit and other parts of the elephant apple tree are used in folk medicines in India and other countries to cure or alleviate many diseases, *e.g.*, hepatitis, dysentery, diarrhoea, peptic ulcer, sore throat, gingivitis, malaria, nephritis, otitis, whooping cough, pruritis, snake bite, cancer and so on.

A review on some medicinal properties of various parts of the tree has been made by Vijayvargia and Vijayvergia (2014) from the University of Rajasthan. Anticancerous activity of fruit extract of elephant apple has been reported by Pradhan *et al.* (2012) from the Utkal University in Odisha.

Phytochemicals

From the fruits, a number of compounds, *viz.*, acidissimin, bergaptin, osthenol, tannin, lupeol, amyrin, sitosterol, stigmasterol, xanthoxine *etc.*, have been reported. Leaves contain compounds, *viz.*, limonin, limonoid, coumarin, bergaptin, lupeol, xanthoxine *etc.* Stem bark of the tree has the presence of marmesin, feronolide, feonones *etc.* and roots contain tannins, alkaloids aurapten, many amino acids *etc.*

Constituents in Developing Fruits

Changes of some physico-chemical constituents in fruit pulp of elephant apple in unripe, half ripe and full ripe stages have been studied by Sharma *et al.* (2014) from the Anand Agricultural University in Gujarat. It was observed by them that the total soluble solids, ratio of brix and acidity, total sugars and the reducing sugars of pulp increased with maturity of the fruits whereas moisture, pectin and the ascorbic acid level had decreased. The *p*H of the pulp juice was found to increase with advancement of maturity. They observed that the ripe fruits were significantly less acidic than the unripe and the half–ripe fruits.

Undesirable Effect

Unripe fruits which are somewhat bitter in taste may produce toxicity in the body. Intake of high amount of the pulp causes flatulence, belching, abdominal pain and in containing much tannin, it may cause costiveness. Many ayurvedas advise pregnant women not to take the fruit. Persons with hypothyroidism are also advised to refrain from taking elephant apple in much quantity.

Post-harvest Storage

Fruits could be stored under ordinary atmosphere for a week or two. Damage of the rind deteriorates shelf-life of the fruits. Sprinkling lime dust on fruits enhances their storage life to a considerable extent.

Utilization

As stated earlier, a stiff jelly could be made from the pulp of the fruit. Pickle made from the pulp has also high popularity. The fruit pulp is much used in India in making chutney. It is done by blending with molasses or sugar or with paste of ginger and green chilli to which salt and spices are added. In West Bengal, there is a practice of making the sweet chutney into small pieces and drying in sun for several days on metallic trays and the product is known as *kath–bael bori*.

Family: Sapindaceae

Species: *Blighia sapida* K. Koenig

[Synonyms: *Cupania sapida* (K. D. Koenig) Voigt; *Cupania akeesia* Cambess ex Spach; *Sapindus obovatus* Wight and Arn. *Cuproma edulis*]

Taxonomic Serial: 506071

Common Name

Akee is the most widely used common name of the species *Blighia sapida* in English. (The word, akee is derived from the west African words, akey fufo). Other names are also used in English, which are Jamaican–akee, Akee–apple *etc.*

In some countries, the fruit is also called as "Vegetable Brain," which is derived from the Spanish words, Seso Vegetal. In India, Akee has not any particular name and where it is grown in this country, it is known by local names such as Aki, Akki, Akiphal, Akkai *etc.*

Botanical Family

The family, Sapindaceae is popularly termed as soap–berry family, which is due to the fact that fruit peel, seed, leaf or root of many plants in this family produce foam, *i.e.*, lather when they come in contact with water. The lather becomes rich when these plant parts are vigorously shaken in water. Production of lather actually takes place due to the presence of a chemical substance, which is known as *saponin*.

Among the different sub-families belonging to the family, Sapindaceae, such as Dodonaeoideae, Hippocastanoideae, Sapindoideae and Xanthoceroideae, the genus, *Blighia* is placed in the sub-family, Sapindoideae.

The number of genera in the Sapindaceae family is around 150 and the number of species ranges from 1500 to 2500. The genera having large number of species are, *Serjania, Paullinia, Acer, Allophylus etc.*

Genus: Blighia

The genus, *Blighia* has its importance in horticulture and is said to be named to honour Captain William Bligh, who had brought the fruit from Jamaica and planted in the Royal Botanical Garden in Kew, England in 1793.

The genus is small, containing 6 or 7 species which are native to tropical Africa, from the east of Guinea to Kenya. The species belonging to the genus are trees or shrubs which are evergreen and their height goes upto 20 metres. Leaves of the species are pinnate and flowers are borne in small panicles. Fruit is oval, capsule, fleshy, 4–8 cm long and contains 3 seeds. Aril is fleshy, cream coloured and surrounds the seed. Peel of the fruit is leathery, reddish or orange coloured and thick. Cotyledons of the seed resemble embryonic leaves they unfurl during germination. Saponin is produced by most of the species.

Importance

The fruit of akee has importance in horticulture for its aril, which surrounds the seed. This is the actual edible part of the fruit like that of lichi and longan. When the fruit is ripe, the aril becomes fleshy, succulent, reddish in colour and sweet in taste. At this stage, *i.e.*, when the fruit is fully ripe, the aril is taken out and is eaten. It is either relished as fresh or is taken after drying, cooking it with other food or after making some products. The aril of akee has fondness by people of many countries and is extremely favoured in Jamaica, in which country it is valued as the national fruit. Akee is grown now in some parts of India, but to limited extent only.

The tree has also other importance. In west African countries, the unripe fruits are used for laundering. This is due to the fact that the peel of the fruit contains saponin which gives profuse foam when mixed with water. This natural lather is preferred particularly to wash delicate clothes which are liable to be damaged if washed with soap. The oil present in the seeds of akee fruit is used as an ingredient to make soap. In Cuba, the fragrance of flower petals is extracted for use in perfumeries. In Gold Coast, the bark of the stem of akee tree is ground and utilized in making skin medicament. The reddish brown and very hard heartwood of the tree trunk is immune to termite and has a great timber value. Akee tree is also planted along avenues for its decorative value and durability.

Leaves of akee tree serve as a good fodder to pet animals. It has been suggested by Aderinola *et al.* (2007) that akee leaves could be a very good feed resource for west African dwarf goats during the dry season in the derived savanna zone in Nigeria. Jimoh *et al.* (2012) in Nigeria observed usefulness of the pods of akee fruits as a low cost adsorbent material to remove heavy metals like lead, cadmium and cobalt from aqueous solutions.

Description of Plant Parts

Akee tree is normally evergreen, rarely deciduous and has a graceful appearance. The height of the tree goes upto 10 metres or more. The trunk is short and measures upto 2 metres in circumference. Crown of the tree is densely branched and has a rounded to broadly spreading canopy. The bark of the stem is greyish in colour and is almost smooth. Root system is fairly strong. Small buttresses are apparent at the base of the trunk.

Leaves are imparipinnately compound, 15–30 cm long with 6–10, elliptical to obovate–oblong leaflets which are leathery. Each leaflet is 8–12 cm long and 5–10 cm wide, entire and petiolate. Leaves are bright green in colour, glossy on the upper surface and paler on the undersurface. Inflorescence is about 20 cm long and is characteristically fragrant.

The tree blooms in warm season. Each flower has 5, whitish or greenish white petals. Flowers are male and bisexual. They are borne together in simple racemes, which measure 8–17 cm in length. Pollination of the flowers takes place by bees and other flying insects.

Fruits are borne in small clusters. They are pear–shaped and on maturity, become 5–10 cm long. Each fruit weighs 150–200 gram. On ripening, fruits turn from green to bright red in colour. A ripe fruit splits open naturally along its seam and reveals 3 shiny black seeds. Each seed is surrounded by soft, spongy, orange or red coloured fleshy material, which is botanically known as *aril* and this constitutes the edible part of the fruit.

Origin

Akee is known to be indigenous to the forests of the Ivory Coast in the western tropical Africa (Morton, 1987) or in the adjoining countries, *viz.*, Cameroon, Gabon, São Tomé, Principe, Benin, Burkino Faso, Ghana, Mali, Guinea, Nigeria, Senegal, Sierra Leone, Tago *etc.* (Anon., 2012).

Geographical Distribution

There are some species of fruit crops which have originated at one place but anchored in other place and there they gained favour. Akee is a classic example of such a fruit, which has originated in western Africa but crossing the Atlantic ocean, it has found its favourite place in the Carribbean and there it covered a vast area. Thus, to speak of akee, one cannot escape mentioning of the country, Jamaica in the West Indies. In that country, the fruit is even given national importance.

When and how the akee tree had migrated from the west African countries to Jamaica is not precisely known but it is assumptive that seeds of the fruit were carried there in the slave ships sometimes in the 18th century. Although it is acceptable that the seeds of the fruit were brought there to raise trees to secure the fruits, it is also conjecturable that another purpose might be to control soil erosion as the roots of the tree are highly efficient to bind and conserve the soil.

At the present time, akee is grown in many tropical and subtropical parts in the world. In its natural habitat, it is extensively grown in the west African countries. In these countries, the trees abound in the forests and are also grown in the non–forest areas, not only for the fruits but also for other uses of the tree and its parts.

The seeds of akee tree were carried to Panama and the Atlantic coast of Guatemala and Costa Rica by the Jamaican slaves. In Trinidad, the akee tree was legally prohibited in 1900 when it was known to cause some fatalities. Akee is now grown in many countries and although not in large scale, some countries where the fruit is considerably grown are Suriname, Venezuela, Columbia, Ecuador and Brazil in south America and in southern Florida in the U.S.A. (Morton, 1987).

Introduction of akee trees has been made in some south Asian and south-east Asian countries as well. Years back, planting of akee trees near Calcutta in India has been mentioned (Morton, 1987). At the present time, planting of akee trees is regularly done in India in the tropical forests, along aveneues and in uncultivated, waste land in some tropical parts.

Climatic Requirement

Akee is a tropical tree but it is also grown in subtropics and even in mild temperate climate, although fruiting performance is less satisfactory in the latter zones. Low temperature causes great harm to the trees, particularly in their mature condition. The young trees have however, much tolerance to chilling and they can withstand low temperature upto about (-) 1 °C for some period. Full sunshine is a very important requirement for the tree for optimum growth, flowering, productivity for fruits, their quality and appearance of colour. The tree is found to grow upto an altitude of 900 metres in Jamaica.

Soil Requirement

Well-textured soil containing copious organic matter is highly suitable for akee. Soil fertility is greatly helpful to ensure faster growth of the tree and its fruiting performance. However, the tree is moderately salt resistant. In the southern Florida and the Bahamas, the tree has been stated to do well in oolitic limestone and on sandy soil (Morton, 1987). By far, drainage of the soil is a very important factor.

Propagation

Akee is propagated by a number of methods. Seed propagation is easy and most common method of propagation. The trees raised from seedlings are robust and they do not show much variation in their characters from that of the mother trees. Direct sowing of seeds in their permanent places is advisable than transplanting in order to avoid shock as may occur in the latter method. As the seeds do not restore viability for long period, fresh seeds should be used.

Propagation by stem–cutting gives good success in akee, if hardwood or semi–hardwood cuttings are used. Success in cuttings is much higher by application of auxin in suitable concentrations. The effects of the auxin, IBA on root formation of akee stem–cuttings containing leaves had been studied by Egbe *et al.* (2012) in Cameroon. Using several concentrations of IBA, it was observed by them that while there had been no root formation in the untreated control, rooting performance was best in the cuttings which were treated with 16,000 microgram concentrations. The same concentration of the auxin was also found to have caused fastest rooting time and largest roots in the cuttings.

The effects of IBA on root formation in air–layering of akee stem had been studied by Maurya *et al.* (2013). They used four concentrations of the auxin and among those, 3500 ppm concentration gave highest rooting response. This concentration of the auxin also ensured maximum number of roots per layer, largest root length as well as maximum fresh and dry weight of the roots.

Budding also gives high success in akee and the method is advised to derive the advantage of rootstock. High success by adopting shield–budding has been brought to light by Morton (1987).

Success with somatic embryogenesis from leaf and zygotic embryo explants in akee has been put forward by Webster *et al.* (2006) from the University of West Indies in Jamaica.

Culture

Where several trees of akee are needed to be set up together in a block or to establish a plantation, spacing the trees at 6–9 metre apart may be considered optimum. However, the actual spacing would depend on the fertility status of the soil, its texture, depth and other conditions.

Akee tree is a heavy feeder and accordingly, it responds well to judicious application of manures and fertilizers. Water management is again, a very important factor for the culture of akee. The soil should never go dry nor should it contain too much moisture in any time of the year. Hence, irrigation practice with lesser amount of water discharge but with greater frequency is advisable. Mulching is a good practice to conserve moisture and control weeds (Mazumdar, 2004 b).

Pests and Diseases

Pests and diseases are not serious problems in akee. However, in a commercial plantation in southern Florida, McMillan *et al.* (2003) observed that akee trees were affected with wilt and showed die–back symptoms in the spring season of 1999. A fungal species which was isolated from the roots onto V–8 agar had been identified as *Verticillum dahlia*e Kelb.

Harvesting

When raised from seeds, akee trees start fruiting after four to six years from sowing. Vegetatively propagated trees produce fruits from two to three years after planting.

There are two bearing seasons of akee, which are January to March and June to August.

In the southern part of Florida in the U.S.A., flowers are produced in spring and the fruits are harvested in the middle of summer but a light flowering and fruiting occurs at that time. In Jamaica, flowering and fruiting continue throughout the year (Morton, 1987).

Akee fruits ripen on the trees *in situ* and the fruits do not ripen any more after harvest. Bright colour is attained by the fruits when they get plenty of sunshine for several hours in the day during their growth.

Maturity of the akee fruits when they should be harvested is a matter of tremendous importance. This is because, an akee fruit which has not fully ripened will have its aril that will have a high level of toxicity in it. Hence, harvesting at proper stage is of utmost importance. Proper stage of harvest of an akee fruit should be determined on examining that the fruit has split open naturally along its seam and thereby, exposed the black coloured shiny seeds and have brought out the bright yellow pulp. The fruit harvested at this stage of ripeness will have the aril that will be almost free from toxic substance. As a matter of fact, the aril will not be completely free from toxicity even when harvested at this stage. But the toxicity will remain only in traces and that amount will not cause any harm on consumption.

Apart from toxicity potential, nutritional deficiency of the aril associated with improper stage of harvest has also been there (Outtara *et al.,* 2011). It may be mentioned that harvesting of akee fruits at improper stage had although been done years back, it is rarely a practice nowadays.

Varieties

Named varieties of akee are not much known, although a number of types of the fruits have been stated from different parts of the world. In Jamaica, some distinct clones are known to be grown. Studies done by Emanuel *et al.* (2012) had mentioned two varieties, *viz.,* Cheese and Butter and they had compared the fatty acids content in their fruits at five stages of their growth.

Mode of Consumption

The aril is the edible part of the akee fruit. When this part is obtained from a fruit which has been fully ripened in the tree, it is fleshy, succulent, reddish, sweet and has a pleasant aroma. Such an aril is highly appealing and is taken in fresh condition.

However, instead of taking as fresh, the aril is also consumed after processing or cooking. Many people in the western part of India and western Africa relish the arils after steaming or boiling in water or cooking with other food. In Jamaica, the arils are sometimes boiled in salted water or in sweetened milk, then lightly fried in butter and taken.

Interestingly, a very popular cuisine is made in Jamaica, using aril of the akee fruit. To prepare this, the fresh arils are at first boiled in water, then they are mixed with pre-salted and dried cod fish, chopped of onion, garlic, ginger and various spices and condiments and half–fried in oil. This is relished as a covetous food of the Jamaicans, especially during their breakfast. The preparation has even been given a high rank by the authorities of the National Geographical Society.

Quality and Nutritive Value

The aril of akee is a nutritious food and has appreciable amount of vitamin C. An analysis done in Mexico revealed the following proximate constituents of the aril per 100 gram of edible part (Morton, 1987).

Moisture: 57.6 g.; Protein: 8.75 g.; Carbohydrate: 9.55 g.; Ash: 1.87 g.; Calcium: 83 mg.; Phosphorus: 98 mg.; Iron: 0.5 mg.; Thiamine: 0.1 mg.; Riboflavin: 0.18 mg.; Niacin: 3.74 mg and vitamin C: 65 mg.

(Index: g = gram, mg = milligram)

Medicinal Value

The aril and other parts of the akee fruit and also other parts of the tree have many medicinal properties. Ripe aril fruits blended with cinnamon and sugar is used by the Cuban people to reduce fever and cure to dysentery. In Brazil, akee seeds are crushed, boiled with water, strained and the juice is taken after taking saline water for expelling intestinal worms.

Bark of the akee stem on mixing with pungent spices in ointment is used by the people of the Ivory Coast to relieve pain. Bark and leaves are used in Cambodia to relieve stomach pain. In the west African countries, paste made from leaves of akee tree is used in migraine and stomach ulcers. In ophthalmia and conjunctivitis, juice made from the leaves is used (Morton, 1987). To treat epilepsy and yellow fever, various medicinal preparations are made in Ghana using parts of the akee tree. While reviewing the anti–fertility properties of higher plants, Mazumdar (1989) mentioned spermicidal effect of the saponins of akee fruits.

Phytochemicals

Saponins are by far, the most significant phytochemicals in akee and also in other members of the Sapindaceae family. These glycosidic compounds are characterized in containing a steroidal or triterpenoid aglycone, attached to one or more sugar side–chains. Distinctive foaming property like soap, *i.e.*, producing lather when added with water and shaken is an attribute of the saponin compounds.

Free–radical scavenging activity and antioxidant properties of akee fruits have been put forward by Sofidiya *et al.* (2012) and Hamzah *et al.* (2013). The latter workers held that alkaloids, tannins, saponins, flavonoids and phenols had their presence in akee fruit. Among fruits of different species tested by them, extract made from akee fruits was found to show highest level of 1, 1–diphenyl–2–picryl–thydrazyl (DPHH) free–radical scavenging activity and reducing potential.

Proximal, mineral and various physico-chemical properties of akee fruits as well as refractive index, specific gravity, saponification value, iodine value, smoke point and other constituents of the fruit oil of akee were determined by Oyeleke *et al.* (2013). Emanuel *et al.* (2012) studied in two varieties of akee that the arils contained much higher amount of fat than the husks of the fruits. According to them, the major fatty acids in the aril of akee fruits were oleic acid and the unsaturated fatty acids in the husk were in higher level than the saturated fatty acids. In both the varieties, oleic acid had been the major fatty acid.

Constituents in Developing Fruits

The variations of the fatty acids in the arils of akee fruits of two variations, *viz.*, Cheese and Butter during their five ripening stages had been assessed by Emanuel *et al.* (2012). Among other observations, the total fatty acids content was found by them to have made a decrease in both the varieties during progress of their ripening. Benkeblia and Lopez (2015) had determined the content of saccharides and potential fructo-oligosaccharides of akee in green and ripe stages. They observed that the stage of ripeness of the fruits had significantly affected the content.

Undesirable Effect

Aril of akee which is the edible part of the fruit has a serious drawback in that, it contains a highly toxic substance. Chemically, the toxic substance is an unusual amino acid, which is known as hypoglycin. This is present in the aril of an unripe or a şemi–ripe fruit but is not present or is not adequately present in the aril of a fruit that has been fully ripened in the tree. Apart from aril, the toxic substance is also present in other parts of the fruit also, much as the outer rind, pinkish membrane underlying the seed and also in the seed.

It may be pointed out that inept consumption of the aril of akee fruit had once been the cause of vomiting and death of about five thousand people in Jamaica. This serious health hazard that had occurred in that country had been referred to in the history as "Jamaican Sickness Syndrome." The cause of this health hazard was confirmed to be the consumption of akee fruit's aril. It was known that when the aril was secured from such fruits which were not fully ripened, they contained the amino acid, hypoglycins (= hypoglycins A and B) and these acted as poison. Symptoms of hypoglycin poisoning are vomiting, drowsiness, convulsion, severe fall of blood pressure and coma. If treatment is delayed, the patient may die.

When an akee fruit is fully ripened in the tree, it splits open naturally along its seam. For such splitting, the aril is exposed to air and light. On such exposure, the toxic substance that is present in it is markedly reduced and such an aril may be considered fit for consumption. However, some quantity of the toxic substance may still remain present in such an aril but that is too small and usually does not cause health hazard. The review made by Olubunmi *et al.* (2009) from Nigeria has provided details of the occurrence of hypoglycin, their increasing and decreasing trend in developing and ripening fruits, possible removal and other aspects.

It may further be mentioned that apart from hypoglycins, other chemical compunds which have anti–nutritional proeprties, *e.g.*, phytate, oxalate and tannin to the amount of 0.928, 5.422 and 0.372 milligram per 100 gram of aril had also been detected by Oyeleke *et al.* (2013).

Post-harvest Storage

Benkeblia (2014) assessed rate of respiration, rate of production of ethylene gas and other biochemical components in the aril of the fruits of the Cheese variety of akee for eight days by storing the fruits at 5°, 10° and 20° C. It was found that respiration rate had decreased while ethylene production tended to increase during storage. Decrease of glucose, fructose, sucrose, short–chain fructo-oligosaccharides and the total phenolic compounds had also occurred and the extent of decrease was higher by storing at 20° C. Arils stored at 5° C were rated best and this was followed by strong at 10° C while those stored at 20° C were spoiled.

Utilization

In Jamaica, large quantity of aril of akee fruits is canned and this is considered to be a major item of export in that country. The product is exported to the U.K. and many other countries. The import of canned arils of akee in the U.S.A. was formerly restricted for the possible danger due to the toxic substance, hypoglycins. Later it had been relaxed and prescribed by the Food and Drug Administration of U.S.A. that import could be permitted, if the hypoglycins content remained within safe limit, which was stated to be 100 parts per million. Apart from Jamaica, export of canned aril is also done from Haiti and other countries of the West Indies.

Canning is conventionally done by putting the fresh or the dried aril into 1.5 per cent salt water and then sealing the cans. The cans are sterilized at 210°C for 15 minutes, taking care that the temperature does not exceed as higher temperature is liable to hamper firmness in the texture of the arils and also lose some flavour from them. The Jamaican immigrants in the U.K. and U.S.A. are particularly fond of canned arils.

Other preserved products are also made from the aril of akee fruits. Wine is also known to be made from the juice extracted from the arils.

Family: Sapindaceae

Species: *Dimocarpus longan* Lour.

[Synonyms: *Euphoria longan* Steud.; *E. longana* Lamk.; *E. morigera* Games.; *E. scandens* Winit and Kerr.; *Nephelium longana* Cambess].

Taxonomic Serial: 506072

Common Name

In English, the most widely used common name of the species, *Dimocarpus longan* Lour is Longan, although other English names on usage are Lungan, Longyan, Dragon's eye *etc.* The name, Dragon's eye is ascribed for the fact that when the shell of the fruit is removed, the black coloured seed that shows the translucent pulp looks like the pupil of eye. The term, longan is in fact, a translated form of the Chinese (Cantonese) word, Dragon's eye.

In other countries and languages, longan is known by the following names.

France: Longanier, Oeil de dragon. Spain: Longan, Longana, Mamoncillo chino. Myanmar: Kyet mouk. Thailand: Lam–yai. Laos: Lam–nhai, Nam–nhai. Vietnam: Nhan, Long. Cambodia: Mien. Malaysia and Indonesia: Leng–keng, Mata kucing, Leng–ker, Lam–yai *etc.*

In India, longan goes by many names in different parts. Some of the names in different languages are Aaansphal, Tokra, Puvatti, Mulay, Poripuna, Oomb, Shempuvan, Shempuna, Malakota, Mulei, Mora, Posakota, Kattupuvam, Naglichi, Murale, Kohala, Porakotta, Konakindel *etc.*

Genus: Dimocarpus

The genus, *Dimocarpus* consists of about 20 species. The trees belonging to the genus are evergreen and are upto about 40 metres in height. Leaves are pinnate and flowers are borne in panicles. Fruit is drupe, oval, 3–5 cm long and the pulp which is botanically recognized as aril is juicy and translucent. Peel of the fruit is hard, thin and orange, reddish or dull brownish in colour.

Importance

The family, Sapindaceae is although very large but only a few species in this family have importance in horticulture. Four species in the family that produce fruits which are consumable to the human beings are of much significance. These are litchi (*Litchi chinensis* Sonn) and longan (*Dimocarpus longan* Lour.) in which the aril (= pulp) remains free and the other two species, *viz.*, rambutan (*Nephelium lappaceum* L.) and pulasan [*Nephelium ramboutan*-ake (Labill.)], in which the aril remains adhered to the seed.

Litchi is by far, the best of all species in producing fruits of excellent quality by taste. Longan could however, be ranked as the next in importance but it could be said that the sourness of longan pulp is much less than that of litchi. Fruits of longan have much importance in oriental medicine. A considerable section of the Chinese people is known to have a preference for longan for its energy giving value and medicinal properties rather than as an edible fruit.

Longan fruits are available when litchis are harvested and hence, they do not compete with litchis in the market.

Adaptability of the longan trees is also much higher than those of litchi as they are less demanding in soil and climatic requirements. Longan trees are more robust than those of litchi and for graceful appearance, they are planted as avenue trees also.

Description of Plant Parts

Longan tree is evergreen, tall and attains a height of 10–15 metres. The height may be greater in favourable climatic condition and when the soil is deep, retentive, well-textured and fertile. Trunk is about 1 metre in diameter and sometimes buttressed. Bark is rough and corky that splits and peels.

New growth of shoots is wine coloured. Leaves are alternate, paripinnately compound, elliptic, ovate to oblong or lanceolate, blunt–tipped, leathery, wavy, glossy, green on upper surface and minutely hairy on the undersurface. They are 10–20 cm long and 3.5–5 cm wide. Petioles measure 0.5–3 cm. Leaflets are in 6–9 pairs. Inflorescence is terminal or axillary panicle, 30–50 cm long, densely tufted tomentose and is borne on new growth. Flowers are staminate, pistillate and bisexual, small, inconspicuous, yellowish brown, leathery, 5–6 petalled and with hairy stalk. Stamens are 6–8 and male flowers have non–functional stamens.

Fruits (Plate-27) are borne in drooping clusters and are spherical to ovoid in shape. Each fruit is 2–4 cm in diameter and weighs 6–19 gram. Pulp is aril, mucilaginous, whitish, translucent and sweet. Peel is thin, rough brown, hard and brittle when the fruit is ripe. In young fruit, it is tender. Seed is small, oblong to avoid, hard, brown to blackish, shining and is marked with a white spot.

Origin

Exact origin of longan is controversial. A great many botanists are of the view that longan has originated in the southern part of China, *i.e.*, somewhere in the provinces of Kwangtung, Kwangsi, Schezwan and Fukien, at an altitude from 150 to 450 metres. Thus, the origin of longan has been stated to be the same or near to the origin of litchi, which is the closely related species of it.

Ke Guan Wu *et al.* (1994) studied morphology of pollen grains in 14 cultivars of longan and also some wild species that were obtained from five zones in China. They observed differences in the veins of the pollen grains. From such characters and also with studies done on botanical geography and evolution, they came to the conclusion that Yunnan might be the primary centre of origin while Hainan, Guangdong and Guanxi provinces, the secondary centre of origin of longan.

Locations other than China as the centre of origin of longan has also been put forward by some. One viewpoint is, the region, in the western part of Myanmar, extending from Assam in India to the western Myanmar which is known as Indo–Burma region is the actual origin of longan, *i.e.*, the same zone where mango is originated. In fact, nativity of longan in the southern part of China or the region between India and Myanmar gains much support. Some botanists consider Assam and Garo hills as the possible origin. Others have opined that it is indigenous to any of the south-east Asian countries. Malaysia has also been speculated to be the origin of longan (Pareek and Sharma, 1993).

The opinions of the different botanists have a consensus in that longan has originated in the eastern part of the world, which may be China in the north to the southern Asia in the south. On contrary to this general opinion, Hayes (1970) quoted that longan might have been a native to the western ghats in India from the Konkan to the Tinnevelly hills. Singh and Saxena (2008) have also held that longan originated in the western ghats where these trees are found to grow as wild upto an altitude of 1000 metres. Hayes (1970) referred to De Candolle's view, according to whom, India might be the origin of longan where the trees could be found to grow as wild from Konkan in the western ghats to Bengal and Myanmar in the east.

Geographical Distribution

Longan is abundantly grown in all countries which are located in the east of 90° longitude. Important longan growing countries in this zone are southern part of China, Thailand, Malaysia, Borneo, Java, Laos, North and South Vietnam, Myanmar, Bangladesh and the eastern and southern parts of India.

China is by far, an important country for growing longan and the fruit tree is largely grown in this country in the central and the southern parts. Provinces in China where it is largely grown are Guangxi, Guangdong, Fujian, Sichuan, Yunnan, Hainai *etc.* In Thailand, longan is much grown in the provinces of Lamphun, Chiang Mai and Chiang Rai.

The tree is grown to a large extent in the South and North Vietnam, where it is a traditional fruit. Main producing areas in the South Vietnam are Tien Giang, Ben Tre, Dong Thap, Vinn Long *etc.* and those in the North Vietnam are Lao Cal, Yen Bai, Thai Nguyen, Phu Thou, Son La, Hung Yen *etc.* Mekong delta has its fame for high production of longan.

All over India, longan is not so popular like the allied fruit litchi but at least in West Bengal, the fruit has high popularity where even many plantations of this fruit tree have been raised in the southern parts. Longan is also much grown in Assam, Odisha and many parts of southern India. Of late, the tree is planted in some parts of northern India where summer and winter are not extreme.

In many countries, longan has been introduced as a new fruit crop. In Florida State in the U.S.A., the tree has been introduced from southern China in 1903 by the United States Department of Agriculture. In some parts of the central and southern America also, longan has been introduced as a new fruit tree. Introduction has been made in other countries as well and the notable ones are Brazil, South Africa, Kenya and Australia.

However, popularity of longan fruit does not appear to increase now because, the allied fruit, litchi is much preferred by the consumers. Nevertheless, it is given importance and grown in many countries for the sweet fruits and their medicinal properties.

Climatic Requirement

Longan and litchi are both subtropical fruit trees and in specific terms, moist subtropical situation is most congenial to them. However, between these two closely related species of fruit trees, litchi is much more specific in its climatic requirement than longan.

For successful growing of these two fruit trees, summer should be warm and humid and winter should be dry and cool but without frost. Tolerance to low temperature is greater in longan and the longan trees are able to grow at higher altitude where litchi trees cannot survive.

Low winter temperature has an impact on flowering of longan trees. The trees come to flowering after the winter is over and it is observed that they bear profuse flowers when low temperature is received by them for 3–4 months in the winter season. If the winter is not very cool or of short duration, the trees do not produce abundant flowers. On the other hand, if the winter temperature goes very low and below 4.5° C, it causes harm to the trees. In severe winter when temperature drops down to (-) 2° to (-) 3° C, the trees are killed. Older trees are more tolerant to chilling and can stand even upto (-) 4° C but not for many days.

Dry atmosphere in winter is another favourable factor, because rain in the winter season produces more vegetative shoots in the trees and less number of inflorescences. On the other hand, dry atmosphere in summer is unfavourable for longan. In prolonged dry atmosphere, there is heavy incidence of flower drop and flower (stigma) dryness. It is also observed that prolonged dry weather in summer when fruits are developing may cause the shell of many fruits to undergo cracking abnormality (Mazumdar, 2004).

As regard wind velocity, longan trees are capable to withstand fairly high wind without causing damage of branches, particularly when the trees are properly trained to open-centre *vase* or modified central leader system at early stages of their growth.

Soil Requirement

Soil is not a very important factor for longan. Nevertheless, for optimum growth, higher yield and better quality of fruits, some types of soil are much helpful. Soil should be sandy loam or may be with high sand but it must be well-drained. High water–holding capacity of the soil is a very important factor. Soil moisture content should be high during fruit development as in dry soil, there is heavy incidence of fruit drop as well as cracking of fruits.

Salinity of soil is stated to be harmful and in saline soil, necrosis results at the margin and tip of the leaves. Trees may be killed where salinity of soil is very high. Lower soil *p*H is found to be favourable for litchi in northern part of Bihar and West Bengal.

Propagation

Longan is propagated by seed and also by a number of vegetative propagation methods. Some methods are in fact, more commonly practised in some countries.

Seed propagation is most common of all methods. Seeds should be sown at 2–3 cm depth from the surface. Light scarification of the seeds by rubbing between two sand–papers facilitates germination. Germination takes place within 10 days. It has been stated by Xia *et al.* (1992 a) that 25°C could be considered as the optimum temperature for germination of longan seeds. Seedlings may be raised in nursery beds, small pots or polythene bags and transplanted after a year or two. Direct sowing *in situ* is no doubt, better as the roots spread out well and freely in all directions and survival rate of the seedlings is higher as the shock due to transplanting of seedlings is avoided.

Longan seeds lose viability within a short period of storage and hence, fresh seeds should be used. It was observed by Xia *et al.* (1992 b) that moist storage of the seeds in perlite at 20 per cent moisture content in sealed polythene bags for 15 days or soaking in 250 ppm of GA_3, solution for 24 hours, 100 per cent germination is achieved at 30° C after 5 days. According to them, seeds lost viability almost completely when their moisture content reduced to around 29 per cent after drying slowly with silica gel. They suggested that storage of the seeds in perlite with 20 per cent moisture was preferable to storage in polythene bags. They cited chemicals which could control fungal infection of seeds. Other aspects of storage of longan seeds in which viability is not much lost have also been put forward by them.

Among vegetative propagation methods, stem–cuttings are not ordinarily successful but good success could be achieved with hardwood or semi–hardwood cuttings by application of auxin in suitable concentrations. In root regeneration of hardwood cuttings of longan, application of bottom heat is much helpful (Mazumdar, 2004). Air–layering gives fairly high degree of success in longan and the success is higher by the use of auxin. The method is largely practised in the southern China.

In some countries, propagation is done by budding but not in large scale. Inarching (= approach grafting) gives high rate of success and is much practised in the southern China. Among grafting by detached methods, whip–grafting is done in Taiwan, whip and tongue grafting in parts of Australia and veneer grafting in the Fujian province in China and in Thailand. As regard temperature for success of grafting, 20°–30° C has been stated to be optimum (Mazumdar, 2004).

Xiang Yan *et al.* (2001) in China studied on the relationship between isoperxidase activity of the scion and the rootstock and their grafting, using five different scions of longan that were grafted onto the same rootstock. The isoperoxidase patterns of the rootstock bark were similar in all the five scion–stock combinations, suggesting that the scion had little influence on the stock. On contrary, the isoperoxidase enzyme patterns of the scion bark were found to be different in different combinations, which indicated that the rootstock had some influence on the scion.

Procedure for micro-propagation of longan using terminal shoots has been described by Chen Jing Ying *et al.* (1999). They stated that culture in liquid medium containing 0.01 mg of 6–BA and 1 mg of IBA per litre was best for rooting. Wang and Bizhu (2000) had studied *in vitro* culture of 0.5 cm shoot tip explants of longan. It has been observed by them that the combination of 0.3 mg of BA along with 0.2 mg of IAA and 0.5 mg of GA_3 per litre was most satisfactory among all treatments.

Culture

Where organized plantation of longan is desired to be set up, a spacing of 5–8 metres is considered satisfactory depending on rainfall in the area and soil condition. In Thailand, around 50 trees are maintained in a hectare area of land. High density planting is also done in Taiwan, maintaining about 300 trees in a hectare. Morton (1987) quoted that in China, if longan is grown on low land, the trees are planted on the edges of raised beds while on high land, they are planted on pre-enriched holes on the surface.

Before planting, *inoculation* of the soil with mycorrhizal fungi should be a must. To do so, a handful soil should be taken from the base of a mature longan or litchi tree. This soil should be thoroughly mixed with the soil around the plant where planting is to be done. A little amount of rotten cow–dung should also be mixed with the soil. Then this part should be watered. When the water leaches down, planting should be done.

Organic manure is much helpful for longan. However, liberal use of fertilizers at the early stage of growth of the trees makes strong framework of them. Analysis of leaves in respect of nitrogen, phosphorus and potash contents in them has been determined by De *et al.* (1996) taking samples from mature longan trees grown near Kolkata in West Bengal. The results of their study revealed that the laminar nitrogen had steadily increased upto the month of April while the other elements tended to increase upto the month of March. In studies done on the range in leaf contents of ten elements in longan, Zhuang *et al.* (1995) in Tongan in Fujian, China observed that the elements were affected by locality and year and the co–efficient of variation were low for nitrogen and successively higher respectively for potassium, magnesium, phosphorus and calcium.

Prolonged drought is harmful for longan and liberal irrigation has its necessity to ensure that the soil should not get dried at any time in the year.

As the longan tree produces inflorescences on the new shoots, annual pruning is very necessary to cut back the shoots which had produced fruits, in order that new shoots appear in optimum. In parts of the tree where there are over–crowding of shoots, some shoots should be thinned–out at their bases so that fruits that would appear may get adequate light and air. Logan fruits do not develop properly and their quality is hampered unless they get sufficient air and light.

The practice of ringing of shoots is much helpful in some years to promote flowering. Wu Ding *et al.* (2000) from China stated that by ringing operation of shoots, carbohydrate, starch and the ratio of carbohydrate and nitrogen in leaves had increased.

In order to enhance flowering in longan, application of potassium chlorate may be helpful but excessive use of the chemical is harmful and may kill the tree. Irregular bearing is sometimes a problem in longan trees. This is difficult to control but heavy pruning of branches, root pruning, girdling *etc.* are effective measures to reduce its intensity. But these practices should not be done in every year.

Pests, Diseases and Physiological Disorder

A range of insect pests attack longan, among which mites, scale insects, mealy bugs, leaf and flower eating caterpillars, elephant beetle and fruit piercing moth cause great damages. Fan Jiquiao *et al.* (2011) in Guangzhou, China, stated that litchi stink bug, (*Tessaratoam apapillosa*) often causes serious damage both in litchi and longan. They tested biological control of the pest with entomo–pathogenic fungi. Fruit-fly by the attack of *Bactrocera dorsalis* is found in longan fruits which mature and ripen in the rainy-season. Maggot feeds on fruit pulp and the adult when it is formed comes out of the fruit, leaving a tunnel. The pest is difficult to control. Methyl eugenol trap may be tried to control the pest.

Apart from insects, some animal pests cause extensive damage to longan fruits. Among animals, bats are by far, most notorious in litchi and longan. These apart, crows and birds also cause great damage. Rats also climb on the trees and feed on the fruits. Suitable mechanical devices should be adopted to scare them away. Smoking is done by some tree owners to scare animal pests and insects. This is done by burning dried leaves and thin branches on the ground under the trees. However, this is not a good practice as smoke hampers photosynthesis, respiration and transpiration of leaves and in containing ethylene gas, smoke causes fruit fall, besides making the fruits to ripen at earlier maturity of them, which hampers their quality.

Among diseases, attack by the fungal species, *viz.*, *Lasiodiplodia theobromae*, *Aspergillus niger*, *Cladosposium* and *Fusarium* species is commonly observed in longan. To control, trees should be sprayed with copper fungicide, *e.g.*, Dithane Z–78 (0.2 per cent), Carbendazim (0.1 per cent), Copper oxychloride (0.3 per cent), Kavach (0.2 per cent), Bordeaux mixture (4: 4: 50) *etc.*

A physiological disorder is met with in longan fruits, going by the name as darkening of skin in which the peel of the fruit turns brown. The disorder appears in fruits mostly during their post-harvest storage. To control the disorder, exposure to sulphur di–oxide gas by burning sulphur to liberate this gas and exposing that to the fruits is done by many but this hampers quality of the fruits as well.

To control weeds in longan plantation, Paraqaut and Glyphosate are applied in Australia.

Harvesting

Longan trees raised from seeds come to bearing maturity after attaining 6–7 years of growth. Vegetatively propagated trees produce fruits from 3 to 4 years of planting.

The trees start blooming from the end of February and continue upto April. Induction of off–season flowering in longan by the use of potassium chlorate has been studied by Suthon *et al.* (2012) in Taichung, Taiwan. They observed that percentage of bud–break in potassium chlorate treated percentage treated plants were 100 and 70 in summer while these were 70 and 6.2 in the autumn. Thus, the chemical could be seen to induce off–season flowering in longan in both seasons. The effects of several plant growth regulations on the flower formation and also fruit setting of longan were studied by Peng *et al.* (2008) in Guangdong, China. They suggested that the reasonable combined treatment of Ethephon, Paclobutrazol and gibberellin could promote flower formation and fruit settings.

Fruits of longan become ready for harvest from the end of May and continue upto the end of July. Some late-maturing varieties are found to produce ripe fruits till the month of September. In south India, flowering and fruiting season, continue over a longer period due to milder winter and high rainfall.

Time of harvest is very important for longan. The fruit is non–climacteric and should be harvested only when it is fully ripe as it ripens best in the tree and does not ripen any more after harvest. Maturation of the fruits for harvest is judged by observing their shape, size and colour. Jiang *et al.* (2011) stated ripeness of longan fruits to harvest could be determined on the basis of fruit weight, skin colour, sugar: acid ratio and flavour of fruits or number of days from anthesis. Harvesting of fruits is done in instalment and along with twigs and leaves and marketed in this conditions. Ripe fruits are soft and much care should be taken in harvest to avoid their damages.

An efficient method of harvest may be done by the use of a special type of pruner. In this method, a sickle–shaped pruner which contains a movable blade with the help of a spring is taken, pruner is tighly fixed at the end of a bamboo pole or a long stick with peg passing through the hole of the pruner. A cloth basket is tied to the pruner. One end of a long rope is tied to the hole of the pruner and other end of the rope reaches at the ground level. The arrangement is such that the harvester, while standing on ground holds the pole and pulls the rope in such a manner that the fruits along with small twigs are cut off by the blade and the cut out materials are collected in the cloth basket. When the bag is filled, it is brought down and the fruits with the twigs attached are collected. The process is continued for harvest of other fruits.

Yield

Yield of longan greatly varies on many factors. Under Indian condition, 30–50 kg of fruits on an average are obtained from a fully mature tree and an individual fruit may weight 5–7 gram. Yield is very high in southern China. In North Vietnam, yield is stated to be 4–5 tonnes per hectare and it is still higher under the condition of South Vietnam.

Varieties

Longan trees that are grown in India are found to have many distinguishing features in their tree and fruit characters. In the South 24–Paraganas district in West Bengal, there are some trees which are found to produce very large and sweet fruits. However, named varieties are not known in India.

In other countries, plentiful varieties of longan are grown. There are at least 300 varieties that are grown in southern China. Largely, 50 of these are cultivated and some of the famous varieties in that zone may be named as Fu Yan, Wu Long Ling, Wu Yuan, Shi Xia, Chuliang *etc.* (Mazumdar, 2004). A large–fruited, late–ripening variety, *viz.*, Baoyuan has been reported by Haung Jinsong *et al.* (1996) in the Fujian province of China and a chance seedling of superior quality has been mentioned by Yang Win Huan (1996) in that province. Xu Xiu Dan *et al.* (1999) referred to an extremely late–maturing variety of Fujian, *viz.*, Lidongben. The variety, Quingshan, which was evolved as a chance seedling has been mentioned by Chen *et al.* (2000). The characteristic features of 42 varieties grown in Taiwan has been mentioned by Chao Cheng Nan *et al.* (1997). Khatun *et al.* (2012) examined physico-chemical composition of 15 promising longan varieties grown in the Research Farms in Bangladesh. A wide range of variations were observed by them among the germplasm, in respect of different characteristics.

Among other varieties that are grown in different parts of the world, some may be mentioned as Biew Kiew, Dai–O–Vien, Duong Phen, Degelman, Diamond River, Edo, Egami, Haew, Huo Kioh, Kay Sweeny, Kohala, Lungan Late, Nhan Long, Nhan Cui, Ponyai, Thach Hiep, Tiger Eye, Tru Luong, Tsao Ho, Xuong Com Vang *etc.*

Mode of Consumption

Longan is most commonly consumed as fresh but the fruit pulp is also served along with salad, snacks, desserts *etc.* as an ingredient. Sweetened and thickened milk boiled with fine rice or sweetened curd, that are blended with arils of longan or litchi is a delicacy in West Bengal which is served as a dessert after lunch or dinner.

Like litchi, pulp of longan is best relished when the fruit has fully ripened in the tree. But in over-ripe condition, the taste deteriorates and a musky odour is emitted by the aril, resembling garlic. Alcoholic smell is also given off from the aril in over-ripen condition. Hence, harvesting before full ripening of the fruits is advisable.

Although longan does not match with litchi in taste, its sourness is less than litchi. Taste of longan is however, highly appealing in some varieties. As compared to litchi, flavour of longan is not so high but its flavour is retained for longer time than litchi. The flavour of longan is stated by many to have a resemblance with grape berries.

Quality and Nutritive Value

Quality of longan pulp in respect of flavour has been studied by Wong *et al.* (1996) in Malaysia and Chang Ching Hui *et al.* (1998) in Taiwan and the said workers identified a large number of volatile compounds.

Mandal and Mazumdar (1997) analyzed longan fruits taking samples from trees growing in the South 24–Paraganas district in West Bengal. Results of their study revealed that weight, specific gravity, size of fruits, total titratable acidity as equivalence of critic acid and the vitamin C content of the aril were respectively 6.81 g, 1.04, 2.38 cm. 0.13 per cent and 10 mg per 100 gram. A large–fruited longan grown in the same zone was found (Mazumdar, B.C. unpublished) to have weight, specific solids of the aril juice, weight and specific gravity of fruits, total soluble solids respectively as 9.65 g, 1.06, 2.1 per cent, 1.7 g and 1.15.

(Index: g = gram, mg = milligram, cm = centimetre).

The proximate composition of 100 gram of longan fruits has been presented by the National Institute of Nutrition (I.C.M.R., Govt. of India) in Hyderabad as follows (Gopalan *et al.,* 2018).

Moisture: 83.09 g.; Protein: 1.4 g.; Fat: 0.3 g.; Minerals: 0.8 g.; Crude fibre: 0.5 g.; Carbohydrate: 13.1 g.; Calcium: 15 g. and Phosphorus: 35 mg. Consumption of 100 g. of pulp gives rise to 61 Kilocalorie (= 255.22 Kilojoule) of energy in the body.

(Index: g = gram. mg = milligram).

Medicinal Value

Longan has much importance in China for its medicinal properties rather than a fruit, dried aril of the fruit is sold in the markets of that country for its preventive and therapeutic value against many ailments.

Nerve toning property of litchi and longan in known since ancient time. Both the fruits are efficacious in insomnia, anxiety and various types of nervous and psychic disorders. Longan pulp is recommended in stomach trouble, vomiting, spleen disorder and to reduce fever. The fruit has quick wound healing property and it improve blood circulation. On making increase in the absorption of iron in the body, the fruit is of great help in anaemia. The fruit is also a rich souce of polyphenols, quercetin and antioxidants.

Phytochemicals

A number of phytochemicals have been reported from longan fruits. Rangkadilok *et al.* (2005) in Thailand extracted phenolic compounds, *viz.*, gallic acid, corilagin, which is an ellagitannin and ellagic acid from the pulp, peel and seed of the fruit using methanolic solvent. The seeds were found to contain the highest level and pulp the lowest. Varietal differences were also observed by them. Phenolic compounds in longan pulp or seed have been detected by other workers (Sudjaroem *et al.*, 2012; Liu Chun Li *et al.*, 2014).

Constituents in Developing Fruits

Bhattacharya *et al.* (1990) determined sugars, acidity and vitamin C contents of the pulp of longan fruits at four stages of their growth. Later, Mandal and Mazumdar (1998) had assessed weight, specific gravity, length and diameter of the fruits and the dry matter, sugars, acidity and the vitamin C contents of their pulp at four stages of growth till harvest–maturity condition of them. Results obtained by them revealed that fruit weight had been higher at the early part of growth and the specific gravity value had increased much during the middle part. At the advanced maturity stage of the fruits, the acidity level decreased with a concomitant increase of sugar content.

Hasan *et al.* (2006 a, b) provided results of the changing pattern of longan fruits during their developmental stages in respect of weight, size and specific gravity of the fruits, total soluble solids, total sugars and acidity of the pulp and the nitrogen, phosphorus, potassium, iron, manganese, zinc and these contents upto the stage of harvest of the fruits.

Undesirable Effect

Harmful effect of longan is not known but too much consumption of the pulp may cause constipation, belching, vomiting and stomach ache. It is believed by many that longan and some other fruits like grape, peach, plum, guava, pineapple *etc.* should not be taken by woman at early stage of pregnancy as these may cause miscarriage in them.

Post-harvest Storage

Shelf-life of longan fruits under ordinary room temperature condition is short but is somewhat greater than litchi. This may be ascribed on the fact that the peel of longan is more tough and less flexible than that of litchi for which moisture and gases are less permeable to each the pulp.

To enhance shelf-life, some growers in Thailand, China and other countries expose the fruits to sulphur di–oxide gas by burning sulphur but such treatment tends to turn the fruits blackish. Many fruit growers in West Bengal prolong shelf-life of the fruits by sprinkling lime dust over them or soaking in lime water for few minutes.

In storage studies of longan, Jiang *et al.* (2002) focused on reducing the major post-harvest problems in order to produce light coloured fruits that would be free from chemicals and attack by pests and diseases. It has been indicated (Jiang *et al.,* 2011) that the main determinants of shelf-life and marketability were microbial decay and pericarp browning. To increase shelf-life of longan fruits, temperature of the storage chamber has however, a market influence (Fang *et al.,* 2010: Wall *et al.,* 2011). Longan fruits are alo canned, *i.e.,* stored in syrup (= sugar solution).

Utilization

To preserve longan fruits, canning in strong syrup is a common practice as is done in litchi. The seeds may be retained or removed in the canned product. Drying of longan fruits is largely done in Thailand and China, either retaining the peels or on removing them. In the latter method, the fruits are at first heated for sometimes to shrink the peels. By this, excoriating the peels becomes easier. On stripping the peels, the pulp along with the seeds are again heated on slow fire. When dried, the seeds are removed or not. The dried pulp becomes brown and leathery. Drink, fruit salad *etc.* are also made with longan pulp. Drying of peeled longan fruits in slow fire and then mixing them with sugar, spices, chilli–dust *etc.* and packing in polythene bags for marketing is done in West Bengal. A liquour is known to be made from the pulp on macerating in alcohol in some countries.

Family: Sapotaceae

Species: *Chrysophyllum cainito* Linn.

[Synonyms: *Chrysophyllum bonplandi* Klots. ex Miq.; *C. monopuyrenum* Spreng.; *C. bicolor* Poir.; *C. eggersii* Pier; *Achras caimito* Ruiz and Pavon] *Cainito pomiferum* Tuss.; *Cynodendron bicolor* (Poir) Baehni].

Taxonomic Serial: 510921

Common Name

Most widely used common name of the species in English is star–apple. It is called so because when the fruit is cut to half, the dark brown seeds in the white pulp form the outline of a star. However, by the common name, star–apple, another species of fruit is also meant, which is *Syzygium samarangense*.

[Mention may be made in this connection that such confusion in naming are there for some other fruits also. Papaya may be cited as an example which is spelt and pronounced as papaw in some countries including most parts of the U.S.A. But papaw is actually a different species, which is *Asimina triloba*. It is grown as a minor fruit in the southern part in the U.S.A. (Hayes, 1970). Similarly, Black plum is referred to in English for the fruit which in northern part of India is called as jamun (*Syzygium cumini*) but Black plum sometimes goes by the variety of the fruit, plum (*Prunus domestica*)].

However, apart from the name, Star–apple, other common names are also applied in English for this fruit species. These are, Golden leaf, West Indian star apple, Cainito, Caimito, Milk–fruit *etc.*

In other countries and languages, the Star–apple goes by other names and some of these are stated in the following:

Java: Sawo, Ijo, Sawo duren. Malyasia: Hnin–thagya, Sawu Duren. Singapore: Chicle durian. Thailand: Sata apoen. Cantonese: Chicle durian. Myanmar: Hnin _ Thagya. Siera Leone: Bobi wata, Breast milk fruit. France: Caimite, Caimitier, Bon Caimite, Caimite franche, Grand caimete. Portugal: Cainito, Ajawa. Spain: Caimito, Estrella. West Indies: Pomme surette, Buis, Pied caimite, Kaimit. Guayabillo. Columbia: Caimo morado (for purple variety), Caimito maduraverade (for green variety). Bolivia: Caimitero, Murucuja. Argentina: Aguay. Olivoa.

Botanical Family

The family, Sapotaceae comes under the Order, Ericales. It includes about 800 species pf evergreen trees and shrubs in around 70 genera. They produce milky latex. Many species produce edible fruits or have other economic use.

Leaves of the tress or shrubs in the family are simple, coriaceous, alternate or opposite though in some species, mostly entire and may be with stipules or not. Flowers are bisexual, calyx consists of 4–12, imbricate, biseriate or spirally arranged sepals. Corolla is gamopetalous and usually has 4–12 lobes. Androecium is adnate to corolla and contains inner whorl of fertile stamens. Gynoecium consists of a single compound pistil, a single style and superior ovary with 4–5 locules. Fruits is a berry.

Genus: Chrysophyllum

The genus, *Chrysophyllum* has been named from Greek words, Chryso meaning gold and phyllos, which means leaf.

About 80 species belonging to this genus are tropical trees and attain a height of 15–20 metres or more. The species of the genus mostly abound in Columbia, Venezuela, Guyana, Suriname, Ecuador and other tropical parts of south America.

The trees of the genus have oval leaves, the upper surface of which is green and the under surface is golden coloured and pubescent. The flowers are bisexual, purplish white, small and measures 4–8 mm. Petals have sweet smell. Fruits are spherical and mostly purplish in colour but greenish brown, greenish yellow and greenish white colour of peel are also found. Seeds are brown and flattened.

Importance

Star–apple has its importance for producing edible fruits, ornamental value of the tree and also for medicinal properties. The fruits are delicious in fresh condition and some varieties are very nice in taste. The tree is planted along avenues for shade and particularly for the striking appearance of the leaves. Upper surface of the leaves is bluish green in colour and the under surface is coppery red and thus, an attractive contrast is created when the leaves are stirred by wind.

It is said that the king Christophe in Haiti was very fond of this fruit and he even held court under the shade of a very large star–apple tree at Milot (Morton, 1987).

Description of Plant Parts

The tree is evergreen and reaches a height of 15 metres or even more in deep rich soil in Cuba. Trunk diameter is about 60 cm. Leaves are alternate, distichorous or spirally arranged, simple, oval or oblong, entire, 7–15 cm long and 4–6 cm wide. Upper surface of the leaves is bluish green and the under surface is coppery red and pubescent. Inflorescence is axillary, ramiflorous or cauliflorous. Flowers are very small, in cluster, axillary, unsexual or bisexual, fasciculate or rarely solitary. They are purplish white and have fragrant smell. Fruit is round, sometimes oblate, 5–10 cm in diameter and the rind is thick, leathery and smooth. Fruit has 8 segments, translucent, whitish and normally one seed is present in each segment which is flattened, light brown, hard and ovate to elliptic in shape.

Origin

Exact provenance of star–apple is not known and is conjectured to have originated in the region from West Indies to the central America. The nativity is also believed to be the Greater Antilles islands in the Caribbean. According to Paul Standley and Louis Williams, the origin of Star–apple is West Indies and not actually central America (Morton, 1987).

Geographical Distribution

Star–apple is grown throughout the tropics. It is grown in Java, Malaysia, South Vietnam, Thailand, Myanmar, Sri Lanka, tropical parts of India, Cameroon, Nigeria, Ivory Coast and other countries in the western Africa. However, it is much grown in Cuba, Jamaica, Haiti and other countries in the West Indies, southern Mexico, Guatemala, Costa Rica, Panama, Venezuela, Suriname and other parts in the north of south America. In many tropical countries, the fruit has been introduces much later.

Climatic Requirement

Star–apple is highly tropical in its climatic requirement and moist tropical situation is most preferred. However, the trees grow well in subtropical climate also but the fruiting performance is less satisfactory than when they are grown in the tropics. In Jamaica, it thrives upto an altitude of 425 metres. Low temperature below 0° C causes serious harm to the trees and younger trees are more sensitive. Even brief exposure to 0° C may kill the trees which are below three years of age. Cold wind also hampers growth and fruiting of the trees. Full sunshine during growth of fruits greatly helps in colour development and quality of the fruits.

Soil Requirements

Soil is not of particular significance for the star–apple tree. It is found to grow well in shallow sandy soil of southern Florida in the U.S.A. as well as deep clay loam in Cuba. In Florida, it is considered to be a moderately salt resistant tree. However, deep soil and high organic matter make the trees highly vigorous and increase their productive capacity. Poor drainage is however, greatly harmful and it seriously hampers growth of the trees.

Propagation

Seed propagation is very common for the tree in different countries but the trees are slow to grow and do not retain all characters of the mother trees. Vegetative propagation hastens bearing maturity and the trees becomes productive within 3 years. Stem–cuttings are successful when taken from mature and well-ripened shoots. Auxins are of great help in root regeneration of cuttings. Air–layering is more successful than stem–cuttings. Budding and grafting are also done on rootstock of the same species but is rarely practised.

Culture

Normally, the star–apple trees grow and fruit satisfactorily without culture but growth and productivity of them greatly increase on receiving manure, if not grown in rich and loamy soil. An important necessity is, frequent irrigation to keep the soil moist throughout the year during first few years of growth of the trees. Mulching the soil with raised mulch is known to be of great benefit to the trees. Drainage is by far, most important as the trees do not grow well or at all in poorly drained soil. Sometimes shy–beating trees are found which give rise to very low production. Pruning some branches of the trees after harvest of fruits is of advantage to increase next yield in such trees in the year. Trimming of some roots sometimes helpful but soil manure mixture must be applied just after root pruning.

Pests and Diseases

The fungal disease, stem–end decay caused by the species, *Pestalotia* and *Diplodia* is often noticed in the trees. Leaf spots due to attack by *Phomopsis, Phyllosticta* and *Cephaleuros* species are also noticed. Mummification of immature fruits is observed in Florida in the U.S.A. Birds, bats and squirrels damage the fruits om ripened condition. Ripe fruits are also attacked by bugs (June bugs).

Harvesting

Star–apple trees raised from seeds start producing after 5 or 6 years of growth while those propagated vegetatively give production within 3 years of planning. Flowering starts from summer to rainy-season and the fruits attain full ripeness from late autumn to summer. In West Indies, fruits are ripened from April to May. In Suriname, the trees are found to produce fruits almost throughout the year.

Star–apple fruits are best ripened in the trees and they do not ripen further if harvested before full ripeness is achieved. Unripe fruits are not sweet, astringent in taste and contain sticky white latex. Full ripeness of the fruits is judged when their colour turns purple in case of the purple varieties and golden colour in case of the green varieties. However, as the fully ripe fruits are greatly lost by the attack of the birds and bats and become very soft, it is advisable to pluck those sightly earlier. Fruits do not fall naturally and hence, clipping the stem with secateur is necessary. Clipping the stems with a tree–pruner fitted with a movable blade and a cloth basket is very suitable for the purpose which may be operated by pulling the rope attached with the spring of the blade without climbing on the trees as stated in the Chapter on longan.

Yield

Yield of star–apple greatly varies in trees depending on soil texture, fertility, climatic condition and the varieties. On an average, 50 to 60 kg of ripe fruits are available from a mature tree.

Varieties

Although peel colour of the star-apple fruits differ, recognized varieties of the fruit are not much known. Most common peel colour of the fruit is deep purple but fruits with greenish brown and greenish yellow coloured peels are also found and sometimes ripe fruits with greenish white peel colour are also met with. Even green colour is apparent around the calyx in the purple-coloured fruits. The fruits belonging to different peel colour are not found to have difference as regard their taste, flavour and quality.

Morton (1987) however, stated an introduced variety in Florida in the U.S.A. According to her, William Whitman of Miami in Florida noticed a star-apple tree growing in Port-au-Prince in Haiti. The trees bore fruits from late January to the end of June. The yield performance and quality of the fruits had been stated to be very high. Whitman brought budwood of the tree in Florida in 1953. The grafted progeny and the trees of this tree grown from air-layers were found to give high performance even before reaching a height of 3 metres. This introduction has been named in Florida as the "Hawaiian Star Apple".

Under condition of Vietnam, the variety of star-apple, named as Lo Ren milk fruit which has come from Vinh Kim commune in Chau Thanh district in Tien Giang province in considered as most famous variety.

Mode of Consumption

Pulp of the star-apple fruit is sweet but the skin and the rind secrete milky and sticky latex which is bitter. Hence, pulp should be taken out with care such that the latex does not come in contact with the pulp. To do so, the best way is to cut the fruit in the half and the pulp which is the edible part and should be scooped out from each half with a spoon.

Fruit is relished as fresh and many consumers prefer to take on chilling the fruits. In Bolivia, the fruit is taken after parboiling also. In Jamacia, pulp of the star-apple is sometimes mixed together with orange juice, nutmeg or pulp of mango, strawberry *etc.* and then taken.

Quality and Nutritive Value

According to Morton (1987), analysis of the star-apple fruits done in Cuba and central America revealed the following components in 100 gram of edible part of the pulp.

Moisture: 78.4–85.7 g.; Protein: 0.72–2.33 g,; Carbohydrate: 14.65 g.; Fibre: 0.55–3.30 g.; Ash: 0.35–0.72 g.; Calcium: 7.4–17.3 mg.; Phosphorus: 15.9–22.0 mg.: Iron: 0.30–0.68 mg.: Carotene " 0.004–0.039 mg.: Thiamine: 0.018–0.08 mg.: Riboflavin: 0.013–0.04 mg.: Niacin: 0.935–1.340 mg.: Ascorbic acid: 3.0–15.2 mg.: Tryptopham: 4 mg.: Methionine: 2 mg.: Lysine: 22 mg and Provides 67.2 Calories of energy in the body on consumption.

(Index: g = gram, mg = milligram)

Analytical results of physical or physico-chemical constituents of star–apple fruits have also been provided by Hernandez Sanchev *et al.* (2009) from Mexico and Adepouju and Akupor *et al.* (2013) from Nigeria.

Medicinal Value

Star–apple tree has many medicinal properties in its leaves, bark, fruits and seeds. Some notable therapeutic value of the fruits may be stated as follows:

Mucilage in the ripe fruit is useful in soothing laryngitis and pneumonia. Mouth wash with infusion of the semi–ripe fruits is also of help in reducing inflammation of throat. Fruits are much used to control haemorrhage and to reduce fever. Polyphenolic antioxidants have also been reported from the fruits (Luo *et al.,* 2002). In Venezuela, unripe fruits are used to overcome intestinal disturbance and seeds are used as diuretic and febrifuge. To reduce blood sugar, fruits are used in Puerto Rico.

Phytochemicals

Volatile compounds in star–apple have been stated by many workers. Pino *et al.* (2002) with star–apple grown in Cuba identified compounds in the aroma concentrate. They stated that the presence of many terpenic compounds were responsible for unique flavour of the fruits. Liu *et al.* (2007) determined constituents from the peel, pulp and juice of the fruits.

Undesirable Effect

Star–apple fruits particularly those which have not fully ripened contain a sticky, milky and astringent latex which is harmful and cause blisters on lips. Therefore, the fruits should be taken carefully to avoid the latex coming in contact with the pulp. Seeds of the fruits also contain toxic substance but seeds are not eaten.

Post-harvest Storage

Star–apple fruits are difficult to be stored for more than few days under ordinary atmosphere but like sapota, they can be stored for 6–8 weeks at a temperature of 1.5–3° C maintaining a humidity of 85–90 per cent. However, Morton (1987) has stated that ripe fruits of star–apple could be stored satisfactorily for 3 weeks at 3°–6°, with a relative humidity of 90 per cent.

Utilization

In most parts of the world where star–apple grows, the fruits are relished in their fresh condition and sometimes by pre-chilling them. In Jamaica however, there is a practice to make preserve from the fruits. To do so, the pulp is carefully scooped out avoiding contact of latex secreted by the rind. Then the extracted pulp is chopped and thoroughly mixed with pulp or juices of other fruits such as mango, pineapple, sapota, orange juice, coconut water *etc.* To this paste, nutmeg is sometimes added for flavour. The preserve is recognised as "Jamaican Fruit Salad Ice"and usually served on chilling.

Family: Solanaceae

Species: *Physalis peruviana* Linn.

[Synonyms: Physalis *edulis* Sims; *P. tomentosa* Medik; *P. esculenta* Salisb.; *P. chenopodifolia* Lam.; *P. latifolia* Lam.; *P. pubescens* L.; *P. puberula* Fer.; *Boberella peruviana* (L.) E.H.L. Krause in Sturm; *Alkekengi pubesens* Moench].

Taxonomic Serial: 30606

Common Name

In English, *Physalis peruviana* is known by the common names as Cape-gooseberry, Golden berry, Husk berry, Poha, Poha berry, Ground cherry, Peruvian ground cherry *etc.* In other countries, the common names are as follows:

France: Capuli, Coqueret du Peru, Groseiller du Cap *etc.* Germany: Andenbeeru, Kapstachelbeere *etc.* Spain: Agaymanto, Passa capule *etc.* Hawaii: Poha, Paina *etc.* South Africa: Apellific, Pompolmoes *etc.* Mexico: Cereza del Peru, Miltomate *etc.* Venezuela: Chuchuva, Topotopo *etc.* Chile: Capuli, Bolsa de Amor *etc.* Columbia: Uchuva, Uvilla *etc.* China: Deng long guo.

In India, the species is known in Hindi as Makowi or Tipari. in Andhra Pradesh and other parts of southern India, commonly used names are Kupantali, Budda, Trippari *etc.* In West Bengal, cape-gooseberry is usually known as Tepari or Tápari. But many people in West Bengal call it as Rasbhari or Rashbhori. It may be mentioned that the names, Rashbhori or Rashbhari are actually misleading terms. This is because, some people thought it to be Raspberry fruit and hence, the name Rasbhari had been given. Some seed stores also sell the seeds writing on the container, Cape-gooseberry seeds–variety Rasbhari or Rashbhori. However, it should be remembered that cape-gooseberry and raspberry are not same fruit. Raspberries are (*Rubus* species) grown in temperate climate while cape-gooseberry (*Physalis peruviana*) is grown in subtropical and tropical climates.

Cape-gooseberry is sometimes confused with another fruit which is gooseberry. But cape-gooseberry and gooseberry are not same fruits. Gooseberry is (*Ribes uva–crispa*) grown in temperate climate only. Again, cape-gooseberry is confused with another fruit which is tomatillo. It should be noted that although tomatillo (*Physalis philadelphica*) is closely related to the fruit of Cape-gooseberry but both are not same.

Botanical Family

Cape-gooseberry comes under the family, Solanaceae which is also known as nightshade- family and is placed under the taxonomic Order, solanales. The family is large and consists of 98 genera and 2700 species. Most of the species grow in tropical and subtropical climates and some thrive in temperate situation. The species has greatest diversity in the south American countries. The plants of the species range from annual and perennial herbs to vines, shrubs and trees. Many of the species are of agricultural and horticultural importance, many are of medicinal importance and some species have toxic alkaloids

Genus: Physalis

The genus, *Physalis* comes under the sub-family, solanoidea of the solanaceae family. It is under the tribe, physaleae and the sub-tribe, physalinae of the family. There are about 100 species of the genus which grow in warm temperate and subtropical countries. The plants are annual or perennial, 0.3 - 1.5 metre tall and have persistent calyx which encloses the fruit completely or partially. Some species are sensitive to frost.

Importance

Cape-gooseberry has a great importance in that, it is ranked as one of the 'quick- growing fruit crops', like pineapple, strawberry, banana, papaya phalsa, water-melon and musk-melon. Most of the fruit crops have a drawback in that, their 'juvenile' period is long, *i.e.*, they take several years of growth to produce flowers and fruits. During the pre-bearing years, manures, fertilizers *etc.* are needed to be provided to them which involve cost but no production is obtained from them. High initial cost in setting up of orchards is another factor. For these cost- involving factors, many crop growers do not prefer establishment of orchards instead of growing agricultural crops. However, such problems do not exist in growing cape-gooseberry and other quick- growing fruit crops referred to above as they give production within a year or two.

Cape-gooseberry and other quick- growing fruit crops do not reach height of more then 1.5 meters. Hence, they are also adequately saved from breaking of branches and fruit fall which occur by high wind that blows at upper atmosphere and appears to be a problem in many of the fruit trees.

In ripe condition, the fruits are rich sources of pectin and this is of high jelly-grade. Hence, excellent quality of jam and jelly could be made from them (Mazumdar, 1979). Removal of the seeds is not necessary in making jam on jelly as they are very small and soft and are taken with the fruits.

Each individual fruit of cape-gooseberry is completely enclosed by an inflated and persistent calyx. This is a contrivance for which no part of the fruit is left exposed. Thus, the calyx acting like a box gives protection to the fruit against attack by birds, fruit fly and other insect. The calyx remains present even after harvest. Thereby, this helps in post-harvest storage of the fruits.

Taste of the fruit is pleasant with a desirable blend of sweetness and sourness and has a nice flavour resembling grape and pineapple. Seeds are very small, soft and are taken with the fruits. Fruits are sometimes used in cooking meat or other culinary preparations. Medicinal value of the fruits is very high and they are rich sources of vitamins B, C and P (= bioflavonoids).

The crop is grown in a wide range of soil types, from *p*H 5 to 8 and does not require special care in its culture. Cropping is done in more than one season. Propagation of the plants is done easily from seeds which are plentiful in number and the resulting plants and fruits retain the characters of the mother plants. Apart from use as fruit plants, cape-gooseberry plants are also sometimes preferred to raise as border plants in ornamental gardens. The crop has also many other merits.

Description of Plant Parts

Cape-gooseberry plant is a herbaceaous (Plate-28) or soft-wooded tree on shrub. It is both annual and perennial and is usually annual in temperate climate. Height of the plant is 1.5–1.8 cm but in rich and moist soil, height may reach upto 2 meters or more.

Roots of the plant are confined within a depth of 20 cm but the tap-root of a plant may go deeper and upto about 80 cm. Branches are spreading, ribbed, often purplish and have dense and soft hairs. Leaves are almost oppositely arranged, heart- shaped, velvety and are randomly toothed. A leaf is 10-15 cm long and 8-12 cm wide. Petiole is 2-4 cm long. Leaves become yellow and fall off when the fruits are ripened.

Flowers are bisexual and arise singly in leaf-axils on 1-2 cm long pedicels. They are bell-shaped, nodding, yellow coloured with 5 dark-purplish marks at the base. They are 10-20 mm wide, 10- lobed and densely pubescent at the base. Calyx is the characteristic features of the flower of cape-gooseberry. It is cup-shaped and is formed of 5 lobes. It is connate at the base, purplish green in colour with tips of triangular shape, distinctly veined and hairy. The calyx is accrescent and remains persistent with the fruit all along. At first, it is of normal size but grows at a faster rate after the petals fall off. It expands and becomes larger in size than the fruit, gets inflated and encloses the fruit completely leaving no part of the fruit exposed. When the fruit ripen, the calyx looses chlorophyll, becomes straw coloured and is much stiffened. Pollination of the flowers takes place by insects.

Fruit (Plate-28) of the plant is a berry. It has smooth and waxy surface, yellowish when ripe and the shape is globose or round. Diameter of the fruit is 1.5-2.5 cm. It has a nice flavour, sweet-sour in taste, highly juicy and a fruit contains 100-300 small and soft seeds. Seeds are taken along with the fruits.

Origin

Amazon basin in south America is considered to be the origin of the species. It has been suggested that higher altitudes in Peru, Columbia and Ecuador are the locations where the plant has originated. Another group of botanists is of the opinion that nativity of the plant is Brazil and from there, it has migrated and naturalized in the highlands of Peru and Chile in very early time.

Geographical Distribution

Cape-gooseberry is grown in many sub-tropical and tropical countries and South Africa is a leading country. The early settlers in the region of Cape of Good Hope introduced the plant and it is conjectured that the name, cape-gooseberry has derived from the word, which is cape. In Africa, the crop is commercially much grown in Namibia, Botswana and Malagasy.

South American countries, notably Venezuela, Ecuador, Peru, Guyana, Brazil, Bolivia have much importance in growing cape-gooseberry crop and in these countries, it is usually grown to a large extent in high altitude regions. Large-scale cropping of the plant is done in many other countries, particularly in New Zealand, Tasmania, part of Australia, Fiji and Hawaii islands in the USA. The crop is also cultivated in Spain, France and other countries in Europe.

In India, the crop is grown in many agro-climatic zones although not to a large extent. However, due to high demand in the market, area under the crop tends to increase in West Bengal, Assam, other eastern states and in south India. Under the condition of south India and particularly in the hilly regions, growing of cape-gooseberry is mostly done as a perennial crop and the plantations are retained for 2-4 years.

Climatic Requirement

Cape-gooseberry grows well in subtropical and warm temperate climates and has naturalized in many tropical countries as well. In warm temperate situation, it grows as an annual plant. Plants are much sensitive to frost. Hence, in north India where it is commercially grown, seeds are sown in summer from May until early July and the seedlings are transplanted when they are less than 25 cm tall. The fruits are harvested during February to March or sometimes earlier. If the seedlings are transplanted late, they are damaged by frost (Hayes, 1970).

Rain is helpful during the growing period of the crop but dry atmosphere during maturation and ripening of the fruits is desirable for better quality of fruits. Plants require bright sunny days for higher productiveness and better quality of the crop.

Soil Requirement

Cape-gooseberry plants grow on a wide range of soil but loam and sandy loam are highly favourable to them. The crop grows in acidic or alkaline soils within a *p*H range of 5 to 8. Top soil should be loose and rich in organic matter as the feeder roots of the plants are superficial. However, in highly rich top soil, the plants make profuse vegetative growth and flowering intensity is reduced. Drainage is very important for the crop as in poorly drained soil, growth of the plants is greatly hampered, leaf fall is greater and fruiting performance is very poor.

Propagation

Seed propagation is conventionally done for propagation of cape-gooseberry plants in all countries where it is cropped. Seeds are usually broadcast directly in the field but in India, raising of seedlings in nursery-beds and transplantation those in the field is a more common practice. Three nursery-beds, each measuring 2 metre long and 1 metre wide appear to be sufficient to raise seedlings to plant one acre area of land. Nursery-beds should be 10-15 cm raised from the ground level and the top soil should be well-pulverized. About 30 gram of seeds would be required to obtain seedlings for planting in an acre of land.

Seeds should be sown from May to July in accordance with climatic condition. As the seeds are very small, these should be mixed with soil or wood ash for uniform sowing in nursery-beds. Between the nursery-beds, 40-60 cm space should be maintained for drainage and path. Nursery-beds should be regularly watered, avoiding over-irrigation.

Within 2-4 weeks of sowing, the seeds would germinate. The emerged out seedlings should be regularly thinned to avoid over-crowding. They would be ready for transplanting when 15-20 cm tall. Transplanting seedlings in a second nursery-bed when they become 5-8 cm tall and then transplanting those in the actual field at 15-20 cm distance results in restriction of vegetative growth. For this, flowers appear earlier and yield becomes higher.

Cape-gooseberry plants may also be vegetatively propagated from stem-cutting and air-layering (Hayes, 1970). However, regeneration of adventitious roots may have difficulty unless the stem which is to be used attains at least one year of growth. Application of auxin at the basal part of a stem-cutting and on the scraped part of the stem surface in case of air-layering quickens root formation and encourages higher production of roots.

It is observed that the plants produced from stem-cuttings and air-layerings make less vegetative growth, for which flowers are produced earlier than the plants raised from seeds.

Micro-propagation in cape-gooseberry has been studied by some workers. Santana and Angarita (1997) observed maximum number of plants by culturing in MS medium supplemented with 1 ppm of BA and GA_3 and 0.5 ppm of 2, 4-D under light. From Tamil Nadu, Raman *et al.* (2014) reported that regeneration was achieved from the node, internode and leaf explants on MS medium with B-5 vitamins and PGRs like BAP, GA_3 and 2, 4-D.

Clonal propagation from nodal segments had been studied by Yucesan *et al.* (2015) from Turkey. An efficient micro-propagation protocol for the cape-gooseberry by the process of *in vitro* culture of nodal, internodal and seed explants has been reported by Singh *et al.* (2016) as well.

Culture

Cropping of cape-gooseberry does not require intensive cultural procedure. The cultivation procedure under Indian condition has been presented by some authors (Hayes, 1970; Mazumdar, 1979; Anon., 1988; Mazumdar, 2004 a) and salient features of cultivation of the crop have been stated in the following.

(i) Land Preparation

Good tilth is necessary for the land for optimum production and hence, repeated ploughing, cross-ploughing and levelling of the land become necessary.

(ii) Planting

In some countries, broadcasting of seeds is done to raise seedlings while in India where it is cultivated, transplanting of seedlings is the usual practice. Transplanting on flat land is most commonly done in India but in some parts, making ridges and transplanting on them is followed.

Cropping season and time of transplanting differ according to agro-climatic zones. Transplanting is usually done in spring. Many growers prefer to transplant the seedlings in early spring for early harvest of fruits, *i.e.*, during the first summer. It has been stated earlier that in north India, seeds are sown in May to early July, seedlings are transplanted at 25 cm height and late transplanting causes damage to the seedlings by frost (Singh and Saxena, 2008). The effect of sowing time and cropping season on the yield behaviour of the crop had been studied by Poi (1989) in the central part of West Bengal.

(iii) Plant Density

Morton (1987) stated that in South Africa, the seedlings are spaced at 0.6-0.9 metre apart in rows. Under the condition of New Zealand, Klinac (1986) stated that yield went higher when spacing was greater than 0.6 metre but yield per plant was greatest when the plants were set at closest spacing.

(iv) Manuring

Cape-gooseberry crop is highly responsive to the application of manures and fertilizers but too much application, particularly nitrogenous fertilizer induces high vegetative growth at the expense of flowering and fruiting. Prasad *et al.* (1985) had studied yield behaviour of the crop by application of N, P and K fertilizers. Effect on yield and quality of the fruits by application of three levels of N, P and K fertilizers singly and in all possible combinations had been studied by Poi (1989).

(v) Irrigation

The crop does not require high irrigation but irrigation in greater frequency is desirable for higher yield. Drip irrigation (Mazumdar, 2004 b) is of much benefit to the crop.

(vi) Weed Control

Weeds become great problem in a cape-gooseberry field and *Physalis minima* is a common weed. In luxuriant growth of weeds, weedicides, *e.g.*, Fluchloralin, Alachlor, Butachlor *etc.* may be applied to the soil at the rate of 2-3 kg per hectare. Mulching (Mazumdar, 2004 b) is much effective to control weeds and conserve soil moisture.

(vii) Pinching

Many growers pinch the apical shoot of cape-gooseberry plants to encourage branching and thus, to raise production. Malla *et al.* (2008) experimented this is Jammu. They pinched the plants at different heights (20, 30, 40 and 50 cm) at which they were pinched. Pinching at 20 cm height was found to be most effective to reduce the plant height and this took lowest number of days to produce flowers.

(viii) Nutrient Solution

It is observed that foliar application of 0.25 per cent aqueous solution of zinc sulphate after a month of transplanting and few days after fruit-set markedly increases yield performance and quality of the berries.

(ix) Growth Regulators

Foliar application of auxin solution at low concentration is observed to make increase in size and quality of the berries. Improvement by application of gibberellin is also observed. Wanyama *et al.* (2006) in Kenya sprayed cape-gooseberry plants with GA_3 solutions of varying concentrations after one week of transplanting. The results of their study revealed that GA_3 treatment increased branching, flowering and fruiting. Growth of the plants and fruit yield were higher with application of higher concentrations. It was observed that number of fruits were maximum and counted as 303 fruits per plant by application of 100 ppm solution of GA_3.

Pest, Diseases and Physiological Disorder

Cape-gooseberry plants are attacked by a number of insect pests, nematodes, bacteria *etc.* and damage by physiological disorder is also observed.

The mite, *Polyphagotarsonemus latus* causes injury to the stem and it sucks the sap. Due to its attack, leaves are deformed and growth of the plant is stunted. To control, affected parts should be clipped off and burnt. Spraying of Dicofol (0.02 per cent) and Dimethoate (0.045 per cent) may be sprayed alternately. Hopper (*Antianthe expansa*), beetles, cut worm, leaf borer, thrips, stem borer, fruit moth, aphids, fruit fly and other insect pests also attack the plant and the fruit. Root knot nematode (*Meloidogyne* species) sometimes causes great damage and to control, Carbofuram may be applied in soil at the rate of 2 kg per hectare.

Symptoms of the fungal disease, powdery mildew, characterized by whitish or greyish spots are often found on leaves. To control, Karathane (0.05 per cent) or wettable sulphur (0.3 per cent) may be sprayed. Fungal infection by *Alternaria* species which produce symptoms like small, purple lesions are noticed on leaves at early stage of growth of the plants and to control, Chlorothalonil or Mancozeb may be applied. Under post-harvest storage condition, berries are often found to be infected by the species of *Penicillium* fungi. Bacterial wilt of the fruits is often observed during rainy-season and aphid vectors should be controlled to prevent the attack of the bacteria (*Pseudomonas* species).

Chadha (2012) reported three types of mosaic symptoms on leaves of cape-gooseberry due to viral infection. It has been mentioned that the mosaic viruses of types- I and II had been responsible for 5-10 per cent loss in yield while the extent of loss might be 15-20 per cent by infection of the mosaic-III type of the virus. According to him, to control the virus, virus-free seeds should be used and the insect vectors which disseminate the virus should be controlled by spraying Dimenthoate (0.05 per cent).

Under West Bengal condition, Poi *et al.* (1987) observed little leaf disease and by spraying tetracycline (60 ppm in water), the symptoms were found to be markedly suppressed. Cracking abnormality of the fruits of cape-gooseberry is sometimes noticed which is a physiological disorder. The symptoms are usually observed in the fruits which are harvested in the first instalment. Oversupply of assimilated products appears to be the cause of the abnormality.

Harvesting

Cape-gooseberry fruits become ripe after 60-100 days from flowering. Harvesting is done in instalment, usually starting from February and continued upto April. Sham Singh *et al.* (1967) stated in their book that in Coonoor in south India, fruits are harvested from January to May. Morton (1987) had mentioned that in the central and southern parts of Africa, the fruits could be harvested from the beginning of April and continued upto the end of June.

The fruits are non-climacteric and they are properly ripened in the plants. They do not ripen any more under post-harvest condition if they are harvested before full ripening. Full ripening of the fruits is best judged on observing colour of the persistent (= accrescent) calyx. That is, they should be plucked when the calices would assume straw colour, which indicates that they have been dried.

Some growers shake the plants and collect the fruits which have fallen. This is however, not a reliable method of harvest because many fruits which have not been fully ripened also fall off by shaking the plants. For harvesting, best method is, to pluck the individual fruits on observing the straw colour of the calices.

Fruits in which calices remain persistent are usually marketed. But in some markets particularly in the western countries, fruits are marketed after removal of husks. Dehusking of the berries manually involves labour cost and Wang (1966) in Hawaii in the USA reported a mechanical husker for dehusking of the berries.

Yield

Yield of cape-gooseberry crop widely varies on many factors such as variety, locality, richness of soil, cultural practices *etc.* In a normal crop, a plant produces 250-300 fruits and a fruit without husk weighs 6-10 gram.

In general, yield per plant is 2-4 kg and per hectare yield is 1650-2250 kg when grown as a rainfed inter-crop and upto 33,600 kg when grown as a pure crop (Anon., 1988). According to Hayes (1970), yield in Uttar Pradesh is 150-250 maunds per acre and in Punjab, 9 maunds per acre (1 maund = 0.37 quintal). He mentioned that at Araku valley at elevation of 3000 feet (1 foot = 0.3048 metre), 250 fruits could be produced by a plant and those would weigh more than 15,000 lb per acre (1 lb = 453.6 gram).

Sham Singh *et al.* (1967) stated that in Coonoor in south India, well-grown cape-gooseberry crop could produce 22,420 kg of fruits per hectare. According to Singh and Saxena (2008), 25 tons of fruits are obtained from one hectare area of land under the condition of south India. Morton (1987) mentioned that a plant could give rise to 300 fruits and 3000 lbs of fruits (= 13,608 kg) might be obtained from one hectare area of land.

Varieties

Some well-known varieties of cape-gooseberry that are grown in other countries may be stated as follows.

Giallo Grosso: Berries of this variety are large, golden yellow coloured and are usually consumed as fresh. The plants grow as perennial in warm temperate climate.

Golden Berry: Berries of this variety are large-sized and have pleasant flavour. These are also dried to make raisin like product.

Gaint: Berries of this variety are very large-sized and have pleasant flavour. Plants are very tall and attain a height of 1-1.5 metres. Profuse branches are produced by the plants and they flower after sufficient growth has been achieved.

Giant Poha: Berry size is very large in this variety. Plants are tall with greyish green leaves and their morphological characters are different from those of other varieties.

In Australia, the berries of the varieties, *viz.*, Golden Nugget and New Sugar Gaint have popularity.

Under Indian condition, any named variety is hardly known but many types are observed which have differences in plant height, branching, fruit size and quality of fruits. Rashbhori or Rashbhari as stated under the Heading on Common name is not regarded to be a variety of cape-gooseberry. (The name, Rashbhari has actually been given by some nurserymen in West Bengal, considering the fruit as raspberry. But this is a misnomer. Raspberry (*Rubus ellipticus*) is a different fruit crop and is grown in temperate climate).

Mode of Consumption

Cape-gooseberry fruits are relished as fresh. They appeal consumers for desirable sweet-sour taste and the flavour. In south-east Asian countries, the berries are used in cooking food also.

Quality and Nutritive Value

Among other fruits, berries of the crop are of great value in quality and nutritive value.

From the National Institute of Nutrition (I.C.M.R., Govt. of India) in Hyderabad, 100 gram of cape-gooseberry fruits containing 82.9 per cent moisture have been stated to have the following nutritive constituents (Gopalan *et al.,* 2018).

Protein: 1.8g.; Fat: 0.2g.; Carbohydrates: 11.1g.; Crude fibre: 3.2g.; Minerals: 0.8g.; Calcium: 10mg.; Phosphorus: 67mg.; Iron: 2mg.; Magnesium: 31mg.; Sodium: 0.9mg.; Potassium: 320mg.; Copper: 0.19mg.; Sulphur: 43mg.; Chlorine: 12mg.; Carotene: 1428mcg.; Thiamine: 0.05mg.; Riboflavin: 0.2mg.; Niacin: 0.3mg.; Vitamin C: 49mg. As regard energy value, 100 gram of fruits has been stated to provide 53 kilocalorie (= 221.75 kilojoule) of energy in the body.

(Index: g = gram; mg = milligram; mcg = microgram)

Vitamin E, *i.e.*, tocopherol has been reported from cape-gooseberry fruits (Barcia *et al.,* 2010). Hayes (1970) stated vitamin P from the fruits. (However, vitamin P which is bioflavonoid is not considered now to be a vitamin).

Medicinal Value

In folk medicine in south and central America, fruit, leaf and root of the plant are prescribed in many diseases *viz.*, asthma, bronchitis, hepatitis, malaria *etc.* Candidiasis caused by the fungal species, *Candida albicans* on skin, cornea, feet, scalp *etc.* is stated to be controlled by cape-gooseberry fruits. Anti-oxidative properties of the fruits have been reported by Wu *et al.* (2005). Ramadan *et al.* (2005) from Egypt reported antioxidants, volatile compounds and anti-cancerous activity from the fruits.

Phytochemicals

Some of the phytochemicals existent in cape-gooseberry fruits are melatonin, sitosterol, myricetin, with a physanolide, chlorogenic acid, flavonoids, alkaloids *etc.* More than 100 volatile compounds including eucalyptol, hexanol, ethyl decanoate *etc.* have been isolated from the fruits. Mohamed Fowzy (2011) stated that cape-gooseberry fruits were rich sources of bioactive compounds for functional foods and pharmaceuticals.

Constituents in Developing Fruits

Changing responses of pectic polysaccharides and activity of the enzymes, pectin methyl esterase and polygalacturonase in cape-gooseberry fruits during their progress of development and ripening after spraying the plants with solution of varied concentrations of IBA and GA_3 have been assayed by Majumder and Mazumdar (2001).

Results of the study revealed that the water-soluble pectic fractions accumulated in greater amount in the auxin-treated fruits and GA_3 application had made increase in the level of the acid-soluble and alkali-soluble pectic substances. Polygalacturonase activity had increased by auxin treatment while it had been highly reduced by GA_3 treatment.

Subsequently, the said authors (Majumder and Mazumdar, 2002) reported that during ripening of the fruits, the water-soluble and the oxalate-soluble pectic substances made increase while the acid-soluble and alkali-soluble fractions decreased. With solubilization of high molecular weight pectic substances, there had been a 5–6 fold increase in the polygalacturonic activity while activity of pectin methyl esterase had not been clearly related to ripening. The enhancement of polygalacturonase activity was highly correlated with evolution of ethylene that had occurred prior to galacturonase synthesis in the fruits.

The authors (Majumder and Mazumdar, 2005) also studied pectic polysaccharides on extraction from fruits at different maturity stages and these were fractioned and analyzed for galacturonate backbone molecules. Pre-harvest spray of the fruits at their 7-day old stage with the ethylene releasing compound, ethephon resulted in a higher accumulation of water-soluble and oxalate-soluble pectic substances than the non-treated fruits while the treatment effectively enhanced the polygalacturonase activity which is necessary for pectic degradation.

It was observed by the authors that treatment with Ethephon not only enhanced the rate of solubilization but also resulted in a net increase in the total pectic polymer content in the developing cape-gooseberry fruits. Influence plant-growth regulators on the crop had been much studied by Majumder (1990).

Undesirable Effect

Morton (1987) stated toxic principles in the cape-gooseberry fruits and quoted that the plant produced illness and death of cattle in Australia. Existence of 1mg of phytin P per 100 gram of fruits had been reported by the National Institute of Nutrition (I.C.M.R., Govt. of India) in Hyderabad (Gopalan *et al.,* 2018).

Post-harvest Storage

Cape-gooseberry fruits may be kept well for two weeks or more if the accrescent calyx of the fruit is not removed. According to Morton (1987), fresh fruits can be stored in a scaled container and maintained in a dry atmosphere for several months. Some growers in India partially dry the fruits with attached calices to increase shelf-life of the fruits.

Utilization

Mazumdar (1979) stated that cape-gooseberries could be best utilized un making fruit jam of excellent quality. Presence of seeds in the berries would not hamper acceptability of the product as they are very small and soft. Canning of the berries is done in some countries. Drying of the berries like raisin is also a practice in some countries.

CHAPTER-3

CONCLUDING PART

Importance of minor fruit crops has been elaborated in the Chapter on Introduction. Keeping in view the various merits of these fruit crops, the horticultural scientists in India strongly hold that with the objective to bring about significant increase in fruit production in the country at the present time, much higher attention is necessary to be paid on the less utilized fruit plants that are conventionally recognized as of minor importance. They draw attention to lay emphasis on to minor fruit crops belonging to all categories, *i.e.*, those which are adaptable to the tropical, subtropical as well as temperate zones in the country. This is brought to notice that apart from securing fruits, making use of these plants for other purposes need also be adequately explored.

Among other considerations, it is felt that much greater attention needs to be paid on the following aspects in the area of minor fruit crops in the country.

(i) In the light of the fact that the species of minor fruit crops and the varieties type etc that exist in them differ according to localities, groups are necessary to be made, specifying such fruit crops having adaptability to each zone. Germplasm collection centres in each zone are also necessary to be set up.

(ii) Making increase in the area of minor fruit crops, has beyond doubt, a great necessity. To achieve this objective, particular attention needs to be paid on exploration of land of low agricultural suitability, uneven physiography, along tanks, streams, rivers and avenues. Minor fruit trees and shrubs that are capable to grow and produce in waste lands needs critical assessment.

(iii) Trees and shrubs belonging to a large number of species of minor fruits are ideal materials for afforestation and hence, mass planting of them in the forests should be ensured.

(iv) Availability of genuine propagating materials is a prerequisite in expanding area under these fruits. Adequate emphasis should therefore, be given to set up reliable nurseries at different zones.

(v) Varietal screening of the minor fruits belonging to each species in respect of morphological, anatomical and physiological characteristics deserve attention.

(vi) Majority of the minor fruit crops being hardy materials are able to withstand adverse edaphic situations. Their effectiveness to use as rootstocks for different species and varieties of fruit crops needs standardization.

(vii) Adequate attention should be paid to explore different species of minor fruit crops as materials for hybridization with species and varieties of principal in other words, major fruit crops.

(viii) Timber value of different species of minor fruit trees for use in various purposes should be critically examined.

(ix) Making preserved food products in small and large scale from the various species of minor fruits leaves an important area of their utilization.

(x) Utilization of juices of many minor fruits or their sap in making fermented, partially fermented or unfermented beverages is an area that demands attention.

References

1. Abdulla, A.T.M.; Hossian, M.A. and Bhuiyan, M.K. (2005). Propagation of latkan (*Baccaurea sapida* Muell. Arg.) by mature stem cuttings. *Research Journal of Agriculture and Biological Sciences*, **1(2):** 129-134.
2. Adepoju, O.T. and Adeniji, P.O. (2012). Nutrient composition and micronutrient potential of three widely grown varieties of African star apple (*Chrysophyllum albidum*) from Nigeria, *African Journal of Food Science*, **6(12):** 344-351.
3. Aderinola, O.A.; Farinu, G.O.; Akinlade, J.A.; Olayeni, T.B.; Ojebiyi, O.O. and Ogunniyi, P.O. (2007). Nutritional potential of *Blighia sapida* K. Konig (Ackee) leaves as a dry season feed resource for West African dwarf goats in the derived savanna zone of Nigeria. *Livestock Research for Rural Development*, **19(6):** 78.
4. Adkar, P.; Dongare, A.; Ambavade, S. and Bhaskar, V.H. (2014). *Trapa bispinosa* Roxb.**:** A review on nutritional and pharmacological aspects. *Advances in Pharmacological Science*, **2014:** 1-13.
5. Agrawal, A. and RAM, H.Y.M. (1995). *In vitro* germination and micro propagation of water chestnut (*Trapa* species). *Aquatic Botany*, **51(1/2):** 135-146.
6. Akinwale, T.O. and Aladesua, O.O. (1999). Comparative study of the physico-chemical properties and the effect of different techniques on the quality of cashew juice from Brazilian and local varieties. *Nigerian Journal of Tree Crop Research*, ***3*(1):** 60-66.
7. Akupor, P.I.; Yusuf, D. and Objegunam, J.E. (2013). Proximate composition and some functional properties of flour from the kernel of African star apple (*Chrysophyllual albidum*). *International Journal of Agricultural Policy and Research*, **1(3):** 62-66.

8. Alam, M.; Siddique, M.B. and Husain, W. (1990). Treatment of diabetes through herbal drugs in rural India. *Filoterapia*, **61(3):** 240-242.

9. Ali, S.L. and Mazumdar, B.C. (1990). Studies on pectin content of some fruits and plant parts. *Science and Culture*, **57(10-11):** 256.

10. Amin, M.N. and Razzaque, M.A. (1993). Regeneration of *Averrhoa carambola* plants *in vitro* from callus cultures of seedling plants. *Journal of Horticultural Science*. **68(4):** 551-556.

11. Amor, E.C.; Villasenor, I.M.; Ghayur, M.N.; Gilani, A.H. and Choudhary, M.I. (2005). Spasmolytic flavonoids from *Syzygium samarangense* (Blume) Merr. and L.M. Perry. *Z. Naturforch* C., **60(1-2):** 67-71.

12. Anandalakshmi, R.; Shivakumar, V.; Warrier, R.R.; Parimalam, R.; Vijayachandran, S.N. and Singh, B.G. (2005). Seed storage studies in *Syzygium cuminii*. *Journal of Tropical Forest Science*, **17(4):** 566-573.

13. Anonymous. (1988). *The Wealth of India–Raw Materials*. Publication and Information Directorate. Council of Scientific and Industrial Research, Govt. of India, New Delhi.

14. Anonymous. (2005). *Handbook of Agriculture*. Indian Council of Agricultural Research, New Delhi.

15. Anonymous. (2012). "Taxon: *Blighia sapida* K.D. Koenig". *Germplasm Resources Information Network, United States Department of Agriculture.*

16. Araujo, E.R.; Alves, L.I.F.; Rego, E.R. do; Castro, J.P. de and Sapucay, M.J.L. da C. (2009). Physico chemical characterization of fruits of Biri-biri (*Averrhoa bilimbi* L.). *Biotemas*, **22(4):** 225-230.

17. Arseculeratni, S.N.; Gunatilanka, A.A.I. and Panabokke, R.G. (1982). Studies on the toxicology of the palmyra palm (*Borassus flabellifer* L.). II. Milk transfer to suckling rats. *Journal of the National Science Council of Sri Lanka*, **10(2):** 277-282.

18. Arumugam, T.; Suthanthirapandian, I.R. and Doraipandian, A. (1994). Effect of organic and inorganic manuring on neera and yield of palymra. *South Indian Horticulture*, **42(4):** 236-238.

19. Avinash, P.; Swapneel, K.; Darshana, P. and Pathak, A. (2012). A comprehensive review of an important medicinal plant–*Averrhoa carambola* L. *Pharmacognosy Communications*, **2(2):** 3.

20. Ayar, K.; Singh, S. and Chavda, J.C. (2011). Effect of postharvest treatments on quality of jamun (*Syzygium cuminii* skeels) fruits during storage. *Asian Journal of Horticulture*, **6(2):** 297-299.

21. Ayyanar, M. and Subash-Babu, P. (2012). *Syzygium cuminii* (L.) Skeels: A review of its phytochemical constituents and traditional uses. *Asian Pacific Journal of Tropical Biomedicine*, **2(3):** 240-246.

22. Babu, M.; Dwivedi, D.H.; Ram, Y.B. and Meena, M.L. (2013). Bioaccumulation and distribution of heavy metals in water chestnut (*Trapa natans* var. *bispinosa* Roxb.) in the Lucknow region. *African Journal of Agricultural Research,* **8(22):** 2765-2768.

23. Bajpai, A.; Singh, A.K. and Ravishankar, H. (2012). Reproductive phenology, flower biology and pollination in jamun (*Syzygium cuminii* L.). *Indian Journal of Horticulture,* **69(3):** 416-419.

24. Bajpai, M.; Pande, A.; Tewari, S.K. and Prakash, D. (2005). Phenolic contents and antioxidant activitiy of some food and medicinal plants. *International Journal of Food Science and Nutrition,* **56(4):** 287-291.

25. Bajpai, P.N. and Chaturvedi, O.P. (1985). Jamun. (In: *Fruits of India–Tropical and Subtropical,* Ed. by T.K. Bose), Naya Prokash, Calcutta.

26. Balakrishnan, K.; Rajendran, C. and Kilandaivelu, G. (2000). Different responses of iron, magnesium and zinc deficiency on pigment composition, nutrient content and photosynthetic activity in tropical crops. *Photosynthetica,* **38(3):** 477-479.

27. Balakumar, S.; Rajan, S.; Thirunalasundari, T. and Jeeva, S. (2011). Antifungal activity of *Aegle marmelos* Correa leaf extract on dermatophytes. *Asian Pacific Journal of Tropical Biomedicine,* **1(4):** 309-312.

28. Balasubramanyan, S. and Bangasuysamy, U. (1998). Maturity standard of emblica (*Emblica officinalis* Gaertn.) under rainfed vertisol. *South Indian Horticulture,* **46(3/6):** 347-348.

29. Balendres, M. and Bengoa, J. (2019). Diseases of dragon fruit: Etiology and current management options. *Crop Protection,* **126:** 104920.

30. Baliga, M.S.; Tilakchand, K.R.; Rai, M.P.; Rao, S. and Venkatesh, P. (2013). *Aegle marmelos* (L.) Correa (Bael) and its phytochemicals in the treatment and prevention of cancer. *Integrated Cancer Therapy,* **12(3):** 187-196.

31. Bandara, B.M.R.; Gunatilake, A.A.E.; Sotheeswaran, S.; Wueratne, E.M.K. and Ranasinghe, M.A.S.K. (1989). Insecticidal properties of an active constituent of *Limonia acidissima*. *Journal of the National Science Council of Sri Lanka,* **17(2):** 237-239.

32. Bandopadhyay, D.P.; Nath, N.; Pandey, H.S. and Yadav, L.P. (1983). Effect of type of rooting in stem cuttings of *Carissa* species. *Plant Science,* **12:** 44-45.

33. Bankeblia, N. (2014). Respiration rate, ethylene production and biochemical variations of ackee fruit and arils (*Blighia sapida* Koenig) stored under three temperature regimes. *Postharvest Biology and Technology,* **97:** 36-43.

34. Bankeblia, N. and Lopez, M.G. (2015). Saccharides and fructo oligosaccharides composition of green and ripe *Averrhoa carambola, Blighia sapida* and *Spondias dulcis* fruits. *Food Chemistry,* **176:** 314-318.

35. Banker, G.J. (1987). A note on influence of gibberellic acid on seed germination and vigour of seedlings in karonda (*Carissa carandas* L.). *Progressive Horticulture,* **19(1-2):** 90-92.

36. Barare, D. and Chretian, L. (1985). Floral anatomy of *Monstera deliciosa* (Araceae). *Canadian Journal of Botany*, **63(8):** 1428.
37. Barcia, M.T.; Jacques, A.C.; Pertuzatti, P.B. and Zambiazi, R.C. (2010). [Determination of HPLC of ascorbic acid and tocopherols in fruits]. *Semina: Ciencias Agrarias* (Londrina), **31(2):** 381-390.
38. Barh, D. and Mazumdar, B.C. (2008). Comparative nutritive value of palm saps before and after their partial fermentation and effective use of wild date (*Phoenix sylvestris* Roxb.) sap in treatment of anaemia. *Research Journal of Medicine and Medical Sciences*, **3(2):** 173-176.
39. Bartoo, A. and Badrie, N. (2005). Physicochemical, nutritional and sensory quality of stirred 'dwarf' golden apple (*Spondias cytherea* Sonn.) yoghurts. *International Journal of Food Science and Nutrition*, **56(6):** 445-454.
40. Begum, M.M.; Haque, M.; Ferdous, M.R.U.; Hasan, M.; Tarek, H.; Alam, M.N.A.; Hossain, M.B.; Uddin, M.N.; Begum, T. and Hasan, C.M. (2015). Screening of antioxidant and antimicrobial properties of the *Syzygium jambos* L. *American Journal of Bioscience*, **3(2-1):** 23-26.
41. Beel, E. (1998). Substrates for greenhouse plants cultivated with several irrigation systems. *Acta Horticulturae* No. **221:** 315-326.
42. Bera, P.K. and Mazumdar, B.C. (1992). Comparative effect of spraying cashew trees with three auxin solutions on the qualitative constituents of their apples and kernels during developmental stages. *Indian Cashew Journal*, **24(3):** 6-10.
43. Beyhan, O.; Bozkurt, M.A. and Boysal, S.C. (2011). Determination of macro and micro nutrient contents in dried fruit and leaves and some pomological characteristics of selected feijoa genotypes (*Feijoa sellowiana* Berg.) from Sakarya provinces in Turkey. *Journal of Animal and Plant Sciences,* **21(2):** 251-255.
44. Bezerra, J.E.F.; Lederman, I.E.; Freitas, E.V. de and Sanitos, V.F. dos. (1999). [Effect of grafting method and rootstock age on propagation of Surinam cherry (*Eugenia uniflora* L.)]. *Revista Brasileira de Fruiticulturu*, **21(3):** 262-265.
45. Bhadauria, S.; Pahari, G.K. and Kumar, S. (2000). Effect of *Azospirillium biofertilizer on seedling growth and seed germination on Emblica officinalis. Indian Journal of Plant Physiology*, **5(2):** 177-179.
46. Bhakare, H.A.; Khotpal, R.R. and Kulkarni, A.S. (1993). Lipid composition of *Withania somnifera, Phoenix sylvestris* and *Indigofera enulaphylla* seeds of central India. *Journal of Food Science and Technology*, **30(5):** 382-384.
47. Bhalla, S.; Verma, M. and Rawal, S. (2012). *In vitro* antioxidant activity of methanolic extract of *Aegle marmelos* leaf. *International Journal of Natural Product Science, Special Issue*, **1:** 185.
48. Bhattacharyya, A.K.; Ali, S.L. and Mazumdar, B.C. (1990). Qualitative constituents of longan (*Euphoria longan* Steud) fruits during their developmental stages. *The Andhra Agricultural Journal*, **37(2):** 192-194.

49. Bhattacharyya, A.K.; Ali, S.L. and Mazumdar, B.C. (1992). Qualitative constituents of some minor fruits during their developmental stages. *Indian Agriculturist,* **36(2):** 125-126.

50. Bhattacharyya, A.K.; Bera, P.K. and Mazumdar, B.C. (1989). Studies on nut-cracking problem of cashewnut (*Anacardium occidentale* L.) in the southern part of West Bengal. *Cashew Bulletin,* **25(8):** 1-5.

51. Bhattacharyya, A.K. and Mazumdar, B.C. (1993). Effect of spraying cashew trees with penicillin solutions on qualitative constituents of their apples and kernels. *Journal of the National Botanical Society,* **47:** 31-36.

52. Bhattacharyya, A.K.; Roy, G.C. and Mazumdar, B.C. (1988). De-greening of carambola fruits by steeping with brine water. *Science and Culture,* **54:** 373-375.

53. Bhattacharya, S.; Das, S. and Mukherjee, K.K. (1993). Biochemical studies on palm pollen. *Grana,* **32(2):** 123-127.

54. Bhattacharya, S.K. (1999). Antioxidant activity of active tannoid principles of *Emblica officinalis* (amla). *Indian Journal of Experimental Biology,* **37(7):** 676-680.

55. Bhaumik, P.K.; Guha, K.P.; Biswas, G.K. and Mukherjee, B. (1987). Flacourtin, a phenolic glycoside ester from *Flacourtia indica. Phytochemistry,* **26(11):** 3090-3091.

56. Bhowmick, N.; Deb, P. and Pradhan, S. (2013). Changes in physico-chemical attributes of Burmese grape (*Baccaurea sapida* Muell. Arg.). *International Journal of Bio-resources and Stress Management,* **4(2):** 179-181.

57. Bicalco, B. and Rezendo, C.M. (2001). Volatile compounds of cashew apple (*A. occidentale* L.). *Zeits Natural Selection, C, Biosciences,* **56(1/2):** 35-39.

58. Biswas, B.; Joshee, N.; Yadav, A. and Yadav, A.K. (2005). *In vitro* propagation and somatic embryogenesis in phalsa. *HortiScience,* **40:** 1104-1104.

59. Biswas, H.G. (1944). Pectin in fruits of the Indian persimmon (*D. embryopteris* Pers.). *Science and Culture,* **9:** 501.

60. Bora, A. and Das, R.P. (1998). Role of indole butyric acid (IBA) on vegetative growth of rooted cuttings of some minor fruits of Assam. *Journal of the Agricultural Science Society of North East India,* **11(2):** 197-201.

61. Bora, N.S.; Kakoti, B.B. and Gogoi, B. (2014). Ethno-medical claims, phytochemistry and pharmacology of *Spondias pinnata*. A Review. *International Journal of Pharmaceutical Sciences and Research,* **5(4):** 1138-1145.

62. Brijesh, S.; Daswani, P.; Tetali, P.; Anita, N. and Birdi, T. (2009). Studies on the antidiarrhoeal activity of *Aegle marmelos* unripe fruit: Validating its traditional urage. *BMC Complementary and Alternative Medicine,* **9(47):** 47.

63. Cartagena, Valenzuela, J.R. and Barreto, Osorio. (1998). [Effect of gibberellic acid and sowing method on seed germination and seedling growth of bullocks heart (*Annona reticulata* L.)]. *Agrnomia Medellin,* **51(2):** 235-244.

64. Chadha, K.L. (2012). *Handbook of Horticulture*. Directorate of Knowledge Management in Agriculture, Indian Council of Agricultural Research, New Delhi.

65. Chalannavar, R.K.; Narayanswamy, V.K.; Baijnath, H. and Odhav, B. (2012). Chemical composition of essential oil of *Psidium cattleianum* var. *lucidum* (Myrtaceae). *African Journal of Biotechnology*, **11(33):** 8341-8347.

66. Chandra, G. (1985). Investigations on essential oils and isolates of potential value at HBTI, Kanpur. *Indian Perfumer*, **29(1/2):** 23-30.

67. Chang Ching Hui; Yu Tunghsi; Lin Liyun and Chang Chiyue. (1998). [Studies on the important volatile flavor compounds of fresh and dried longan fruits]. *Journal of the Chinese Agricultural Chemical Society*, **36(5):** 521-532.

68. Chantaranothai, P. and Parnell, J.A.N. (1994). The breeding biology of some Thai *Syzygium* species. *Tropical Ecology*, **35(2):** 199-208.

69. Chao Cheng Nan; Chang Jerway and Yen Chungruy. (1997). [Variety improvement of longan (*Dimocarpus longan* Lour.) in Taiwan]. In: *Proceedings of a Symposium on Enhancing Competitiveness of Fruit Industry, Taichung, Taiwan*, March 20-21, 1997. (Edited by: Yungwu and Chang LinRen).

70. Chatterjee, D. and Randhawa, G.S. (1952). Standardized names of cultivated plants in India. I. Fruits. *Indian Journal of Horticulture*, **9(2):** 24-36.

71. Chattopadhyay, P.K. and Ghosh, A. (1996). Mineral composition of inflorescence and fruits at different stages of growth in rose apple. *Odisha Journal of Horticulture*, **20(2):** 28-31.

72. Chattopadhyay, P.K. and Mahanta, S.K. (1989). Bael: media requirements for seed germination and seedling establishment. *Indian Horticulture*, **36(3):** 27.

73. Chen Jing Ying; Chen Jing Yao and Chen Xi. (1999). [Shoot apex culture of longan (*Dimocarpus longan Lour.*) trees]. *Journal of Tropical and Subtropical Botany*, **7(2):** 159-164.

74. Chen Qixuan; Chen Lel Jin; Yu Desheng and Li Chun. (2000). [Qingshan O, a promising late variety of longan]. *South China Fruits,* (2000), **29(1):** 22.

75. Chintala, S.; Kandula, A.; Janapathi, Y.K.; Khan Md, F. and Vani, P.D. (2012). Pharmacogenistic studies on *Diospyros melanoxylon*. *International Journal of Pharmaceutical Sciences and Research*, **3(9):** 3438-3443.

76. Chovatia, R.S. and Singh, S.P. (2000). Success of air-layering in custard apple (*Annona squamosa* L.) as influenced by ringing of shoots and growth regulators. *Odisha Journal of Horticulture*, **28(2):** 61-65.

77. Chovatia, R.S. and Singh, S.P. (2006). Effect of time on ringing of shoots and applications of root promoters of air layering in jamun (*Syzygium cuminii* Skeel). *Scientific Horticulture*, **10:** 165-170.

78. Chowdhary, S.M.; Kadam, A.S.; Ramniwas and Chavan, I. (2016). Success and growth of softwood grafts of different aonla (*Emblica officinalis* Gaertn.) varieties at nursery stage. *The Bioscan*, **11(4):** 2533-2536.

79. Chowdhury, S. (1947). A *Glomerella* rot of *nuna. Current Science,* **16:** 384.

80. Chowdhury, F.; Pal, S.; Sharmin, T.; Rashid, R.B.; Sikder, M.; Kabir, S.; Rahman, M.S. and Rashid, M.A. (2013). Bioactivities of *A. chaplasha* R. and *Bougainvillea spectabilis* Willd. *Bangladesh Pharmaceutical Journal,* **16(1):** 33-37.

81. Cozza, R.; Galant, L.G.; Bitonti, M.B. and Innocenti, A.M. (1994). Effect of storage on the germination of the water chestnut (*Trapa natans* L.). *Phyton (Horn),* **34(2):** 315-320.

82. Dahiya, K.K. and Lakra, R.K. (2001). Population build up and seasonal activity of thrips. *Rhipiphorothrips cruentatus* Hood in jamun, *Syzygium cuminii* Skeels. *National Journal of Plant Improvement,* **3:** 113-114.

83. Darwesh, M.A.; El-Shiaty, O.H.; Farahat, M.M. and Ibrahim, A.S. (2011). Response of *Monstera deliciosa* Liebm. to plant spacing and nitrogen sources fertilizer. *Bulletin of the Faculty of Agriculture, Cairo University,* **62(1):** 81-92.

84. Das, M. and Sarma, B.P. (2013). Evaluation of hypoglycemic effect of an Indian fruit: *Dillenia indica* L. *International Journal of Research, Ayurveda Pharma,* **4(4):** 545-546.

85. Das, N. (2014). The effect of seed sources variation and processing treatments on the seed germination of *Acacia catechu* and *Elaeocarpus Floribundus* species in Bangladesh. *International Journal of Forestry Research,* Article ID 984194.

86. Das, S.C.; Prakash, J.; Deb, A.K. and Biswas, T. (2013). Medicinal value of underutilized fruits in hilly Tripura. *Acta Horticulturae,* **972:** 135-141.

87. De, Rupa; Mitra, Jaya; Majumder, K. and Mazumdar, B.C. (1996). Laminar nitrogen, phosphorus and potassium content during winter to summer seasons in some minor fruit crops grown in West Bengal. *Indian Biologist,* **28(2):** 37-39.

88. Deb, P. and Bhowmick, N. (2013). Physico-chemical properties of Burmese grape (*Baccaurea sapida* Muell, Arg.)–An underutilized fruit crop of West Bengal. *International Journal of Agriculture and Food Science Technology,* **4(5):** 415-420.

89. Devmurari, V.; Shivanand, P. and Jivani, N.P. (2010). A Review on *Carissa congesta*: Phytochemical constituents, traditional use and pharmacological properties. *International Journal of Chemical Science,* **8(1):** 81-87.

90. Devi, V.C.; Naveen, J. and Kannan, P.V. (2015). Process optimization of bio-ethanol production through cashew apple juice fermentation. *International Journal of ChemTech Research,* **8(1):** 79-81.

91. Dewanjee, S.; Das, A.K.; Sahu, R. and Gangopadhyay, M. (2009). Antidiabetic activity of *Diospyros peregrina* fruit: Effect on hyperglycemia, hyperlipidemia and augmented oxidative stress in experimental type 2 diabetes. *Food Chemistry and Toxicology,* **47(10):** 2679-2685.

92. Dhandar, D.G. and Bhargava, B.S. (1993). Leaf sampling technique for nutritional diagnosis in custard apple. *Indian Journal of Horticulture*, **50(1):** 1-4.

93. Dhankhar, D.S. and Kumar, S. (1996). Effect of bio-regulator on seed germination and seedling growth in aonla (*Phyllanthus emblica* L.) cv. Anand-2. *Recent Horticulture*, **3(1):** 45-48.

94. Dinesh, M.R. and Reddy, B.M.C. (2001). Evaluation of *Psidium guajava* accessions and some other *Psidium* species for fruit characters. *Journal of Applied Horticulture*, **3(1):** 41-43.

95. Ducroquet, J.P.H.J.; Santos, K.L. Dos; Andrade, E.R. De; Boneti, J.A. Da S; Bonin, V. and Nodari, R.O. (2007). The first Brazilian cultivars of feijoa: SCS 411 Alcantara and SCS 412 Helena. *Agropecuaria Catarinense*, **20(2):** 77-80.

96. Dutta, K.; Basumatary, S. and Deka, D.C. (2013). Physiochemical properties of *Baccaurea ramiflora* (Letuk) seed oil. *Journal of Chemical and Pharmaceutical Research*, **5(9):** 210-218.

97. Dutra, L.F.; Mendez, M.E.Z.; Schwengber, J.E. and Kersten, E. (2000). Effect of shading, indole butyric acid (IBA) and phloroglucinol on rooting of *Psidium cattleianum* Sabine cuttings. *Revista Clentificia Rural*, **5(1):** 95-100.

98. Egbe, E.A., Chuyong, G.B.; Fonge, A.B.; Tata, B.L. and Tabot, P.T. (2012). The effects of different concentrations of Indole-3-Butyric Acid (IBA) on leafy stem cuttings of four timber species. *Journal of Horticulture and Forestry.* **4(5):** 85-91.

99. Egbekun, M.K. and Otiri, A.O. (1999). Changes in ascorbic acid contents in oranges and cashew apples with maturity. *Ecology of Food and Nutrition,* **38(3):** 275-284.

100. Ekta Singh; Sharma, S.; Pareek, A.; Dwivedi, J.; Yadav, S. and Sharma, S. (2011). Phytochemistry, traditional uses and cancer chemopreventive activity of amla (*Phyllanthus emblica*): The sustainer. *Journal of Applied Science*, **2(1):** 176-183.

101. El-Kadder, E.M.A. and Hammad, H.H. (2012). *In vitro* propagation of *Dillenia indica. Australian Journal of Basic and Applied Sciences*, **6(7):** 452-457.

102. El Siddig, K.; Ludders, P.; Ehert, G. and Adiku, S.G.K. (1998). Response of rose apple (*Eugenia jambos* L.) to water and nitrogen supply. *Angewandte Botanik*, **72(5/6):** 203-206.

103. Emanul, M.A.; Gutierrez-Orozco, F.; Yahia, E.M. and Bankblia, N. (2012). Assessment and profiling of the fatty acids in two ackee fruit (*Blighia sapida* Koenig) varieties during different ripening stages. *Journal of Science Food and Agriculture*, **93(4):** 722-726.

104. Fan Ji Qiao; An Xin Cheng; Xia Jin Ming and Hu Qiong Bo. (2011). Bioactivity of soil dwelling entomopathogenic fungi against *Tessaratom apapillosa* (*Hemintera:* Peptatomidae). *Chinese Journal of Biological Control*, **27(2):** 197-201.

105. Fang-Rong Chang; Jien-Lin Chen; Hui-Fen Chiu; Ming-Jung Wu and Yang Chang Wu. (1998). Acetogenins from seeds of *Annona reticulata. Phytochemistry,* **47(6):** 1057-1061.

106. Fang, T.; Chen, J.Q. and Peppelenbos, H. (2010). Quality changes of fresh longan (*Dimocarpus longan*) under different storage condition. *Acta Horticulturae,* **863:** 593-598.

107. Fawzy, M.R.H. (2011). *Physalis peruviana*: A rich source of bioactive phytochemicals for functional foods and pharmaceuticals, *Food Reviews,* **27:** 259-273.

108. Ferro, E.; Schinini, A.; Maldonado, M.; Rosner, J. and Schmeda Hiraschmann, G. (1988). *Eugenia uniflora* leaf extract and lipid metabolism in *Cobs apella* monkeys. *Journal of Ethnopharmacology,* **24(2-3):** 321-325.

109. Figueiredo, R.W. de; Lajolo, F.M.; Ailves, R.E. and Filgueiras, H.A.C. (2002). Physico-chemical changes in early dwarf cashew pseudofruits during development and maturation. *Food Chemistry,* No.**485:** 155-160.

110. Figueiredo, R.W.; Lajolo, F.M.; Alves, R.E.; Figueiras, H.A.S. and Araujo, N.C.C. (2001). [Change of firmness, pectins and pectolytic enzymes during the development and maturation of fruits of early dwarf cashew (*A. occidentale* L. var. *nanum*) CCP-76]. *Proceedings of the Interamerican Society for Tropical Horticulture,* **43:** 82-86.

111. Filgueiras, H.A.C.; Alvers, R.E.; Mosca, J.L. and Menezes, J.B. (1999). Cashew apple for fresh consumption. *Acta Horticulturae,* No.**485:** 155-160.

112. Flores, G.; Dastmalchi, K.; Wu, S.B.; Whalen, K.; Dabo, A.J.; Reynertson, K.A.; Foronjy, R.F.; D. Armiento, J.M. and Kennelly, E.J. (2013). Phenolic-rich extract from the Costa Rican guava (*Psidium friedrichsthalianum*) pulp with antioxidant and anti-inflammatory activity. Potential for COPD therapy. *Food Chemistry,* **141(2):** 889-895.

113. Fu Ping, Z.; Xiao Chun L.; Manxia, Du; Huanbin, C.; Yuanqing, W. and Xuedong, C. (2012). Effects of plant growth regulators on cutting propagation. *Guizhou Agricultural Sciences,* **8:** 176-178.

114. Gao, Y.; Hu, Q. and Li, X. (2012). Chemical composition and antioxidant activity of essential oil from *Syzygium samarangense* (Bl.) Merry *et* Perry flower-bud. *Spatula* DD, **2(1):** 23-33.

115. Garnde, V.K.; Joshi, G.D.; Magdum, M.B. and Waskar, D.P. (1998). Studies on physical changes during growth and development of jamun (*Syzygium cuminii* Skeels) fruit. *Agricultural Science Digest,* **18(3):** 206-208.

116. Gatho, A.S.; Lopes, N.F.; Raseira, A. and Bacarin, M.A. (2000). Growth of *Psidium cattleianum* Sabine fruits. *Revista Brasileira de Fruiticulturae,* **22(2):** 223-225.

117. Ghayur, M.N.; Gilani, A.H.; Khan, A.; Amor, E.C.; Villasenor, I.M. and Choudhary, M.I. (2006). Presence of calcium antagonist activity explains the use of *Syzygium samarangense* in diarrhoea. *Phytotherapy Research,* **20(1):** 49-52.

118. Gholap, S.V.; Dod, V.N.; Bhuyar, S.A. and Bharad, S.G. (2000 a). Effect of plant growth regulators on seed germination and seedling growth in aonla (*Phyllanthus emblica* L. under climatic condition of Akola. *Crop Research* (Hissar), **20(3):** 546-548.

119. Gholap, S.V.; Dod, V.N.; Bharad, S.G. and Wankar, A.M. (2000 b). Studies on vegetative propagation in bullock's heart (*Annona reticulata* L.). *Crop Research* (Hisar), **20(1):** 158-159.

120. Ghosh, A.K.; Nagar, P.K. and Sircar, P.K. (1985). Endogenous gibberellins in developing fruits of *Ziziphus mauritiana* Lam. and *Aegle marmelos* Correa. *Journal of Plant Physiology*, **120(5):** 381-388.

121. Ghosh, P.; Bandopaddhyay, A.K.; Thakur, S.; Akihisa, T.; Simizu, N.; Tamura, T. and Matsumoto, T. (1989). Acidissimin, a new tyramine derivative from the fruit of *Limonia acidissima*. *Journal of Natural Products*, **52:** 1323-1326.

122. Ghosh, S.N.; Bera, B. and Roy, S. (2011). Propagation studies in karonda (*Carissa carandas*) under Paschim Midnapore condition of West Bengal. *Journal of Crop and Weed*, **7(2):** 247-248.

123. Gomes, J.; Hossain, G.; Awal, A. and Haq, Q.N. (1986). Polysaccharide components in the pulp and seeds of *Eugenia jambolana* L. (*Syzygium cumini*). *Bangladesh Journal of Scientific and Industrial Research*, **3/4:** 186-194.

124. Gopalan, G.: Ramasastri, B.V. and Balasubramanian, S.C. [Revised and updated by B.S. Narasinga Rao; Deosthale, Y.G. and Pant, K.C.]. (2018). *Nutritive Value of Indian Foods*, National Institute of Nutrition, Indian Council of Medical Research, Hyderabad.

125. Gopalan, I. and Mohanram, M. (2000). *Fruits*. National Institute of Nutrition, Indian Council of Agricultural Research, Hyderabad.

126. Gordon, A.; Friedrich, M.; Matta, V.M. da; Moura, C.F.H. and Marx, F. (2012). Changes in phenolic composition, ascorbic acid and antioxidant activity in cashew apple (*Anacardium occidentale* L.) during ripening. *Fruits (Paris)*, **67(4):** 267-276.

127. Goyal, A.K.; Mishra, T. and Sen, A. (2013). Antioxidant profiling of latkan. *Indian Journal of Biotechnology*, **12:** 137-139.

128. Goyal, D. and Bhadauria, S. (2007). *In vitro* proliferation in *Emblica officinalis* Gaertn. *var.* Balwant from nodal explants. *Indian Journal of Biotechnology*, **17:** 394-397.

129. Guerra, M.P.; Cangahualainocente, G.C.; Vesco, L.L.; Pescador, R. and Caprestano, C.A. (2013). Micropropagation systems of Feijoa. [*A. sellowiana* (O. Berg) Burret]. *Methods of Molecular Biology*, **11013;** 45-62.

130. Guha, Gopa and Mazumdar, B.C. (1996). Vitamin C content in different layers of fruits. *Applied Nutrition*, **21(2):** 40.

131. Gummadi, V.P.; Battu, G.R.; Kirthana, D.M.S. and Manda, K. (2016). A review on palmyra palm (*Borassus flabellifer*). *International Journal of Pharmaceutical Research*, **8(2):** 17-20.

132. Gupta, P.; Patni, V. and Kant, U. (1997). *In vitro* shoot differentiation in *Emblica officinalis* Gaertn. *Journal of Physiological Research,* **7(2):** 171-172.

133. Gupta, R. and Saxena, A.M. (2011). Hypoglycemic and anti-hyperglycemic activities of *Syzygium cumini* (Linn.) Skeels whole fruit, in normal and streptozoetin-induced diabetic rats. *Asian Journal of Pharmaceutical and Biological Research,* **1(3):** 262-272.

134. Gurbachan Singh; Bhist, J.K.; Singh, N.T. and Dagar, J.C. (1998). Evaluation of fruit plants for sodicity tolerance. *Range Management and Agroforestry,* **19(1):** 87-91.

135. Hamzah, R.U.; Egwim, E.C.; Kabiru, A.Y. and Muazu, M.B. (2013). Phytochemical and *in vitro* antioxidant peroperties of the methanolic extract of fruits of *Blighia sapida, Vitellaria paradoxa* and *Vitex doniana. Oxidation and Antioxidation Medical Science,* **2(3):** 217-223.

136. Handique, P.J. and Bhattacharjee, S. (2000). *In vitro* shoot proliferation of wood apple (*Feronia elephantum*). *Advances in Plant Sciences,* **13(1):** 241-243.

137. Haq, M.Z.U.; Stancovic, M.S.; Rizwan, K. and Feo, V.D. (2013). *Grewia asiatica* L., a food plant with multiple uses. *Molecules,* **18(3):** 2663-2682.

138. Harish, C.A. and Vijay, K.P. (2010). Nutritive and mineral value of ripe fruits: Evaluation of nutritive and mineral value in ripe fruits of *Spondias pinnata* from two locations of Western Himalaya, India. *International Journal of Phytomedicines and Related Industries,* **2(3):** 233-236.

139. Hasan, M.A.; Ghosh, A.; Chowdhury, R.R.; Sarkar, S. and Chattopadhyay, P.K. (2006 a). Studies on the physiology of fruit growth and development of longan (*Euphoria longan*). *Proceedings of the National Symposium on Production, Utilization and Export of Underutilized Fruits with Commercial Potentialities, Kalyani, West Bengal,* 22-24 November, 2006**:** 184-187.

140. Hasan, M.A.; Ghosh, A.; Chowdhury, R.R.; Sarkar, S. and Chattopadhyay, P.K. (2006 b). Changes in mineral composition during growth and development of fruits of longan (*Euphoria longan*). *Proceedings of the National Symposium on Production, Utilization and Export of Underutilized Fruits with Commercial Potentialities, Kalyani, West Bengal,* 22-24 November, 2006**:** 188-191.

141. Hasmah, S.N.; Bhatt, A. and Keng, C.L. (2013). Micropropagation of Asam karanda (*Carissa carandas* Linn.). *Pertanika Journal of Tropical Agricultural Science,* **36(1):** 89-98.

142. Haung, Jinsong; Xu Xiudan; Zhang, Shaoquan; Zhu Tangming and Li Chun. [A new large fruited, late ripening longan cultivar–Quingke Baoyuan]. *China Fruits,* No.**4:** 4, 5, 7.

143. Hayes, W.B. (1970 Reprinted). *Fruit Growing in India*. Kitabistan, Allahabad, India.

144. Heredia, J.B.; Siller, J.H.; Barz, M.A.; Araiza, E.; Portillo, T.; Garcia, R. and Muy, M.D. (1998). [Changes in the quality and content of carbohydrates in tropical and sub-tropical fruits at the supermarket level]. *Proceedings of the Intraamerican Society for Tropical Horticulture*, **41:** 104-109.

145. Hernandez Sanchev, M.; M. de la L. Hernandez, Fluentes, A.D.; Elorja Martinez, P.; Lopez Herrera, M. and Lopez Jimenez, M.A. (2009). Fruit characteristics of star apple (*Chrysophyllum* cainito L.) at Varacruz State, Mexico. *Revista Cientifica Udo Agricola*, **9(1):** 70-73.

146. Hiregowder, L.V.; Kumar, H.G.A. and Murthy, H.N. (2005). *In vitro* culture of *Feronia limonia* (L.) Swingle from hypocotyl and internodal explants. *Biologia Plantarum*, **49(1):** 41-45.

147. Hongyang, L.; Yongchun, K.; Zengtog, X.; Hong, R. and Jun, X. (2013). Advances in research on processing technology of *Syzygium samarangense*. *Food Research and Development*, **34(7):** 131-134.

148. Hoque, A.; Nahar, A.; Razvy, M.A.; Biswas, M.K. and Kabir, A.H. (2006). Micropropagation of water chestnut (*Trapa* sp.) through local varieties of Rajshahi division. *Asian Journal of Plant Sciences*, **5:** 409-413.

149. Hore, J.K. and Sen, S.K. (1994). Effect of non-auxinic compounds and IBA on root formation in stem cuttings of rose apple (*Syzygium jambos* Alston). *Crop Research* (Hissar), **7(1):** 44-48.

150. Hore, J.K. and Sen, S.K. (1995). Effect of seed treatment on germination, seedling growth and longevity of bael (*Aegle marmelos* C.) seeds. *Agricultural Science Digest*, **15(1/2):** 26-30.

151. Hossain, M.; Islam, R.; Karim, M.R. and Rahman, S.M. (1994). Production of plantlets from *Aegle marmelos* nucellar callus. *Plant Cell Reports*, **13(10):** 570-573.

152. Hussain, M.D.; Hussain, M.I. and Alam, M. (1992). Juice harvesting from date and palmyra palm tree in Bangladesh. *Indian Journal of Agricultural Engineering*, **2(1):** 17-24.

153. Ibrahim, M.A.; Jasim, A.M. and Abbas, M.F. (2012). *In vitro* plant regeneration of Indian jujube (*Ziziphus mauritiana* Lamk.) *ev.* Zaytoni via indirect organogenesis. *Akta Agriculturae Slovenica*, **99-1:** 65-67.

154. Ishak, S.A.; Ismail, N.; Noor, M.A.M. and Ahmad, H. (2005). Some physical and chemical properties of ambarella (*Spondias cytherea* Sonn.) at three different stages of maturity. *Journal of Food Composition and Analysis*, **18(8):** 819-827.

155. Islam, R.; Hossain, M.; Joarder, O.I. and Karim, M.R. (1993). Adventitious shoot formation on excised leaf explants of *in vitro* grown seedlings of *Aegle marmelos* Corr. *Journal of Horticultural Science*, **68(4):** 495-498.

156. Itankar, P.R.; Lokhande, S.J.; Verma, P.R.; Arora, S.K.; Sahu, A. and Patil, A.T. (2011). Antidiabetic potential of unripe *Carissa carandas* Linn. fruit extract. *Journal of Ethnopharmacology*, **135(2):** 430-433.

157. Jacob, A.; Pandey, M.; Kapoor, S. and Saroja, R. (1988). Effect of the Indian gooseberry (amla) on serum cholesterol levels in men aged 35-55 years. *European Journal of Clinical Nutrition*, **42(11):** 939-944.

158. Jain, N. and Babbar, S.B. (2000). Recurrent production of plants of black plum, *Syzygium cuminii* (L.) Skeels, a myrtaceous fruit tree, from *in vitro* cultured seedling explants. *Plant Cell Reports*, **19(5):** 519-524.

159. Jain, N.L. and Lal, G. (1954). Studies on the preparation of amla syrup. *Bulletin of the Central Food Technological Research Institute*, **3:** 297-300.

160. Jana, H. and Jana, S. (2017). Palmyra palm: Importance in Indian Agriculture, *Rashtriya Krishi*, **12(2):** 35-40.

161. Jaiswal, H.R.; Misra, K.K. and Singh, V. (1999). Clonal variation in growth and yield of bael (*Aegle marmelos* Correa). *Scientific Horticulture*, **6:** 31-36.

162. Jauhari, O.S. and Singh, R.D. (1971). Bael–Valuable fruit. *Indian Horticulture*, **16:** 9-10.

163. Jauhari, O.S.; Singh, R.D.; and Awasthi, R.K. (1969). Survey of some important varieties of bael (*Aegle marmelos* Corr.). *Punjab Horticultural Journal*, **9:** 48-53.

164. Jayant Kumar and Parmar, C. (2000). Standardization of sexual and asexual propagation techniques for some wild fruits of sub-himalayan region. *Indian Forester*, **26(8):** 870-873.

165. Jena, B.S.; Jayaprakasha, G.K. and Sakariah, K.K. (2002). Organic acids from leaves, fruits and rinds of *Garcinia cowa. Journal of Agricultural and Food Chemistry*, **50(12):** 3431-3434.

166. Jerry, A. (2018). A comprehensive review on the medicinal properties of *Borassus flabellifer. Journal of Academia and Industrial Research*, **7(7):** 93-97.

167. Jiang, J.; Joyce, D. and Lin, H. (2011). *Longan: Postharvest Biology and Technology of Tropical and Subtropical Fruits.* Vol.3: Cocona to Mango.**:** 408-423.

168. Jiang, Y.; Zhang, Z.; Joice, D.C. and Ketsa, S.I. (2002). Postharvest biology and handling of longan fruit (*Dimocarpus longan* Lour.). *Postharvest Biology and Technology*, **26:** 241-252.

169. Jimoh, T.O.; Buoro, A.T. and Muriana, M. (2012). Utilization of *Blighia sapida* (ackee apple) pod in the removal of lead, cadmium and cobalt ions from aqueous solution. *Journal of Environmental Chemistry and Ectotoxicology*, **4(10):** 178-182.

170. Joglekar, S.N. and Gaitonde, B.B. (1970). Histamine releasing activity of *Carissa carandas* roots (*Apocynaceae*). Japanese Journal of Pharmacy, **20:** 367-372.

171. Johar, D.S. and Anand, J.C. (2003). Brining and pricklings of fruits and vegetables. In: *Fruit and Vegetable Preservation* (Ed. by Bhutani, R.C.). Biotech Books, Delhi.

172. Johnson, L.A.S. and Briggs, B.G. (1984). Myrtales and Myrtaceae–a phylogenetic analysis. *Annals of the Missouri Botanic Garden*, **71:** 700-756.

173. Joshee, N.; Bastola, D.R.; Agrawal, V.P. and Yadav, A.K. (2002). Lakoocha: A multipurpose tree of warm climate. In: Janick, A. and Whipkey, A. (eds.) *Trends in New Crops and New Uses*. ASHS Press, Alexandria, V.A.

174. Kadono, Y. and Schneider, E.L. (1986). Floral biology of *Trapa natans* var. *japonica. Botanical Magazine*, **99(1056):** 435-439.

175. Kaladhar, D.S.V. and Rayavarapu, K.A. (2014). Phytochemical analysis, antioxidant and antimicrobial activities of *Annona reticulata new fruit* peel extracts. *World Journal of Pharmacy and Pharmaceutical Sciences*, **3(11):** 1226-1226.

176. Kamaluddin, M.; Nath, T.K. and Jashimuddin, M. (1998). Indigenous practice of khejur palm (*Phoenix sylvestris*) husbandry in rural Bangladesh. *Journal of Tropical Forest Science*, **10(3):** 357-366.

177. Kandiah, S. and Kokudathasan, S. (1987). EDTA–stimulation of sap (toddy) flow from inflorescence during tapping in palmyra palm (*Borassus flabellifer* L.). *Vinganam Journal of Science*, **2(1-2):** 58-61.

178. Kandiah, S. and Mahendran, S. (1986). A new method for culturing palmyrah palm (*Borassus flabellifer* L.). *Vinganam Journal of Science*, **1(1):** 40-43.

179. Kant, U. and Verma, B. (1997). Propagation of *Emblica officinalis* Gaertn. through tissue culture. [In: *Plant Tissue Culture and Biotechnology: Emerging Trends*]. *Proceedings of Symposium*, January, 1997.

180. Kantharajah, A.S.; Richards, G.D. and Dodd, W.A. (1992). Roots as a source of explants for the successful micropropagation of carambola (*Averrhoa carambola* L.). *Scientia Horticulturae*, **51(1-2):** 169-177.

181. Kapoor, N.; Bedi, K.L.; Tikoo, M.K.; Kapoor, S.K. and Sarin, Y.K. (1990). Some lesser known oilseeds of India. *Journal of the Oil Technologists' Association of India*, **22(3):** 65-66.

182. Kapoor, S.; Ranote, P.S. and Sharma, S. (2015). Antioxidant potentials and quality aspects of jamun (*Syzygium cumini* L.) supplemental unleavened flat bread (Indian chapatti). *Journal of Applied and Natural Science*, **7(1):** 309-315.

183. Karale, A.R.; Keskar, B.G.; Dhawale, B.C.; Shete, M.B. and Kale, P.N. (1991). Flowering, sex expression and sex ratio in Banarasi aonla seedling trees. *Journal of Maharastra Agricultural Universities*, **16(2):** 270-271.

184. Karale, A.R.; Keskar, B.G.; Shete, M.B.; Dhawale, B.C.; Kale, P.N. and Choudhari, K.G. (1989). Seedling selection in karonda (*Carissa carandas* L.). *Maharashtra Journal of Horticulture*, **4:** 125-129.

185. Karami, M.; Saeidnia, S. and Nosrati, A. (2013). Study of the hepatoprotective activity of methanolic extract of *Feijoa sellowiana* fruits against MDMA using the isolated rat liver perfusion system. *Iranian Journal of Pharmaceutical Research*, **12(1):** 85-91.

186. Karami, M.; Salehi, H. and Naghshvar, F. (2008). Study of histopathology and antioxidant activity of methanolic extract of *Feijoa sellowiana* against dosage induced by MDMA in mouse liver. *Pharmacology online*, **3:** 315-321.

187. Ke Guanwu; Wang Changchun and Tang Zifa. (1994). Palynological studies on the origin of longan cultivation. *Acta Horticulturae Sinica*, **21(4):** 323-328.
188. Keskar, B.C.; Dhawale, B.C.; Karale, A.R. and Kale, P.N. (1991). Propagation techniques in aonla. *Journal of Maharashtra Agricultural University*, **6(2):** 282.
189. Keskar, B.C.; Karale, A.R.; Dhawale, B.C. and Choudhary, K.G. (1989). Improvement of jamum (*Syzygium cumini* Skeels) by selection. *Maharashtra Journal of Horticulture*, **4:** 117-120.
190. Khan, K.H. (2009). Roles of *Emblica officinalis* in medicine–A Review. *Botany Research International*, **2(4):** 2221-3635.
191. Khattab, M.; Kamel, H. and Yacout, M. (1987). Nitrogen and potassium nutrition of *Monstera delicios* Liebm. *Alexandria Journal of Agricultural Research*, **32(2):** 277-288.
192. Khatun, M.M.; Karim, M.R.; Molla, M.M.; Khatun, M.M. and Rahman, M.J. (2012). Study on the physico-chemical characteristics of longan (*Euphoria longana*) germplasm. *Bangladesh Journal of Agricultural Research*, **37(3):** 441-447.
193. Klinac, D.J. (1986). Cape-gooseberry (*Physalis peruviana*) production systems. *New Zealand Journal of Experimental Agriculture*, **14:** 425-430.
194. Krishnaveni, M. and Mirunalini, S. (2010). Therapeutic potential of *Phyllanthus emblica* (amla) the ayurvedic wonder. *Journal of Basic Clinical Physiology and Pharmacology*, **21(1):** 93-105.
195. Kubo, I.; Masamitsu, O.; Vieira, P.C. and Komatsu, S. (1993). Anti-tumour agents from the cashew (*Anacardium occidentale*) apple. *Journal of Agriculture and Food Chemistry*, **41(6):** 1012-1015.
196. Kubo, J.; Lee, J.R. and Kubo, I. (1999). Anti-*Helicobacter pylori* agents from the cashew apple. *Journal of Agriculture and Food Chemistry*, **47(2):** 533-537.
197. Kumar, D.; Mallick, S.; Vedasiromoni, J.R. and Pal, B.C. (2010). Anti-leukemic activity of *Dillenia indica* L. fruit extract and quantification of betulinic acid by HPLC. *Phytomedicine*, **17(6):** 431-435.
198. Kumar, N.A.; Lakshmana; Jagadeesh, S.L. and Laxmankuknor. (2015). Effect of postharvest treatments of calcium on cashew apple (*Anacardium occidentale* L.) 'Ullal-2' during storage. *Acta Horticulturae*, **1080:** 377-382.
199. Kumar, P.H. and Sreedharan, C. (1987). Correlation studies between leaf nutrients and fruit quality characters in cashew (*Anacardium occidentale* L.). *Indian Cashew Journal*, **18(3):** 15-16.
200. Kumar, R.; Tiwari, J.P. and Misra, K.K. (2003). *In vitro* cloning of Chinese guava (*Psidium friedrichsthalianum*) (Berg.)(Nierdz). *Advances in Horticulture and Forestry*, **9:** 59-69.
201. Kumar, S.; Gupta, P. and Virupaksha Gupta, K.L. (2013). A critical review on karamarda (*Carissa carandas* Linn.). *Journal of Pharmaceutical and Biological Archives*, **4(4):** 637-642.

202. Kumar, S. and Tandon, M.P. (1982). A new fruit rot of *Carissa*. *Current Science*, **51:** 1119.

203. Kumar, V. and Patel, P.N. (1968). Symptoms and host range in a bacterial disease on *Carissa carandas* L. *Indian Phytopathology*, **21:** 315-315.

204. Kumar, V.K. (2001). *Tendu* (*Diospyros* species). Daya Publishing House, Delhi.

205. Kumri, M.; Majumder, K. and Mazumdar, B.C. (1997). Qualitative constituents of some types of bael (*Aegle marmelos* Cor.) fruits growing in West Bengal. *Indian Biologist*, **29(2):** 52-54.

206. Lakshmi, R. (2013). Postharvest storage and value addition of jamun fruits (*Syzygium cuminii* L.). *Mysore Journal of Agricultural Science*, **47(2):** 456.

207. Lal, A. and Singh, R.P. (1953). Anthracnose disease of *Carissa carandas* Linn. caused by *Colletotrichum inamdarii*. *Journal of Indian Botanical Society*, **32(1,2):** 54-63.

208. Landrum, L.R.; Clark, W.D.; Sharp, W.P. and Brendecke, J. (1995). Hybridization between *Psidium guajava* and *P. guineense* (Myrtaceae). *Economic Botany*, **49(2):** 153-161.

209. Leite, T.S.; Augusto, P.E.D. and Christianini, M. (2015). Using high pressure homogenization (HPH) to change the physical properties of cashew apple juice. *Food Biophysics*. **10(2):** 169-180.

210. Li Ji Hong; Shao Hanshuang; Huang Guixiu and Fu Shaop. (1999). [Experiment of fast propagation of carambola *in vitro*]. *China Fruits*, No.**4:** 37.

211. Linnaeus, C.V. (1753). *Species Plantarum*, **1:** 470.

212. Lira, D.N.; Uddin, M.A.; Uddin, M. and Rouf, A.S.S. (2014). Assessment of cytotoxic activities of *Phyllanthus amarus* and *Monstera deliciosa*. *Journal of Applied Pharmaceutical Science*, **4(7):** 110-113.

213. Litz, R.E. (1984). *In vitro* responses of adventitious embryos of two polyembryonic *Eugenia* species. *HortScience*, **19(5):** 720-722.

214. Liu Chun Li; Guan Xiao Li; Li Dian Peng and Huang Yong Lin. (2014). Study on the chemical constituents of *Dimocarpus longan* pericarp. (Chinese). *Guangxi Zhiwu Guihaia*, **34(2):** 167-169, 173.

215. Li Yu Qi Ao and Zhou Cuiping. (1997). [Study on the tissue culture for *Monstera deliciosa* Liebn.]. *Journal of Jiangsu Forestry Science and Technology*, **24(1):** 42-45.

216. Lo Jo, J. de and Benedetto, A. di. (2014). Biomass accumulation and leaf shape can be modulated by an exogenous spray of 6-benzyl amino purine in the ornamental foliage plant, *Monstera deliciosa* (Liebm.). *Journal of Horticultural Science and Biotechnology*, **89(2):** 136-140.

217. Lozano Veronica, V. and Luis Fernando, O.V. (2013). Changes in the antioxidant activity during the development of fruits of *Psidium cattleianum* Sabine (Myrtaceae). *Revista Faculted Nacional de Agronomia Medellin*, **66(1):** 6939-6947.

218. Lu, P.L. and Chinho, L. (2011). Physiology of fruit caracking in wax apple (*Syzygium samarangense*). *Journal of Plant Science*, **8:** 70-76.

219. Luo, X.D.; Basile, M.J. and Kennelly, E.J. (2000). Polyphenolic antioxidants from the fruits of *Chrysophyllum cainito* L. (Star apple). *Journal of Agriculture and Food Chemistry*, **50(1):** 1379-1382.

220. Luo, X.D.; Basile, M.J. and Kennelly, E.J. (2007). Chemical composition and physical properties of African star apple (*Chrysophyllum albidum*). *Agriculture and Enviornment*, **7(1):** 37-42.

221. Macleod, J.K.; Moeller, P.D.R.; Bandara, B.M.R.; Gunatilaka, A.A.I. and Wijeratne, E.M.K. (1989). *Acidissimin*, a new limonoid from *Limonia acidissima*. *Journal of Natural Products*, **52(4):** 882-885.

222. Madhava Rao, V.N. (1967). *Cashewnut Cultivation in India*. Ministry of Food, Agriculture, Community Development and Co-operation, Govt. of India, New Delhi.

223. Maharachchi Kumbura, S.S.N.; Liang Dong, G.; Chukeatirote; McKenzie, E.H.C. and Hyde, K.D. (2013). A destructive new disease of *Syzygium samarangense* in Thailand caused by the new species, *Pestalotiopsis samarangensis*. *Tropical Plant Pathology*, **38(3):** 227-235.

224. Mahato, S.B.; Banerjee, S.K. and Chakravarti, R.N. (1971). Triterpenes of the stem bark of *Artocarpus chaplasha*. *Phytochemistry*, **10(6):** 1351-1354.

225. Maikup, M.; Majumder, K. and Mazumdar, B.C. (1997). Important constituents in apple and kernel of three types of cashew grown in southern part of West Bengal. *South Indian Horticulture*, **45(5 and 6):** 299-300.

226. Maiti, A.; Cuendet, M.; Kondratyuk, T.; Croy, V.L.; Pezzuto, J.M. and Cushman, M. (2007). Synthesis and cancer chemopreventive activity of zapotin, a natural product from *Casimiroa edulis*. *Journal of Medicinal Chemistry*, **50(2):** 350-355.

227. Maithi, C.S.; Nath, A., and Sen, S.K. (1999). Studies on the propagation of bael (*Aegle marmelos* Correa) by different grafting methods in West Bengal. *Journal of Applied Horticulture*, **1(2):** 131-132.

228. Majumder, K. (1990). *Involvement of Ethylene on Changing Pattern of Pectic Polysaccharides and Related Enzymes in Developing Fruits of Physalis peruviana* L. Ph.D. thesis submitted to the Calcutta University under supervision of research by Dr B.C. Mazumdar.

229. Majumder, K. and Mazumdar, B.C. (2001). A note on pectin content in some underutilized minor fruits of India. *Science and Culture*, **67(3-4):** 125.

230. Majumder, K. and Mazumdar, B.C. (2001). Effects of auxin and gibberellin on pectic substances and their degrading enzymes in developing fruits of cape-gooseberry (*Physalis peruviana* L.). *Journal of Horticultural Science and Biotechnology*, **76(3):** 276-279.

231. Majumder, K. and Mazumdar, B.C. (2002). Changes of pectic substances in developing fruits of cape-gooseberry in relation to the enzyme activity and evolution of ethylene. *Scientia Horticulturae*, **96:** 91-101.

232. Majumder, K. and Mazumdar, B.C. (2005). Ethephon-induced fractional changes of pectic polysaccharides in developing cape-gooseberry (*Physalis peruviana* L.) fruits. *Journal of the Science of Food and Agriculture*, **85:** 1222-1226.

233. Malla, A.A.; Sharma, R.M.; Singh, A.K. and Masoodi, F.A. (2008). Propagation and pinching studies in cape-gooseberry (*Physalis peruviana* L.). *Journal of Research, Skuast-J*, **7(1):** 89-98.

234. Manda, H.; Vyas, K.; Pandya, A. and Singhal, G. (2012). A complete review on: *Averrhoa carambola. World Journal of Pharmacy and Pharmaceutical Sciences*, **1(1):** 17-33.

235. Mandal, U. (1997). *Quantitative Changes of Important Constituents in Some Minor Fruits During Development and Improvement of the Constituents in a Minor Fruit of Agro-industrial Importance*. Ph.D. thesis submitted to the Calcutta University, under supervision of research by Dr. B.C. Mazumdar.

236. Mandal, U.; Majumder, K. and Mazumdar, B.C. (1997). Comparative study of some qualitative constituents of water-chestnuts. *Indian Biologist*, **24(2):** 37-38.

237. Mandal, U. and Mazumdar, B.C. (1995). Physico-chemical constituents of some tropical and sub-tropical minor fruits of India. *Proceedings of the National Symposium on Sustainable Agriculture in Sub-humid Zone*, March 3-5, 1995. Institute of Agriculture, Sriniketan, West Bengal.

238. Mandal, U. and Mazumdar, B.C. (1997). Qualitative characteristics of some summer season tropical and sub-tropical minor fruits grown in West Bengal. *Indian Agriculturist*, **41(4):** 291-293.

239. Mandal, U. and Mazumdar, B.C. (1998). Qualitative constituents in some tropical and sub-tropical minor fruits during development. *Indian Biologist*, **30(1):** 70-80.

240. Mandal, U. and Mazumdar, B.C. (2000). Qualitative improvement of some minor fruits by foliar application of plant nutrients and growth regulants. *Indian Agriculturist*, **44(1 and 2):** 93-96.

241. Mandal, U. and Mazumdar, B.C. (2008). Effect of preharvest growth regulator sprays on sugar content of karonda (*Carissa carandas* L.) fruits. *Haryana Journal of Horticultural Science*, **37(1 and 2):** 91-92.

242. Mandal, U.; Sinha Roy, S. and Mazumdar, B.C. (1992). A recently developed agro-industry in the southern suburb of Calcutta city, using a bramble fruit. *Indian Journal of Landscape Systems and Ecological Studies*, **15(1):** 100-102.

243. Marcelino Ponce, J. (1986). [Performance of *Annona reticulata* grafted on various rootstocks]. *Proceedings of the Tropical Region, American Society for Horticultural Science*, **23:** 119-121.

244. Martins, A.B.G. and Antunes, E.C. (2000). Propagation of rose apples (*Syzygium jambos* L. Alston) by air layering. *Revista Brasileira de Fruiticultura*, **22(2):** 205-207.

245. Mathur, N.; Ramawat, K.G. and Nandwani, D. (1995). Rapid *in vitro* multiplication of jujube through mature stem explants, *Plant Cell Tissue and Organ Culture*, **43:** 75-77.
246. Maurya, R.P.; Lewis, D.M. and Chandler, J.A. (2013). Studies on the propagation of Jamaican Ackee (*Blighia sapida* L.) by air-layering. *HortScience*, **48(No.10):** 1298-1300.
247. Mazumdar, B.C. (1975). Physico-chemical analyses of some types of bael (*Aegle marmelos* Correa) fruits growing in West Bengal. *Indian Agriculturist*, **19(3):** 295-298.
248. Mazumdar, B.C. (1979). Cape-gooseberry–the Jam fruit: Cultivation in India. *World Crops* (U.K.), **31(1):** 19, 23.
249. Mazumdar, B.C. (1982). [*Less known water-chestnut.*] *Basundhara*, Govt. of West Bengal, **34:** 21-24.
250. Mazumdar, B.C. (1983). [Cultivation of bael–Improved method.] *Basundhara*, Govt. of West Bengal, **35:** 25-27.
251. Mazumdar, B.C. (1984 a). [Emblica: The fruit of winter season] *Basundhara*, Govt. of West Bengal, **36:** 23-25.
252. Mazumdar, B.C. (1984 b). [Palmyra and wild date in afforestation.] *Dhanadhanney*, Ministry of Information and Broadcasting, Govt. of India, **16(6):** 20-21.
253. Mazumdar, B.C. (1985 a). Water-chestnut–the aquatic fruit: Cultivation in India. *World Crops* (U.K.), **37:** 42-44.
254. Mazumdar, B.C. (1985 b). Studies on quality of fruits and yield of sweet lime (*Citrus limettioides* Tan.) due to spraying with 2,4-D solutions. *Plant and Nature*, **3(1):** 45-50.
255. Mazumdar, B.C. (1986). Qualitative behaviour of different types of emblica (*Phyllanthus emblica* L.) fruits and loss of their vitamin C by steeping with brine water. *Indian Biologist*, **18(1):** 34-36.
256. Mazumdar, B.C. (1989). A review on anti-fertility properties of higher plants. *Indian Journal of Landscape Systems and Ecological Studies*, **12(2):** 53-60.
257. Mazumdar, B.C. (1990). Polyembryony in seeds of some fruit trees. *The Andhra Agricultural Journal*, **37(3):** 311-312.
258. Mazumdar, B.C. (2004 a). *Minor Fruit Crops of India: Tropical and Sub-tropical.* Daya Publishing House, Darya Ganj, Delhi.
259. Mazumdar, B.C. (2004 b). *Orchard Irrigation and Soild Management Practices.* Daya Publishing House, Darya Ganj, Delhi.
260. Mazumdar, B.C. (2004 c). *Principles and Methods of Orchard Establishment.* Daya Publishing House, Darya Ganj, Delhi.
261. Mazumdar, B.C. (2007 a). Possible attempts on making improvements of some horticultural nursery practices. *Indian Journal of Landscape Systems and Ecological Studies*, **30(1):** 17-26.

262. Mazumdar, B.C. (2007 b). Productivity and kernel quality of water-chestnut [*Trapa natans* L. var. *bispinos* (Roxb.) Makino]. *Journal of Research, Birsa Agricultural University,* Ranchi, Jharkhand, **19(1):** 99-101.

263. Mazumdar, B.C. (2007 c). Need to thrust on minor fruit crops in western tableland of West Bengal. *Indian Journal of Landscape Systems and Ecological Studies.* **30(2):** 35-36.

264. Mazumdar, B.C. (2009). Pectic substances: Use in industries and survey of studies in some horticultural fruits. *Indian Journal of Landscape Systems and Ecological Studies,* **33(1):** 11-14.

265. Mazumdar, B.C. (2014). Palmyra–The invaluable palm of India. *Kurukshetra,* Ministry of Rural Development, Govt. of India, New Delhi, **62(7):** 43-45.

266. Mazumdar, B.C. (2016). Quality of an inferior type of black plum [*Syzygium cuminii* (Linn.) Skeels] of West Bengal claimed to have very high medicinal value. *Indian Journal of Landscape Systems and Ecological Studies,* **39(2):** 247.

267. Mazumdar, B.C. (2018). Cattley guava–A minor fruit for making jelly and a tree for ornamental garden. *Agriculture World,* **4(12):** 78-79.

268. Mazumdar, B.C. (2019 a). [*Improved Method of Preparation of Jelly, Jam and Marmalade from Fruits*]. Dey's Publishing, College Street, Kolkata.

269. Mazumdar, B.C. (2019 b). [*Improved Method of Water-chestnut* (*Trapa natans* L. var. *bispinosa* and var. *quadrispinosa*) Cultivation]. Dey's Publishing, College Street, Kolkata.

270. Mazumdar, B.C. and Ali, S.L. (1998). *Production Technology of Papaya Latex.* New Age International (P) Ltd., Darya Ganj, Delhi.

271. Mazumdar, B.C. and Majumder, K. (2003). *Methods on Physico-chemical Analysis of Fruits.* Daya Publishing House, Darya Ganj, Delhi.

272. Mazumdar, B.C. and Mukhopadhya, Paramita. [*Fruit Processing and Preservation–Scientific and Improved Methods*]. Sreebhumi Publishing Co., Mahatma Gandhi Road, Kolkata.

273. McMillan, R.T. Jr.; Graves, W.R. and Wood, T.F. (2003). Diabetes caused by *Verticillium dahiliae on Blighia sapida. Proceedings of the Florida State Horticultural Society,* **116:** 6-8.

274. Medina, A.L.; Haas, L.I.R.; Chaves, F.C.; Salvador, M.; Zambiazi, R.C.; Silva, W.P. da; Nora, L. and Rambaldi, C.V. (2011). Araca (*Psidium cattleianum* Sabine) fruit extracts with antioxidant and antimicrobial activities and antiproliferative effect on human cancer cells. *Food Chemistry,* **128(4):** 916-922.

275. Mehmood, M.H.; Anila, N.; Begum, S.; Syad, S.A.; Siddiqui, B.S. and Gilani, A.H. (2014). Pharmacological basis for the medicinal use of *Carissa carandas* in constipation and diarrhea. *Journal of Ethnopharmacology,* **153(2):** 359-367.

276. Mehra, K.L. and Arora, R.K. (1982). *Plant Genetic Resources of India: Their Diversity and Conservation.* National Bureau of Plant Genetic Resources, New Delhi.

277. Merai, A.H. and Jadhav, A.G. (2014). Anti-uler activity of *Carissa carandas* using root extract in albino rats. *World Journal of Pharmacy and Pharmaceutical Sciences*, **3(4):** 1314-1326.

278. Miller, W.R. and McDonald, R.E. (1997). Carambola quality after ethylene and cold treatments and storage. *HortScience*, **32(5):** 897-899.

279. Minn, N.P.; Nhan, N.P.T.; Tha, D.T.; Thuy, L.K.; Khai L.Q. and TU, L.N. (2019). Different aspects affecting to production of dragon fruit (*Hylocereus undatus*) nectar. *Journal of Pharmaceutical Science and Research*, **11(3):** 1040-1043.

280. Mishra, D.; Lal, B. and Pandey, D. (2007). Clonal multiplication of *Psidium* species with mound layering. *Acta Hoticulturae*, **735:** 339-342.

281. Mishra, M.; Candra, R.; Tiwari, R.K.; Pati, R. and Pathak, R.K. (2005). Micropropagation of certain underutilized fruit crops: A review. *Small Fruits Review*, **4(4):** 7-18.

282. Misra, K.K. and Jaiswal, H.R. (1993). Effect of growth regulators on rooting of stool layers of karonda (*Carissa carandas* L.). *Indian Journal of Forestry*, **16(2):** 181-182.

283. Misra, K.K. and Jaiswal, H.R. (1998). Effect of foliar sprays of gibberellic acid on the growth of caronda (*Carissa carandas* L.) seedlings. *Indian Journal of Forestry*, **21(1):** 70-71.

284. Misra, K.K. and Singh, R. (1990). Effect of growth regulators on rooting and survival of air layers of karonda (*Carissa carandas* L.). *Annals of Agricultural Research*, **11(2):** 208-210.

285. Misra, K.K.; Singh, R. and Jaiswal, H.R. (2000). Performance of bael (*Aegle marmelos*) genotypes under foot-hills region of Uttar Pradesh. *Indian Journal of Agricultural Sciences*, **70(10):** 682-683.

286. Misra, R.S. and Bajpai, P.N. (1975). Studies on the floral biology of jamun [*Syzygium cumini* (L.) Skeels]. *Indian Journal of Horticulture*, **32:** 15-24.

287. Monforte, M.T.; Lanuzza, F.; Mondello, F.; Naccari, C.; Pergolizzi, S. and Gulati, E.M. (2014). Phytochemical composition and gastroprotective effect of *Feijoa sellowiana* Berg. Fruits from Sicily. *Journal of Coastal Life Medicine*, **2(1):** 14-21.

288. Mortensen, L.M. (1991). The effect of air temperature on the growth of foliage plants. *Norwegian Journal of Agricultural Sciences*, **5(3):** 289-294.

289. Morton, J. (1987). *Fruits of Warm Climates*. Julia F. Morton, Miami, Florida, U.S.A.

290. Morton, J.F. (1964). The jambolan (*Syzygium cumini* Skeels)–Its food, medicinal, ornamental and other uses. *Proceedings of the Florida Society of Horticultural Science*, **76:** 328-338.

291. Moschini, F.; Tosi, D. and Graifenberg, A. (1985). [The effect of type of fertilization on the raising of *Monstera deliciosa* Leib. And *Ficus benjamina* L. in hydroculture]. *Revista Della Ortofloro*–Fruiticoltura Italiana, **69(6):** 397-411.

292. Moura, C.F.H.; Alves, R.E.; Inneco, R.; Filgueiras, H.A.C.; Mosca, J.L. and Pinto, S.A.A. (2001). [Physical characteristics of cashew apples for fresh fruit market]. *Revista Brasileira de Fruiticultura,* **23(3):** 537-540.

293. Mukherji, H. (1974). *Plant Groups.* Books and Allied Private Ltd., College Street, Kolkata.

294. Mukherjee, S.K.; Rao, D.P.; Chakladar, B.P. and Chatterjee, B.K. (1986). Effect of growth regulators, invigoration and etiolation on rooting of air layers of bael (*Aegle marmelos* Correa). *Indian Journal of Horticulture,* **43(1/2):** 9-12.

295. Mukhopadhya, P.; Gangopadhyay, D. and Mazumdar, B.C. (2002). Further studies on qualitative constituents of bael (*Aegle marmelos* Correa) fruits grown in the *Sunderbans* area of West Bengal. *Indian Agriculturist,* **46(1 and 2):** 97-102.

296. Nahar, N.; Rahman, S. and Mosihuz Zaman, M. (1990). Analysis of carbohydrates in seven edible fruits of Bangladesh. *Journal of the Science of Food and Agriculture,* **51(2):** 185-192.

297. Naidu, T.C.M.; Sreekanth, B. and Narayanan, A. (1998). Growth and development of cashew fruit. I. Dry matter accumulation in different components of fruits. *Journal of Plantation Crops,* **26(1):** 82-92.

298. Naim, Z.; Khan, M.A. and Nizami, S.S. (1988). Isolation of a new isomer of ursolic acid from fruits and leaves of *Carissa carandas. Pakistan Journal of Scientific and Industrial Research,* **31(11):** 753-755.

299. Nand, D. (1970). Flowering and bearing behaviour of carambola (*Averrhoa carambola* L.). *Indian Journal of Horticulture,* **27:** 145-152.

300. Narain, N.; Bora, P.S.; Holschuh, H.J. and Vasconcelos, M.A. Da S. (2001). Physical and chemical composition of carambola fruit (*Averrhoa carambola* L.) at three stages of maturity. *Cienc Technology Aliment,* **3(3):** 144-148.

301. Nayak, G. and Sen, S.K. (1999). Effect of growth regulators, acid and mechanical scarification on germination of bael (*Aegle marmelos* Correa). *Environment and Ecology,* **17(3):** 768-769.

302. Nayak, D.R.G. (2006). Propagational studies in aonla. M.Sc. (*Agri.*) *Thesis,* University of Agricultural Sciences, Dharwad, Maharastra.

303. Neog, M. and Mohan, N.K. (1993). Physico-chemical changes during growth and development of dillenia (*Dillenia indica* L.). *South Indian Horticulture,* **41(2):** 115-116.

304. Ngamkiti Dechakul, C.; Jaijoy, K.; Hansakul, P.; Soonthornchareonnon, N. and Sireertawong, G. (2010). Anti-tumour effects of *Phyllanthus emblica* L.: Induction of cancer cell apoptosis and inhibition of *in vivo* tumour promotion and *in vitro* invasion of human cancer cells. *Phytotherapy Research,* **24(9):** 1405-1413.

305. Nikawela, J.K.; Abeysekara, A.M. and Jansz, E.R. (1998). Flabelliferin–steroidal saponins from palmyrah (*Borassus flabellifer* L.) fruit pulp. I. Isolation by flash chromatography, quantification and saponin related [foaming, haemolytic] activity. *Journal of the National Science Council of Sri Lanka*, **26(1):** 9-18.

306. Nisha, P.; Singhal, R.S. and Pandit, A.B. (2004). A study on degradation kinetics of ascorbic acid in amla (*Phyllanthus emblica* L.) during cooking. *International Journal of Food Science and Nutrition*, **55(5):** 415-422.

307. Oguntimein, B.O. and Elakovich, S.D. (1991). Allelopathic activity of the essential oils of Nigerian medicinal plants. *International Journal of Pharmacognosy*, **29(1):** 39-44.

308. O'Hare, T.J. (1993). Postharvest physiology and storage of carambola (star fruit): A review. *Postharvest Biology and Technology*, **2(4):** 257-267.

309. Oh Minseok; Jiyeon, Y.; Mingi, K. and Hoiseon, L. (2014). Acaricidal activities of beta-caryophyllene oxide and structural analogues derived from *Psidium cattleianum* oil against house dust mites. *Pest Management Science*, **70(5):** 757-762.

310. Oliveira, A.L. de; Almeida, E. de; Silva, F.B.R. da and Nascimento, Filho, V.F. (2006). Elemental contents in exotic Brazilian tropical fruits evaluated by Energy Dispersive X-Ray Fluorescence. *Scientia Agricola*, **63(1):** 82-84.

311. Olveira, M.E.B. de; Olveira, G.S. de; Maia, G.A.; Moreira, R. de A. and Monteiro, A.C. de O. (2002). [Main free amino acids in cashew fruit juice: variation during harvest]. *Revista Brasileira de Fruiticultura*, **24(1):** 133-137.

312. Oltramari, A.C.; Vesco, L.L.D.; Pedrotti, E.L.; Ducroquet, J.P.H.J.; Nodari, R.O. and Guerra, M.P. (2000). Protocol for micropropagation of feijoa [*Acca sellowiana* (Berg) Burret]. *Cliencia Rural*, **30(1):** 61-68.

313. Olubunmi, A.; Ademola, G.A. and Adenike, F.A. (2009). *Blighia sapida*: The plant and its hypoglycins–An overview. *Journal of Scientific Research*, **39(2):** 15-25.

314. Outtara, H.; Meite, A.; Amonkan, A.K.; Kouame, K.G. and Kati-Coulibaly, S. (2011). [Aril of *Blilghia sapida* (K. Koenig, 1778). Potential nutritional deficiency and toxic effects associated with its use, according to the harvesting stage and the degree of consumption]. *International Journal of Biological and Chemical Sciences*, **5(1):** (unpaginated).

315. Oyeleke, G.O.; Oyetede, O.E.; Fatai, A. and Adegoke, B.M. (2013). Nutrients, antinutrients and physicochemical composition of *Blighia sapida* pulp and pulp oil (ackee apple). *IOSR Journal of Applied Chemistry*, **4(1):** 5-8.

316. Padmaja, A.P.; Jose, R.K.J. and Hisham, A. (2013). Diuretic activity of the roots of *Flacourtia indica*. *Hygeia Journal for Druga and Medicine*, **5(1):** 79-83.

317. Padmavathi, D.; Deshpande, N. and Sarala, A. (2011). *Dillenia indica*: A review on morphology, phytochemistry and pharmacological aspects. *Research Journal of Pharmacy and Technology*, **4(7):** 1037-1039.

318. Panchal, J.B.; Gaikwad, R.S.; Dhemre, J.K. and Chavan, U.D. (2018). Studies on preparation and storage of jelly from dragon fruit (*Hylocereus undatus*). *Journal of Pharmacognosy and Phytochemistry*, **7(4):** 2648-2655.

319. Pande, N.C.; Singh, A.R.; Maurya, V.N. and Katiyar, R.N. (1986). Studies on the biochemical changes in bael (*Aegle marmelos* Correa) fruit. *Progressive Horticulture*, **18(1/2):** 29-34.

320. Pareek, O.P. (1983). *The* Ber, I.C.A.R., New Delhi.

321. Pareek, O.P. and Sharma, S. (1993). Genetic resources of under-exploited fruits. [In: *Advances in Horticulture, Vol.1 Fruit Crops, Part 1*: Ed. by K.L. Chadha and O.P. Pareek]. Malhotra Publishing House, New Delhi.

322. Parmar, C. and Kaushal, M.K. (1982). *Aegle marmelos* (In: Wild Fruits). Kalyani Publishers, New Delhi.

323. Patel, V.B.; Pandey, S.N.; Singh, S.K. and Das, B. (2005). Variability in jamun (*Syzygium cuminii* Skeels) accessions from Uttar Pradesh and Jharkhand. *Indian Journal of Horticulture*, **62(3):** 244-247.

324. Pathak, R.K. (2003). *Status Report on Genetic Resources of Indian Gooseberry–Aonla* (*Emblica officinalis* Gaertn.) *in South and Southeast Asia*. National Agriculture Science Centre, Pusa, New Delhi.

325. Pathak, R.K.; Ojha, C.M.; Dwivedi, R. and Hari Om. (1991). Studies on the effects of methods and duration of budding in aonla. *Indian Journal of Horticulture*, **48(3):** 207-212.

326. Pathak, R.K. and Pathak, R.A. (1993). Improvement of Minor Fruits. In: *Advances in Horticulture, Vol. 1.–Fruit Crops, Part I.* (Ed. by K.L. Chadha and O.P. Pareek). Malhotra Publishing House, New Delhi.

327. Peng, J.; Gu, M.; Xu, B.Q.; Dang, J.Z.; Li, Y.H.; Zhang, S.A.; Lu, Z.H.; Yu, Z.Y.; Bai, X.Q.; Cai, Z.F.; Deng, H.G.; Zeng, S.Y.; Xi, J.B. and Huang, W.D. (2010). Study of the flowering and fruiting phenological law of 'Sizihua' longan (*Dimocarpus longan*) and its regulation. *Acta Horticulturae*, **863:** 259-266.

328. Peralta-Bohorquezo, A.F., Parada, F.; Quijano, C.E. and Pino, J.A. (2010). Analysis of volatile compounds of sour guava (*Psidium guineense* Swartz) fruit. *Journal of Essential Oil Research*, **22(6):** 493-498.

329. Peng Sheng. (1999). [Xiangmi Yangtao, a high quality carambola variety]. *South China Fruits*, **28(6):** 35.

330. Pino, J.A. (1997 a). [The volatile constituents of tropical fruits. III. Feijoa and cashew]. *Alimentaria*, **35(286):** 41-45.

331. Pino, J.A. (1997 b). [The volatile constituents of tropical fruits. IV. Kiwi fruit, carambola and mangosteen]. *Alimentaria*, **35(286):** 47-50.

332. Pino, J.; Marbot, R. and Rosado, A. (2000). [Volatile constituents of star apple (*Chrysophyllum cainito* L.) from Cuba]. *Flavour and Fragrance Journal*, **17(5):** 401-403.

333. Pino, J.A.; Marbot, R. and Fuentes, V. (2003). Characterization of volatiles in bullock's heart (*Annona reticulata* L.) fruit cultivars from Cuba. *Journal of Agriculture and Food Chemistry*, **51(13):** 3836-3839.

334. Pino, J.A.; Marbot, R. and Vazques, C. (2002). Characterization of volatiles in Costa Rican guava [*Psidium friedrichsthalianum* (Berg) Nied.] fruit. *Journal of Agriculture and Food Chemistry*, **50(21):** 6023-6026.

335. Pinto, D.S.; Freitas, R.M.O.D.; Araujo, P.C. and Praxedes, S.C. (2012). Propagation of *Spondias* species by cutting of different lengths treated with sugars. *Genetics and Plant Physiology*, **2(3-4):** 171-176.

336. Poi, A.K. (1989). *Studies on the Improvement of Cape-gooseberry Crop (Physalis peruviana* L.) *by Standardization of Some Agro-practices and Application of Plant Growth Regulants.* Ph.D. thesis, submitted to the B.C. Krishi Viswavidyalaya, Mohanpur, under supervision of research by Dr. B.C. Mazumdar and Mr. S.K. Sen.

337. Poi, A.K. and Mazumdar, B.C. (1989). Auxin-induced root regeneration in stem cuttings of kath bael (*Limonia acidissima*). *Indian Biologist*, **21(1):** 51-52.

338. Poi, A.K.; Mazumdar, B.C. and Sengupta, P.K. (1987). Suppersion of little leaf disease of cape-gooseberry by tetracycline treatment. *Indian Journal of Mycological Research*, **25(2):** 153-155.

339. Pradhan, D.; Tripathy, G. and Patnaik, S. (2012). Anticancer activity of *Limonia acidissima* Linn. (Rutaceae) fruit extracts on human breast cancer cell lines. *Tropical Journal of Pharmaceutical Research*, **11(3):** 413-419.

340. Prasad, I.D.; Sengupta, B.N.; Singh, R.K. and Singh, S.P. (1985). Effect of NPK on yield attributes and quality of cape-gooseberry (*Pghysalis peruviana*). *Haryana Journal of Horticultural Science*, **14(3/4):** 151-155.

341. Qureshi, S.A.; Asad, W. and Sultana, V. (2009). The effect of *Phyllanthus emblica* L. on type-II diabetes, triglycerides and liver-specific enzyme. *Pakistan Journal of Nutrition*, **8(2):** 125-128.

342. Rahamat, M. (2014). *In vitro* techniques for the proliferation of axillary shoots to establish complete plants of *Artocarpus chaplasha* Roxb. *American Eurasion Journal of Sustainable Agriculture*, **8(9):** 60-65.

343. Rahman, M.; Amin, M.N.; Ishiguri, F.; Yokota, S.; Sultana, R.S.; Takashima, Y.; Lizuka, K. and Yoshizawa, N. (2009). Effects of sucrose and pH on shoot multiplication of Indian olive (*Elaeocarpus robustus* Roxb.). *Bulletin Utsunomiya University Forests*, **45:** 5-8.

344. Rahman, M.M.; Amin, M.N.; Ishiguri, F.; Yokota, S.; Sultana, R.S.; Takashiman, Y.; Lizuka, K. and Yoshizawa, N. (2009). *In vitro* plantlet regeneration of dwarf Indian olive (*Elaeocarpus robustus* Roxb.): A fruit plant of Bangladesh. *Plant Biotechnology Reports*, **3(3):** 259-266.

345. Rahman, S.M.I.; Bhuyan, M.H.M.B.; Mustakim, A.A.M.M.; Sarker, J.C. and Hussian, M.A. (2014). Fruit characteristics, yield contributing characters and yield of some lotkon genotypes under north eastern region of Bangladesh. *International Journal of Sustainable Crop Production*, **9(1):** 8-10.

346. Rai, R. and Misra, K.K. (2005). Micropropagation of karonda (*Carissa carandas*) through stool multiplication. *Scientia Horticulture*, **103(2):** 227-232.

347. Rakesh, L. and Shivanna, H. (2015). Variation in important traits of jamun (*Syzygium cuminii* Skeels). *Karnataka Journal of Agricultural Sciences*, **28(1):** 120-122.

348. Ramachandran, V.S.; Swarupanandan, K. and Renuka, C.A. (2004). Traditional irrigation system using palmyra palm (*Borassus flabellifer*) in Kerala, India. *Palms*, **48(4):** 175-181.

349. Ramadan, M.M.; El-Ghorab, A.H. and Ghanem, K.Z. (2015). Volatile compounds, antioxidants and anticancer activities of cape-gooseberry fruit (*Physalis peruviana* L.). An *in vitro* study. *Journal of the Anab Society for Medical Research*, **10(2):** 56-64.

350. Ramar, K.; Ayyadurai, V. and Arulprakash, T. (2014). *In vitro* shoot multiplication and plant regeneration of *Physalis peruviana* L., an important medicinal plant. *International Journal of Current Microbiology and Applied Sciences*, **3(3):** 456-464.

351. Ramrirez, J.M.; Galvis, J.A. and Fischer, G. (2005). Postharvest ripening of feijoa (*Acca sellowiana* Berg.) treated with $CaCl_2$ at three storage temperatures. *Agronomia Columbiana*, **23(1):** 117-127.

352. Rangkadilok, N.; Worasuttayangkurn, L.; Bennett, K.N. and Satayavivad, J. (2005). Identification and quantification of polyphenolic compounds in longan (*Euphoria longana* Lam.) fruit. *Journal of Agriculture and Food Chemistry*, **53(5):** 1387-1392.

353. Rao,' P.R.; Babu, M.D.; Rao, R.V.K. and Nammi, S. (2002). Comparative study of sweet and bitter varieties of *Aegle marmelos* (L.) Correa on hypoglycemic activity in rabbits. *Journal of Natural Remedies*, **2(1):** 88-91.

354. Rao, S.S.R. and Charyulu, N.V.N. (1989). Physiological changes during development and ripening of *Syzygium cuminii* (L.) Skeels (Syn. *Eugenia jambolana* Lamk.) fruits. *Indian Botanical Reporter*, **8(1):** 31-34.

355. Rathore, T.S.; Singh, R.P.; Dora, N.S. and Shekhawat, N.S. (1992). Clonal propagation of *Ziziphus* species through tissue culture. *Scientia Horticulturae*, **51:** 165-168.

356. Reddy, P.V.R. and Vasugi, C. (2008). Resistance to fruit fly, *Bactrocera dorsalis* (Hendel) and tea mosquito bug, *Helopeltis antonii* in certain wild *Psidium* species. *Journal of Horticultural Sciences*, **3(1):** 79-81.

357. Rema, J.; Krishnamoorthy, B. and Mathew, P.W. (1997). Vegetative propagation of major tree species–A review. *Journal of Species and Aromatic Crops*, **6(2):** 87-105.

358. Ramashree, A.B.; Thomas, T.V.; Nabeesa, E.; Neelakandan, N. and Nandakumar, S. (2003). *In vitro* plant regeneration from nodal explants of *Syzygium cuminii* L. *Plant Cell Biotechnology and Molecular Biology*, **4(3/4):** 197-200.

359. Reynertson, K.A.; Basile, M.J. and Kennelly, E.J. (2005); Antioxidant potential of seven myrtaceous fruits. *Ethnobotany Research and Applications*, **3(1):** 25-35.

360. Rodia, M.H.; Jamila, B.; Sharifah, K.S.M. and Russly, A.R. (2019). Physico-chemical and anti-oxidant properties of mesocarp and exocarp from *Borassus flabellifer*. *International Food Research Journal*, **26(5):** 1469-1479.

361. Romeiras, M.M.; Duarte, M.C.; Indjal, B. and Catarino, L. (2012). Medicinal plants used to treat neurological disorders of west Africa: A case study with Guinea-Bissau flora. *American Journal of Plant Sciences*, **3:** 1028-1038.

362. Roy, G.C. and Mazumdar, B.C. (1989). Qualitative improvement of cashew apples and nuts by spraying with water and zinc sulphate solutions, in the coastal area of West Bengal. *Indian Journal of Landscape Systems and Ecological Studies*, **12(2):** 95-98.

363. Roy, P. and Mazumdar, B.C. (1989). Pectin content in some minor fruits. *Science and Culture*, **55(3):** 110-111.

364. Roy, S.K. (1953). Embryology of *Eugenia jambos* L. *Current Science*, **22:** 249-250.

365. Roy, S.K. and Singh, R.N. (1978). Studies on utilization of bael fruit (*Aegle marmelos* Correa) for processing. I. Physico-chemical characteristics in different cultivars. *Indian Food Packer*, **32:** 3-8.

366. Roy, S.K. and Singh, R.N. (1979). Bael fruit (*Aegle marmelos*)–A potential fruit for processing. *Economic Botany*, **33(2):** 203-212.

367. Rudriamampionona, D.; Rafamantanana, M.; Rabemanantsoa, C.; Rakotoniriana, F.; Cheuk, F.; Corbisier, A.M.; Mahilon, J.; Ratsima Manga, S. and El-Jaziri, M. (2008). *Ex situ* conservation and clonal propagation of the Malagasy *Syzygium cuminii*, an antidiabetic plant. *Belgium Journal of Botany*, **141(1):** 14-20.

368. Rupal, A.V. and Narasimhachary, A.V.R.L. (2011). Antioxidant potential of seasonal fruits in fluoride toxicity: An *in vitro* study. *Journal of Cell and Tissue Research*, **11(2):** 2783-2794.

369. Sabour, A.M.T.; Ebeed, S.S. and El Habid, F.A. (2002). *In vitro* propagation of *Opuntia tuna* and *Casimiroa edulis* fruit trees. *Annals of Agricultural Science*, Moshtohor, **40(3):** 1739-1753.

370. Saeed, S. and Tariq, P. (2007). Antibacterial activities of *Emblica officinalis* and *Coriandrum sativum* against gram-negative urinary pathogens. *Pakistan Journal of Pharmaceutical Science*, **20(1):** 32-35.

371. Sah, A.K. and Verma, V.K. (2011). *Syzygium cuminii*: An overview. *Journal of Chemical and Pharmaceutical Research,* **3(3):** 108-113.

372. Sahu, U.C.; Panda, S.K.; Mohapatra, U.B. and Ray, R.C. (2012). Preparation and evaluation of wine from tendu (*Diospyros melanoxylon* L.) fruits with antioxidants. *International Journal of Food and Fermentation Technology,* **2(2):** 167-178.

373. Saini, J.P. and Dube, S.D. (1992). Effect of storage on germination, tube elongation and fertilization ability of wild date pollen grains. *Progressive Horticulture,* **21(3-4):** 305-307.

374. Sajib, M.A.M.; Haque, M.M.; Yeasmin, S. and Khatun, M.H.A. (2014). Minerals and heavy metals concentration in selected tropical fruits of Bangladsh. *International Food Research Journal,* **21(5):** 1731-1736.

375. Sanabria, C.M.E.; Cammino, A.J.M.; Garcia, G. and Renaud, J. (1999). [Leaf monphology of *Monstera deliciosa* Liebm. (Araceae) during ontogeny]. *Ern.,* **9(2):** 37-43.

376. Sanjay Tyagi; Mishra, K.K. and Jaiswal, H.R. (1999). Effect of auxins on rooting of softwood stem cuttings of *Carissa* under mist. *Scientific Horticulture,* **6:** 37-43.

377. Sankat, C.K. and Balkisson, F. (1992). The effect of packaging and refrigeration on the shelf life of the carambola. *ASEAN Food Journal,* **7(2):** 114-117.

378. Santana, G.E. and Angarita, A. (1997). [Adventitious regeneration in somaclones of cape gooseberry (*Physalis peruviana*)]. *Agronomia Columbiana,* **14(1):** 59-65.

379. Saraju Devi, S. and Paul, S.B. (2011). An overview on *Cicea acida* (*Phyllanthus acidus*). *Assam University Journal of Science and Technology,* **7(1):** 156-160.

380. Saravanan, S.; Chander, P.; Kumar, R. and Singh, J. (2013). Influence of different levels of NPK on growth, yield and quality of phalsa (*Grewia subinaequalis* L.). *Asian Journal of Horticulture,* **8(2):** 433-435.

381. Sarkar, A.K.; Dhua, R.S. and Sen, S.K. (1984). Interaction of phenolic compounds with IBA and NAA in the regeneration of roots in water-apple (*Syzygium javanicum* L.) stem-cuttings. *Progressive Horticulture,* **16(1/2):** 41-46.

382. Sarkar, G.K.; Singh, M.M. and Misra, R.S. (1985). Nutritional studies in aonla (*Emblica officinalis* Gaertn.) cv. Banarasi. *Progressive Horticulture,* **17(1):** 41-46.

383. Saroj, P.L. (2012). Prepetuation of *Psidium* species through stooling and assessment of growth behaviour of clonal rootstocks under subtropical condition. *Indian Journal of Agricultural Sciences,* **82(3):** 230-235.

384. Sauls, J.W.; Jackson, L.K. and Crocker, T.E. (1986). Fruit crops fact sheet–Dooryard fruit varieties. *Florida Co-operative Extension Service, Institute of Food and Agricultural Sciences, University of Florida.*

385. Sawant, L. (2010). Determination of gallic acid in *Phyllanthus emblica* Linn. dried fruit powder by HPTLC. *Journal of Pharmacy and Bioallied Sciences,* **2(2):** 105-108.

386. Schotsmans, W.C.; East, A.; Thorp, G. and Woolf, A.B. (2011). *Postharvest Biology and Technology of Tropical and Subtropical Fruits. Volume 3*: *Cocona to Mango.* Woodhead Publishing Ltd.

387. Sehgal, C.B. and Khurana, S. (1985). Morphogenesis and plant regeneration from cultured endosperm of *Emblica officinalis* Gaertn. *Plant Cell Reports,* **4(5):** 263-266.

388. Sekar, D.K.; Kumar, G.; Karthik, L. and Rao, K.V. Bhaskara. (2011). A review on pharmacological and phytochemical properties of *Aegle marmelos* (L.) Corr. Serr. (Rutaceae). *Asian Journal of Plant Science and Research,* **1(2):** 8-17.

389. Sekar, T. and Francis, K. (1998). Some plant species screened for energy, hydrocarbons and phytochemicals. *Bioresource Technology,* **65(3):** 257-259.

390. Sen, S.K.; De, R.K. and Bandopadhyay, A. (1990). Effect of preconditioning stock plants and exogenous application of growth regulators on rooting of semi-hardwood cuttings of aonla (*Emblica officinalis* Gaertn.). *Advances in Plant Science,* **3(2):** 195-199.

391. Sham Singh; Krishnamurthy, S. and Katyal, S.L. (1967). *Fruit Culture in India.* I.C.A.R. Publications, New Delhi.

392. Shankar, K.B.; Raghavathatham, D.; Chezhian, N.; Vijayakumar, R.M. and Balasubramanian, A. (1999). BSR-I, a high yield amla variety for different agro-climatic regions of Tamil Nadu. *South Indian Horticulture,* **47(1/6):** 143-144.

393. Sharma, G.N.; Dubey, S.K.; Sharma, P. and Sati, Nitin. (2011). Medicinal value of bael: A Review. *International Journal of Current Pharmaceutical Review and Research,* **1(3):** 12-22.

394. Sharma, H.P.; Patel, H.; Sharma, S. and Vaishali. (2014). Study of physico-chemical changes during wood apple (*Feronia limonia*) maturation. *Journal of Food Research and Technology,* **2(4):** 148-152.

395. Sharma, J.; Bandyopadhyay, A. and Sen, S.K. (1989). Effect of auxinic and non-auxinic chemicals on rooting of rose apple (*Syzygium jambos* Alston) stem cuttings. *South Indian Horticulture,* **37(2):** 108-111.

396. Sharma, R.K.; Bhoi, S.K.; Pandey, N.; Shinde, S. and Pandey, V.K. (2012). *Agriculture at a Glance*. Daya Publishing House, Delhi.

397. Sharma, S.S.; Sharma, R.K. and Yamdagni, R. (1989). Studies on the yield and quality of fruit of aonla cultivars (*Emblica officinalis* Gaertn.) under rainfed conditions of Haryana. *Research and Development Reporter,* **6(1):** 41-43.

398. Sidhu, S.; Pandhi, P.; Malhotra, S.; Vaiphei, K. and Khanduja, K.L. (2011). Beneficial effects of *Emblica officinalis* in arginine-induced acute pancreatitis in rats. *Journal of Medicinal Food,* **14(1-2):** 147-155.

399. Silveira, M.S.; Fontes, C.P.M.L.; Guilherme, A.A.; Fernandes, F.A.N. and Rodrigues, S. (2012). Cashew apple juice as substrate for lactic acid production. *Food and Bioprocess Technology*, **5(3):** 947-953.

400. Singh, A.K. and Singh, P. (1998). Power of significance of difference among fruit and seed size parameters of karonda (*Carissa carandas* Linn.). *Annals of Agricultural Research*, **19(1):** 66-71.

401. Singh, D. (1963). Kamrakh–Sweet and Vitamin rich. *Indian Horticulture*, **7:** 8.

402. Singh, G.D.; Singh, S.; Jindal, N.; Bawa, A.S. and Saxena, D.C. (2010). Physico-chemical characteristics and sensory quality of signhara (*Trapa natans* L.): An Indian water-chestnut under commercial and industrial storage conditions. *African Journal of Food Science*, **4(11):** 693-702.

403. Singh, G.K. and Parmar, A.S. (1998). Studies on the effects of methods and dates of budding in aonla (*Emblica officinalis* Gaertn.). *Annals of Arid Zone*, **37(2):** 199-201.

404. Singh, I.S. and Pathak, R.K. (1987). Evaluation of aonla (*Emblica officinalis* Gaertn.) varieties for processing. *Acta Horticulture*, No.**208:** 173-177.

405. Singh, J.P. and Dhuria, H.S. (1960). Studies on floral biology of sweet lime (*Citrus limettioides* Tanaka). *Indian Journal of Horticulture*, **17:** 9-20.

406. Singh, P.; Singh, S.P.; Shalitra, R.; Singh, S. and Vishwakarma, P.K. (2016). *In vitro* clonal propagation of cape-gooseberry (*Physalis peruviana* L.). *Ecology, Environment and Conservation Paper*, **22(2):** 859-863.

407. Singh, R. and Saxena, S.K. (2008). *Fruits*. National Book Trust, Govt. of India, New Delhi.

408. Singh, R.D. (1986). Studies on the bearing habit and fruit development in bael (*Aegle marmelos* Correa) cv. Mirzapuri Kagzi. *Progressive Horticulture*, **18(3-4):** 277-284.

409. Singh, R.N. and Roy, S.K. (1984). *The Bael, Cultivation and Processing*. I.C.A.R., New Delhi.

410. Singh, R.R.; Joon, M.S. and Daulta, B.S. (1984). A note on physico-chemical characteristics of fruit in two cvs. of aonla (*Phyllanthus emblica* Linn.). *Haryana Journal of Horticultural Sciences*, **13(3/4):** 133-134.

411. Sinha Roy, S. (1996). *Assessment of Some Agricultural Waste Materials as Sources of Pectin and Changes of Pectic Compounds in Some Minor Fruits During Development*. Ph.D. thesis, submitted to the Calcutta University, under supervision of research by Dr. B.C. Mazumdar.

412. Sinha, Roy, S. and Mazumdar, B.C. (1996). Potentiality of the berries of *Carissa carandas* L. in making human consumable jelly. *Indian Biologist*, **28(2):** 19-20.

413. Shu Zen Hong; Chu Chengchung; Hwang Leejuan and Sheieh Chingshung. (2001). Light, temperature and sucrose affect colour, diameter and soluble solids of disks of wax apple fruit skin. *HortScience*, **36(2):** 279-281.

414. Siti Halijah Ali and Md. Yunus Jaafar. (1992). Effect of harvest maturity on physical and chemical characteristics of carambola (*Averrhoa carambola* L.). *New Zealand Journal of Crop and Horticultural Science,* **20(2):** 133-136.

415. Sobhana, A. and Mathew, J. (2015). Novel products from cashew apple–jelly and cake. **6(1):** 45-49.

416. Sofidiya, M.O.; Jimoh, F.O.; Aliro, A.A.; Afolayan, A.J.; Odukoya, O.A. and Familoni, O.B. (2012). Evaluation of antioxidant and antibacterial properties of six Sapindaceae members. *Journal of Medicinal Plants Research,* **6(1):** 154-160.

417. Souza, S.N. De. (2013). Grafting technique for the propagation of feijoa plants [*Acca sellowiana* (Berg.) Burret]. *Revista de Ciencias Agroveterinarias,* **12(3):** 314-316.

418. Spinola, V.; Perestrelo, R.; Camara, J.S. and Castilho, P.C. (2015). Establishment of *Monstera deliciosa* fruit volatile metabolomic profile at different ripening stages using solid-phase micro-extraction combined with gas chromatography–mass spectrometry. *Food Research International,* **67:** 409-417.

419. Srivastava, K.K. and Singh, H.K. (2000). Floral biology of bael (*Aegle marmelos*) cultivars. *Indian Journal of Agricultural Sciences,* **70(11):** 797-798.

420. Srivastava, K.K. and Singh, H.K. (2004). Physico-chemical quality of bael (*Aegle marmelos* Correa) cultivars. *Agricultural Science Digest,* **24(1):** 65-66.

421. Subramanyam, M.D. and Iyer, C.P.A. (1993). Improvement of guava. [In: *Advances in Horticulture, Vol.1*: Fruit Crops, Part 1: Ed. by K.L. Chadha and O.P. Pareek]. Marathon Publishing House, New Delhi.

422. Sudjaroen, Y.; Hull, W.E.; Erben, G.; Wurtele, G.; Chang Bum Rung, S.; Ulrich, C.M. and Owen, R.W. (2012). Isolation of characterization of ellagitannins as the major polyphenolic components of longn (*Dimocarpus longan* Lour.) seeds. *Phytochemistry,* **77:** 226-237.

423. Suthon, W.; Lee Ching Lung and Yang Yau Shiang. (2012). Effect of potassium chlorate on off-season flowering in longan (*Dimocarpus longan*). *Journal of the Taiwan Society for Horticultural Sciences,* **58(1):** 61-70.

424. Susumu Arima; Daigoho, M. and Hoque, W.A. (1999). Flower development and anthesis behaviour in the water chestnut (*Trapa* species). *Bulletin of the Faculty of Agriculture, Saga University, Japan,* No.**84:** 83-92.

425. Swami, S.; Thakor, N.; Patil, M. and Haldankar, P. (2012). Jamun [*Syzygium cumini* (L.)]: A Review of its food and medicinal uses. *Food and Nutrition Sciences,* **3(8):** 1100-1117.

426. Takeuchi, J.; Ono, T.; Sano, M.; Kagiwada, S.; Kanagawa, T.; Nishlo, T. and Horie, H. (2011). [First report of anthracnose of Swiss cheese plant and phoma rot of leather leaf fern in Japan]. *Annual Report of the Kanto-Tosan Plant Protection Society,* **58:** 69-72.

427. Tangkijchote, P.; Thaensa, M. and Wanichkul, K. (2012). Belete (*Boletus colossus* Helm.) spawn affecting air-layering rooting of Java apple [(*Syzygium samarangense* (Blume) (Merr. And Perry)] 'Phetchsairung'. *Proceedings of the 50th Kasetsart University Annual Conference, Kasetsart University, Thailand,* 31 January–2 February, 2012: 272-279.

428. Thimmapaiah and Shirley, Raichal Samuel. (1999). *In vitro* regeneration of cashew (*Anacardium occidentale* L.). *Indian Journal of Experimental Biology,* **37:** 384-390.

429. Thiripurasundari, G. and Usharani, G. (2011). Comparative production of vinegar using cashew apple juice by different immobilization techniques. *Current Botany,* **2(3):** 31-33.

430. Thiyam, B. and Sharma, G.D. (2013). Isolation and identification of fungi associated with local fruits of Barak Valley, Assam. *Current World Environment,* **8(2):** 319-322.

431. Tomar, A.; Kumar, A. and Tirpathi, S. (201). Seed germination behaviour in *Artocarpus lakoocha* Roxb. *Indian Forester,* **140(3):** 325-326.

432. Utami, R.; Khalid, N.; Sukari, M.A.; Rahmani, M.; Abdul, AB and Dachriyanus. (2013). Phenolic contents, antioxidant and cytotoxic activities of *Elaeocarpus floribundus* Blume. *Pakistan Journal of Pharmaceutical Science,* **26(2):** 245-250.

433. Uthaiah, B.C. (1988). Studies on changes in physical parameters of the developing wild edible fruit, *Carissa carandas. Myforest,* **24(4):** 256-258.

434. Varghese, S.K.; Inamdar, J.A. and Kalia, K. Subramanian, R.B. and Nataraj, M. (1993). Micropropagation of *Aegle marmelos* (L.) Corr. *Phytomorphology,* **43(1/2):** 87-92.

435. Vasudeva, R.S. and Raychaudhury, S.P. (1994). Guava disease in Pushkar Valley and its control. *Indian Phytopathology,* **7:** 78-81.

436. Vasugi, C. and Dinesh, M.R. (2007). Genetic variability in some *Psidium* species. *Indian Journal of Agricultural Science,* **77(7):** 420-423.

437. Venkitakrishnan, M.; Panjatcharam, V.; Kumaravelu, G. and Ramanujam, M.P. (1997). Physico-chemical changes during maturation and ripening of jambolan fruit. *Indian Journal of Plant Physiology,* **2(4):** 267-270.

438. Verma, B. and Kant, U. (1999). Propagation of *Emblica officinalis* Gaertn. through tissue culture. *Advances in Plant Sciences,* **12(1):** 21-25.

439. Vitale, A.; Aiello, D.; Castello, I.; Parlavecchio, G. and Polizzi, G. (2008). First report of crown rot and root rot caused by *Cylindrocladium pauciramosum* on feijoa (*Feijoa sellowiana*) in Italy. *Plant Disease,* **92(11):** 1590.

440. Voltolini, J.V. and Fachinello, J.C. (1997). Effect of shading cattley guava stock plant (*Psidium cattleyanum* Sabina) on propagation by cuttings. *Acta Horticulturae,* No.**452:** 59-62.

441. Vriesmann, L.C.; Petkowiez, C.L. de O; Carneiro, P.I.V.; Costa, M.E. and Beleski-Carneiro. (2009). Acidic polysaccharides from *Psidium cattleianum* (Araca). *Brazilian Archives of Biology and Technology*, **52(2):** 259-264.

442. Wahyuni, F.S.; Shaari, K.; Stanslas, J.; Lajis, N.H. and Dachriyanus. (2015). Cytotoxic xanthones from the stem bark of *Garcinia cowa* Roxb. *Journal of Chemical and Pharmaceutical Research,* **7(1):** 227-236.

443. Wall, M.M.; Nishijima, K.A.; Keith, L.M. and Nagao, M.A. (2011). Influence of packaging on quality retention of longans (*Dimocarpus longan*) under constant and fluctuating postharvest temperatures. *HortScience*, **46(6):** 917-923.

444. Wang Dan; Liu Ren Dao; Zhang Dong Xue; Chen Jing and Wu Jie. (2007). Study on the tissue culture for *Feijoa sellowiana,* a new fruit crop. *South China Fruits,* **2:** 21-23.

445. Wang Jia Fu and He Bizhu. (2000). [*In vitro* culture of longan shoot tips]. *Journal of Fujian Agricultural University*, **29(1):** 23-26.

446. Wang, J.K. (1966). Equipment for husking poha berries. *Technical Bulletin of the Hawaii Agricultural Experiment Station*, **60:** 18.

447. Wang, P.; Chen, Y.; Lin, M.; Tsou, Y. and Ko, W. (2010). Severe decline of wax apple trees caused by *Fusarium solani* in northern Taiwan. *Botanical Studies,* **51:** 75-80.

448. Wani, R.A.; Prasad, V.M.; Hakeem, S.A.; Sheema, S.; Angchuk, S. and Dixit, A. (2013). Shelf life of karonda jams (*Carissa carandas* L.) under ambient temperature. *African Journal of Agricultural Research*, **8(21):** 2447-2449.

449. Wani, T.A.; Rana, S.; Bhat, W.W.; Pandith, S.A.; Dhar, N.; Razdan, S.; Chandra, S.; Sharma, N. and Lattoo, S.K. (2016). Efficient *in vitro* regeneration, analysis of molecular fidelity and *Agrobacterium tumifaciens*- mediated genetic transformation of *Grewia asiatica* L. *Journal of Plant Biochemistry and Physiology*, **4(2):** 167.

450. Wanichkul, K. and Natheung, M. (2011). Effect of hand pollination with certain pollinizers on fruit set and fruit quality of Java apple [(*Syzygium samarangense* (Blume) (Merr. and Perry)]. *Proceedings of the 49th Kasetsart University Annual Conference, Volume 1, Plants*: 241-248.

451. Wanyama, D.O.; Wamocha, L.S.; Ngamau, K. and Ssonkko, R.N. (2006). Effect of gibberellic acid on growth and fruit yield of greenhouse grown cape gooseberry. *African Crop Science Journal,* **14(4):** 319-323.

452. Wattanawikkit, P.; Tantiwiwat, S.; Surawatananon, S. and Sangthongprow, S. (1999). [*In vitro* culture of Malay apple (*Eugenia malaccensis* Linn.)]. In: *The 37th Kasetsart University Annual Conference,* 3-5 February, 1999. (Edited by Outes, C.G.) Bangkok, Thailand, Text and Journal Publication Co. Ltd.: 235-240.

453. Webster, S.A.; Mitchell, S.A.; Reid, W.A. and Ahmad, M.H. (2006). Somatic embryogenesis from leaf and zygotic embryo explants of *Blighia sapida* 'Cheese' Ackee. *In vitro Cellular and Developmental Biology. Plant,* **42(5):** 467-472.

454. Wilson, P.G.; O'Brien, M.M.; Gadek, P.A. and Quinn, C.J. (2001). Myrtaceae revisited: A reassessment of infrafamilial groups. *American Journal of Botany*, **88(11):** 2013-2025.

455. Wilson, P.G.; O'Brien, M.M.; Helswood, M.M. and Quinn, C.J. (2005). Relationships within Myrtaceae sensu lato based on matk phylogeny. *Plant Systematics and Evolution*, **251:** 3-19.

456. Wang, K.C.; Wong, S.N.; Loi, H.K. and Lim, C.L. (1996). Volatile constituents from the fruits of four Sapindaceae: rambutan (*Nephelium lappaceum* L.), pulasan [*N. ramboutanake* (Labill.)], longan (*Dimocarpus longan* Lour.) and mata kucing (D. *longan* spp. *Malesianus* Leenh.). *Flavour and Fragrance Journal*, **11(4):** 223-229.

457. Wu Ding Yao; Qui Jindan; Zhang Hailan and Luo Xiaozheng. (2000). [A study on flowering promotion by ringing in longan (*Dimocarpus longana* Lour.)] *Scientia Agricultural Sinica*, **33(6):** 40-43.

458. Wu, S.J.; Ng, L.T.; Huang, Y.M.; Lin, D.L.; Wang, S.S.; Huang, S.N. and Lin, C.C. (2005). Antioxidant activities of *Physalis peruviana*. *Biological and Pharmaceutical Bulletin*, **28(6):** 963-966.

459. Venkataratnam, L. (1963). *It is Easy to Grow Phalsa. Farm Information Bulletin No.61,* Directorate of Extension, Ministry of Food and Agriculture, Govt. of India, New Delhi.

460. Vijayvargia, P. and Vijayvergia, R. (2014). A review on *Limonia acidissima* L.: Multipotential medicinal plant. *International Journal of Pharmaceutical Science Review Research*, **28(1):** 191-195.

461. Xiaohui, Zheng,; Peikun, Oi and Zi De Jiang. (2001). [Identification of the fungal diseases of ornamental plant (Araceae) in Guangzhou region-II]. *Journal of South China Agricultural University*, **22(2):** 39-41.

462. Xia, Q.H.; Chen, R.Z. and Fu, J.R. (1992 a). Effects of desication, temperature and other factors on the germination of lychee (*Litchi chinensis* Sonn.) and longan (*Euphoria longan* Steud.) seeds. *Seed Science and Technology*, **20(1):** 119-127.

463. Xia, Q.H.; Chen, R.Z. and Fu, J.R. (1992 b). Moist storage of lychee (*Litchi chinensis* Sonn.) and longan (*Euphoria longan* Steud.) seeds. *Seed Science and Technology*, **20(2):** 269-270.

464. Xiang Yan; Huang Jian Chang; Gao Ping Fei; Huang Zi Hai and Wang Jiang Bio. (2001). Relationship between isoperoxidase of the scion and the rootstock and their grafting affinity. *Journal of Southwest Agricultural University*, **23(1):** 70-73.

465. Xu Xiudan; Huang Jinson; Zhen Shaoquan; Lu Xiumin and Xu Jiahui. (1999). [An extremely late-maturing longan variety 'Lidongben']. *Acta Horticulturae Sinica*, **26(2):** 136.

466. Yadav, V.K. and Singh, H.K. (1999). Studies on the preharvest application of chemicals on shelf-life of aonla (*Emblica officinalis* Gaertn.) fruits at ambient temperature. *Journal of Applied Horticulture*, **1(2):** 118-121.

467. Yamada, M.; Hidaka, T. and Fukamachi, H. (1996). Heat tolerance in leaves of tropical fruit crops as measured by chlorophyll fluorescence. *Scientia Horticulturae*, **67(1/2):** 39-48.

468. Yang Win Huan. (1996). [Songfengben, an uncommon and promising late longan variety]. *China Fruits*, **1:** 52.

469. Yasseen, M.; Yasseen. (2002). Micropropagation of pitaya (*H. undatus*). *In vitro Cellular and Developmental Biology-Plant*, **38:** 427-429.

470. Yeh Der Ming. (1998). [Growth of six foliage plants in sub-irrigation systems]. *Journal of the Chinese Society for Horticulture Science*, **44(1):** 81-92.

471. Yonemoto, Y.; Nomura, K.; Ide, M.; Inoue, H.; Majikina, M. and Okuda, H. (2006). Index for harvesting time of white sapote (*Casimiroa edulis* Llave and Lex.) cv. Cuccio. *Journal of Horticultural Science and Biotechnology*, **81(1):** 18-22.

472. Yucesan, B.B.; Mohammed, A.; Arslan, M. and Gurel, E. (2015). Clonal propagation and synthetic seed production from nodal segments of cape gooseberry (*Physalis peruviana* L.), a tropical fruit plant. *Turkish Journal of Agriculture and Forestry*, **39:** 797-806.

473. Yumei, H.; Menqjaiu, T. and Chinho, L. (1999). The fluctuation of carbohydrates and nitrogen compounds in flooded wax-apple trees. *Botanical Bulletin of Academia Sinica*, **40(3):** 193-198.

474. Veerasamy, S.; Rajarathinam, K. and Jayabalan, M. (1995). Regeneration of shoots from split shoot apical meristem in *Borassus flabellifer*. *Journal of Ecotoxicology and Environmental Monitoring*, **5(1):** 39-43.

475. Velu, G. (1989). Impact of organic manuring of palmyrah. *Madras Agricultural Journal*. **76(10):** 592-598.

476. Vengaih, P.C.; Rabindrababu, D.; Murthy, G.N. and Prasad, K.R. (2013). Jaggery from palmyrah palm (*Borassus flabellifer* L.). *Indian Journal of Traditional Knowledge*, **12(4):** 714-717.

477. Zang Xiaoping and Xu Nengkudn. (2000). [On the fertilizing for custard apple trees]. *South China Fruits*, **29(3):** 29-30.

478. Zarga, M.H.A. (1986). Three new simple indole alkaloids from *Limonia acidissima*. *Journal of Natural Products*, **49(5):** 901-904.

479. Zhang Meng; Tang Hao Ru; Wang Dan; Ren Shao Xiong and Liu Ren Dao. (2009). Studies on rooting characteristics and anatomical structure of feijoa cuttings. *Journal of Fruit Science*, **26(4):** 498-501.

480. Zhang Meng; Wang Dan; Ren Shao Xiong and Liu Ren Dao. (2009). Effect of different substratum and plant growth regulators on the rooting ability of feijoa soft cuttings. *South China Fruits*, **4:** 47-49.

481. Zhang, Y.J.; Abe, T.; Tanaka, T.; Yang, C.R. and Kouno, I. (2001). Phyllanemblinins A-F, new ellagitannins from *Phyllanthus emblica*. *Journal of Natural Products*, **64(12):** 1527-1532.

482. Zhou Li Juan; Wang Dan; Huang Hai Tao and Zhang Xiao Xue. (2008). Sterilization of the explant and callus induction of *Feijoa sellowiana* Berg. *in vitro* culture. *Journal of Tropical and Subtropical Botany,* **16(2):** 179-183.

483. Zhuang Yimei; Wang Renji; Xie Zhinan and Xi Wenbao. (1995). [Optimum range of mineral element contents in the leaves of Shuizhang longan]. *Journal of Fujian Agricultural University*. **24(3):** 281-286.

Index

M

N

O

P

Colour Plates

PLATE 1

Tree and Plantation of *Anacardium occidentale*

PLATE 2

Fruits of *Annona reticulata*

PLATE 3

Fruits of *Carissa carandas*

PLATE 4

Preparing 'Imitation Cherries' from the berries of *Carissa carandas*

PLATE 5

Palms of *Borassus flabellifer*

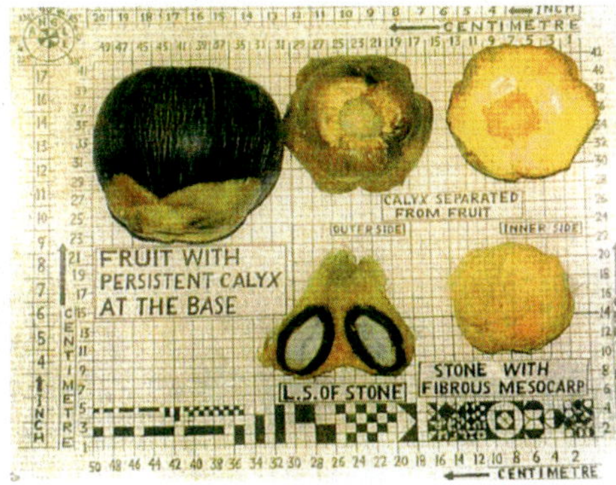

PLATE 6

Fruit of *Borassus flabellifer*

PLATE 7

Palms of *Borassus flabellifer* grown around pond

PLATE 8

Tender, succulent endosperm of immature seeds of *Borassus flabellifer*

PLATE 9

Fruits of *Phoenix sylvestris*

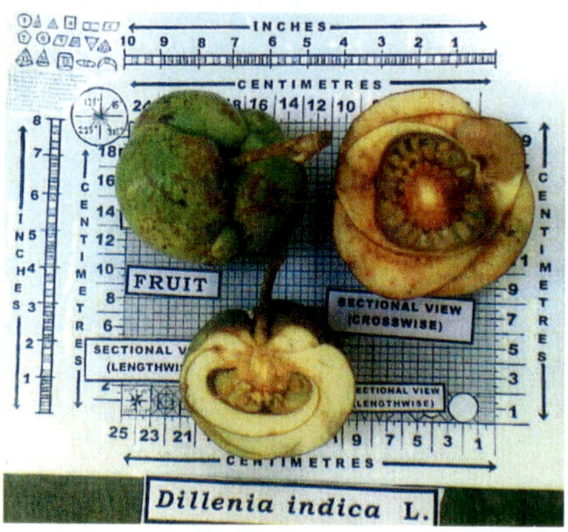

PLATE 10

Fruits of *Dillenia indica*

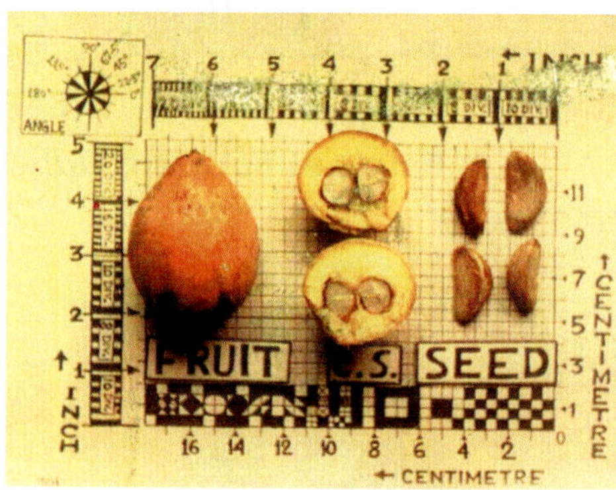

PLATE 11

Fruits of *Diospyros melanoxylon*

PLATE 12

Plant of *Trapa natans*

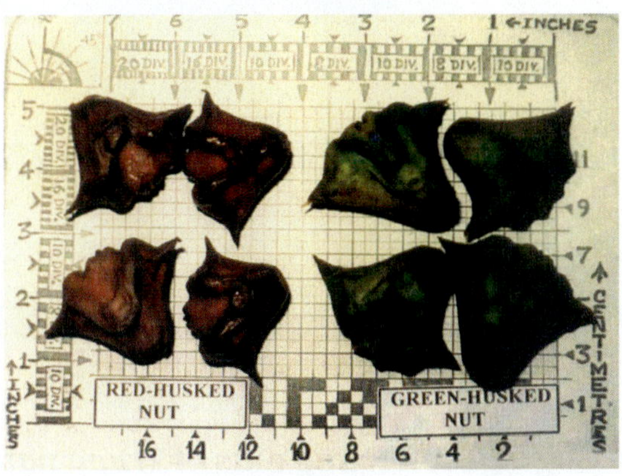

PLATE 13

Red and Green husked nuts of *Trapa natans*

PLATE 14

Raft made for harvest of nuts of *Trapa natans*

PLATE 15

Plucking and collecting nuts of *Trapa natans*

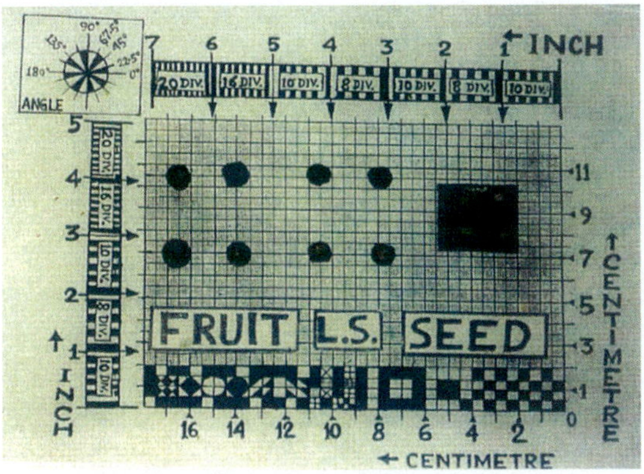

PLATE 16

Fruits of *Grewia subinaequalis*

PLATE 17

Fruits of *Artocarpus lacucha*

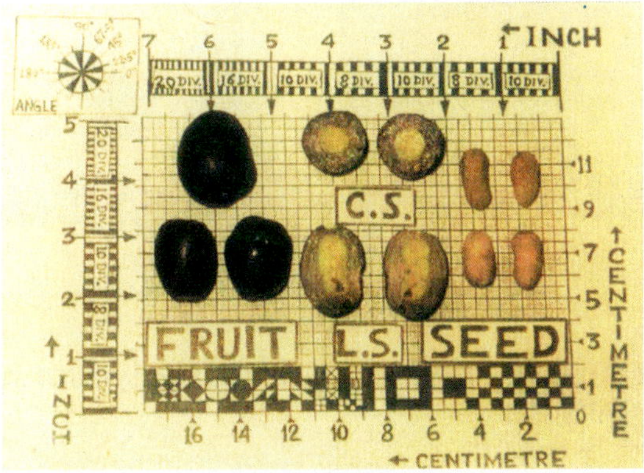

PLATE 18

Fruits of *Syzygium cumini*

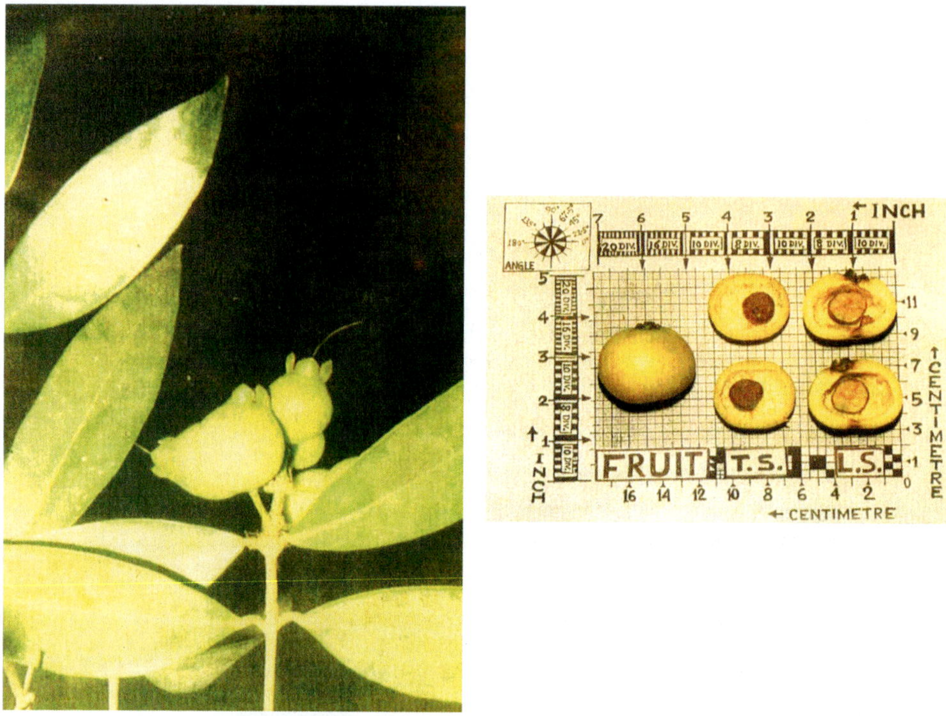

PLATE 19

Leaves *and Fruits of Syzygium jambos*

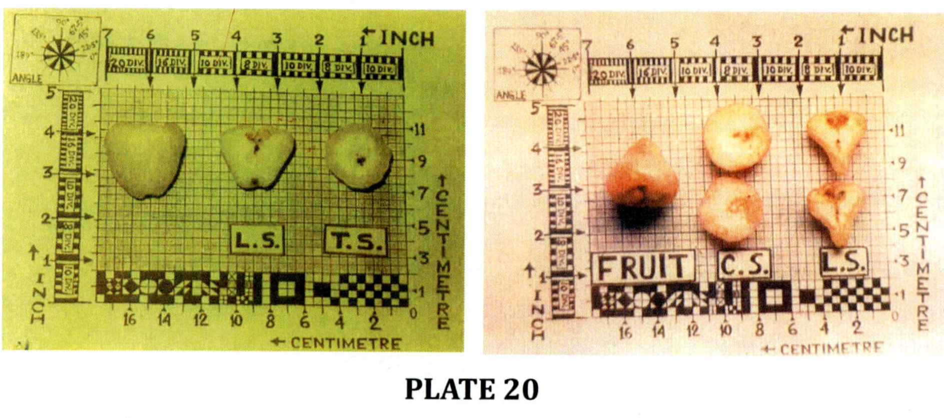

PLATE 20

Greenish white and red peeled Fruits of *Syzygium samarangense*

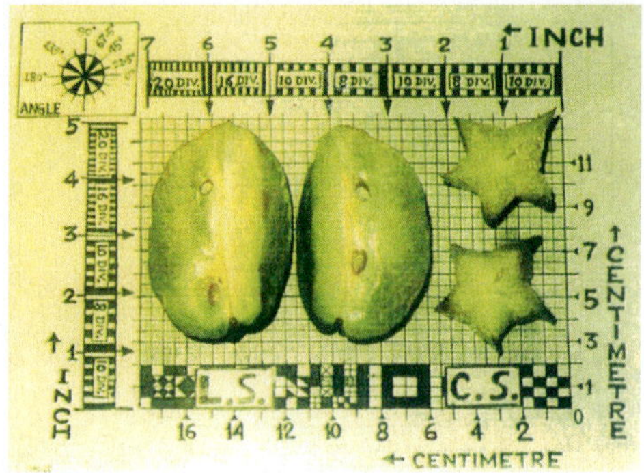

PLATE 21

Fruits of *Averrhoa carambola*

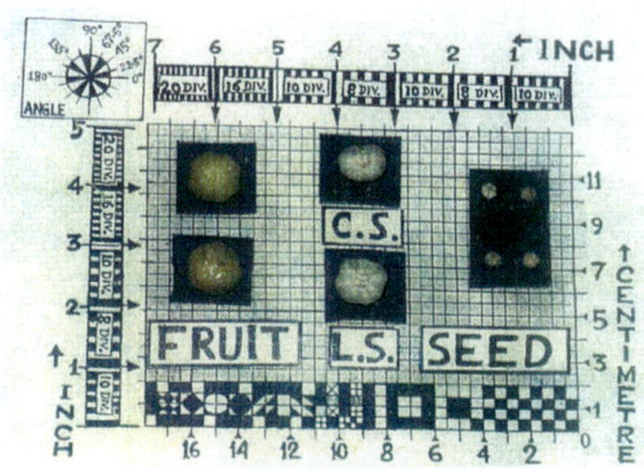

PLATE 22

Fruits of *Phyllanthus acidus*

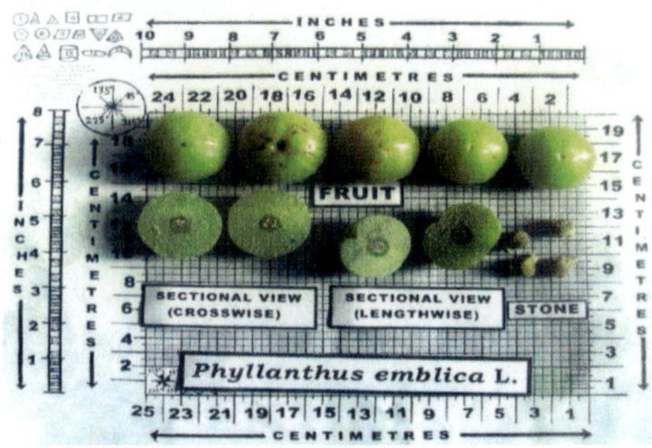

PLATE 23

Fruits of *Phyllanthus emblica*

PLATE 24

Fruits of *Ziziphus mauritiana*

PLATE 25

Fruits of *Aegle marmelos*

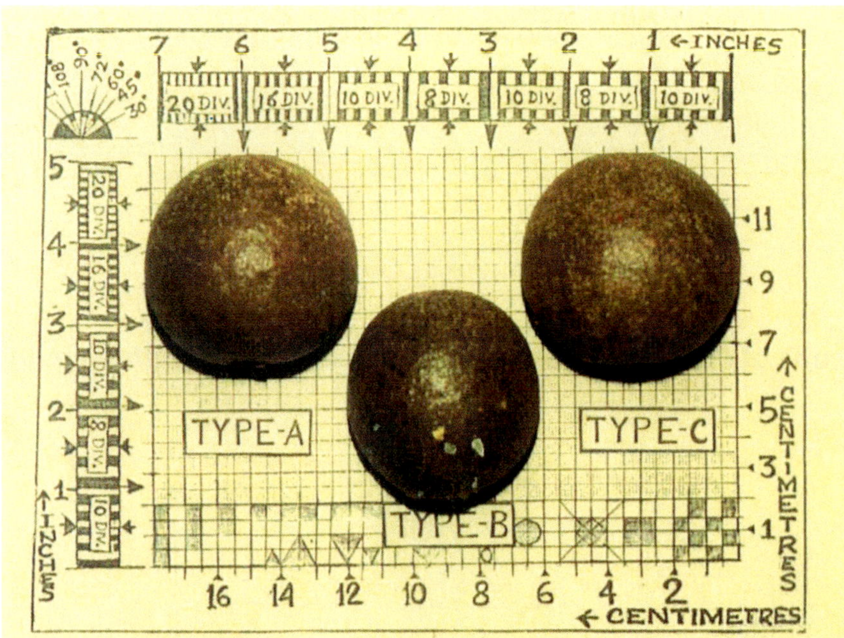

PLATE 26

Fruits of *Limonia acidissima*

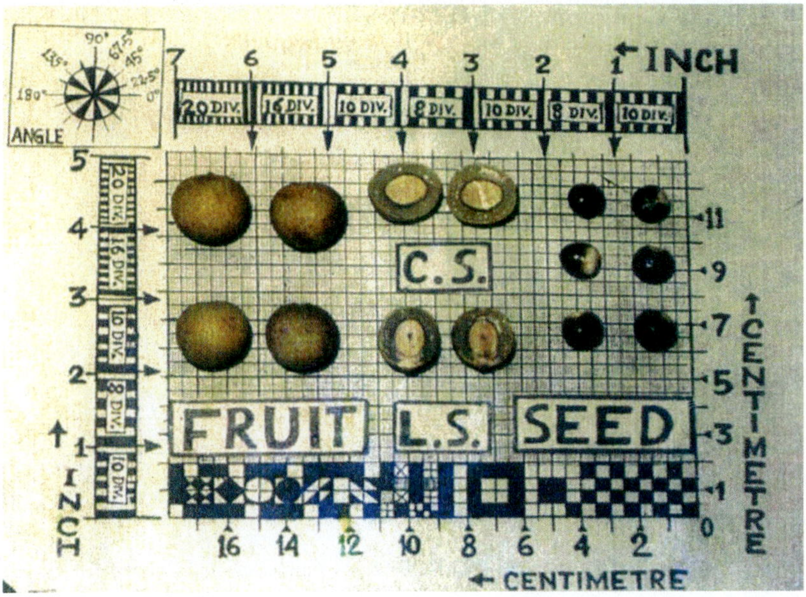

PLATE 27

Fruits of *Dimocarpus longan*

PLATE 28

Plant, Twig and Fruits of *Physalis peruviana*